Solid State Astrochemistry

T0135171

NATO Science Series

A Series presenting the results of scientific meetings supported under the NATO Science Programme.

The Series is published by IOS Press, Amsterdam, and Kluwer Academic Publishers in conjunction with the NATO Scientific Affairs Division

Sub-Series

I. Life and Behavioural Sciences	IOS Press
II. Mathematics, Physics and Chemistry	Kluwer Academic Publishers
III. Computer and Systems Science	IOS Press
IV. Earth and Environmental Sciences	Kluwer Academic Publishers
V. Science and Technology Policy	IOS Press

The NATO Science Series continues the series of books published formerly as the NATO ASI Series.

The NATO Science Programme offers support for collaboration in civil science between scientists of countries of the Euro-Atlantic Partnership Council. The types of scientific meeting generally supported are "Advanced Study Institutes" and "Advanced Research Workshops", although other types of meeting are supported from time to time. The NATO Science Series collects together the results of these meetings. The meetings are co-organized bij scientists from NATO countries and scientists from NATO's Partner countries – countries of the CIS and Central and Eastern Europe.

Advanced Study Institutes are high-level tutorial courses offering in-depth study of latest advances in a field.
Advanced Research Workshops are expert meetings aimed at critical assessment of a field, and identification of directions for future action.

As a consequence of the restructuring of the NATO Science Programme in 1999, the NATO Science Series has been re-organised and there are currently Five Sub-series as noted above. Please consult the following web sites for information on previous volumes published in the Series, as well as details of earlier Sub-series.

http://www.nato.int/science
http://www.wkap.nl
http://www.iospress.nl
http://www.wtv-books.de/nato-pco.htm

Proceedings of the NATO Advanced Study Institute on
Solid State Astrochemistry
Erice, Sicily, Italy
5–15 June 2000

A C.I.P. Catalogue record for this book is available from the Library of Congress.

ISBN 1-4020-1558-5 (HB)
ISBN 1-4020-1559-3 (PB)

Published by Kluwer Academic Publishers,
P.O. Box 17, 3300 AA Dordrecht, The Netherlands.

Sold and distributed in North, Central and South America
by Kluwer Academic Publishers,
101 Philip Drive, Norwell, MA 02061, U.S.A.

In all other countries, sold and distributed
by Kluwer Academic Publishers,
P.O. Box 322, 3300 AH Dordrecht, The Netherlands.

Printed on acid-free paper

Solid State Astrochemistry

edited by

Valerio Pirronello
DMFCI Università di Catania,
Catania, Sicily, Italy

Jacek Krelowski
Center for Astronomy,
N. Copernicus University, Torun, Poland

and

Giulio Manicò
DMFCI Università di Catania,
Catania, Sicily, Italy

Kluwer Academic Publishers

Dordrecht / Boston / London

Published in cooperation with NATO Scientific Affairs Division

Contents

Preface

The fundamental role that Astrochemistry plays into regulating the processes that in interstellar clouds lead to the formation of stars, and how these processes concur into affecting the shape and the dynamics of galaxies and hence into showing the Universe in the way it appears to us is well established.

Together with those occurring in the gas phase a special relevance is recognized to processes that involve interstellar dust grains, the solid component of matter diffused among stars.

The school on "Solid State Astrochemistry", held at the Ettore Majorana Centre for Scientific Culture in Erice (Sicily) from the 5th to the 15th of June 2000, was the fifth course of the International School of Space Chemistry.

In spite of its very focused aim it was attended by 66 participants from 17 different countries, that in the very special environment provided by the Majorana Centre, discussed in great details the various aspects of the subject.

During the course it has been presented the state of the art of the processes of interaction of grains with the gas, they are mixed with; of the catalytic role of dust in allowing the formation on its surface of key molecular species (like H_2, but also many complex, possibly prebiotic species) that cannot be obtained efficiently by other mechanisms; of the interaction between solids (dust grains and icy mantles, but also cometary nuclei, satellites of the giant planets and minor bodies of our Solar System) in space and energetic agents like UV photons and fast particles.

The presence and the importance of Polycyclic Aromatic Hydrocarbons, that may represent the smallest component of grains, their chemical role have been considered together with possible paths toward astrobiology and the ever standing mystery of the ubiquitous presence of Diffuse Interstellar Bands and their carriers - also very likely some complex organic molecular species.

All the mentioned issues have been tackled from both the observational and the theoretical point of view, but a special attention has been paid to experimental simulations of such processes, performed in the laboratory to put the whole matter on a firm Galilean ground.

We would like to dedicate this volume with the Proceedings of the Erice School on Solid State Astrochemistry to J. Mayo Greenberg, who has been a great Scientist and a close friend who inspired all of us and unfortunately passed away on the 29th of November 2001.

We acknowledge gratefully support from NATO under the framework of the Advanced Study Institute, from the United States National Science Foundation and from NASA. We like to thank gratefully also the administration and the staff members of the Ettore Majorana Centre for Scientific Culture that with their efficiency and personal support allowed to render particularly agreeable the fruition of the discussions and the exchange of ideas that pervaded the course.

Valerio Pirronello, Jacek Krelowski, Giulio Manicò
Catania, Sicily, Italy
26 September, 2002

THE INTERSTELLAR MEDIUM: AN OVERVIEW

David A. Williams
*Department of Physics & Astronomy, University College London, Gower Street, LONDON
WC1E 6BT*

Keywords: Interstellar medium - clouds - dust - chemistry

Abstract: An overview of the Interstellar medium is presented. Properties of interstellar
dust together with the most relevant locations of solid state interstellar and
circumstellar astrochemical processes are reviewed.

INTRODUCTION

The interstellar medium is the matter filling the space not occupied by
stars in a galaxy. We often describe the interstellar medium in terms of our
own galaxy, the Milky Way. In fact, all galaxies have an interstellar
medium. In some galaxies, the average interstellar gas density is relatively
high, as seems to be the case in so-called star-burst galaxies, for example.
Other galaxies, such as elliptical galaxies, have very little matter in their
interstellar space.

I take an inclusive view of the interstellar medium to mean all matter, gas
and solid-phase, at high and low temperatures, in interstellar and
circumstellar space. From the point of view of astronomy, the interstellar
medium is an important component in the evolution of any galaxy: it is the
reservoir of mass from which new stars form, and into which the ashes of
stars - in the form of elements heavier than H_2, and dust - are ejected. The
interstellar medium therefore is continually consumed as stars are formed,
and is continually enriched in elements particularly oxygen, carbon, and
nitrogen, and in dust, as the stars evolve and die, sometimes explosively, in
supernovae.   From the point of view of chemistry, the discovery of

1

V. Pironello et al. (eds.), Solid State Astrochemistry, 1–20.
© 2003 *Kluwer Academic Publishers. Printed in the Netherlands.*

molecules in the interstellar medium has been a great stimulus and source of new areas of study. The interstellar medium presents a range of physical conditions that are difficult to reproduce in the laboratory. The pressures are very low, $\sim < 10^{-10}$ Torr, the temperatures range from less than 10 K to about one million K, the medium is pervaded by UV and other electromagnetic radiation, and by fast protons and electrons (mostly with energies $\sim$ 1MeV, but ranging to very high energies). Mixed in the gas are dust grains of sizes from nanometres to micrometres, at such a low number density that the grains rarely collide. The discovery of a rich interstellar chemistry in the 1970s and 1980s stimulated an intense programme of research into ion-molecule and neutral-neutral gas-phase reactions. The recognition of the importance of relatively high density regions, where star formation is occurring and in which the gas-dust interaction dominates any purely gas-phase process, is now driving a similarly intense study of the interaction of atoms and molecules with surfaces, of heterogeneous chemistry, of the formation and processing of ices, and of thermal and non-thermal processes that can desorb molecules.

This paper gives a brief description of the various regions of interstellar space in the Milky Way Galaxy, discusses some recent work on interstellar dust in our galaxy, and then identifies interstellar regions where the gas-dust interaction plays a significant role.

1.   REGIONS OF INTERSTELLAR SPACE

Table 1 summarizes our present knowledge of the various regions in the interstellar medium of the Milky Way Galaxy. It should be recognised that much of the data remains uncertain, though the broad picture is well established. The table indicates that much of the interstellar medium is either hot (the radiatively excited regions, heated by the radiation from hot stars) or very hot (so-called coronal gas), much too hot for dust grains to survive or molecules to form. The cooler regions (diffuse clouds and molecular clouds) which are of interest to solid-state astrochemistry, occupy only a few percent of the volume. However, they contain nearly all the mass. Therefore, if the interstellar medium is considered as a reservoir of mass from which new stars can form, it is the cooler, denser and mainly neutral parts which are the most significant.

Table 2 gives information about pressures in the interstellar medium. Radiatively ionized regions and molecular clouds are overpressured compared to the other regions which are in approximate pressure equilibrium. Radiatively ionized regions are in fact expanding as a consequence of this overpressure; the molecular clouds are gravitationally bound and may be collapsing in the process of star formation.

Table 1: Regions in the interstellar medium of the Galaxy - typical parameters

	Coronal gas	HII Regions	Intercloud gas	Diffuse Clouds	Molecular Clouds
Temperature (K)	10^6	10^4	10^3-10^4	10^2	10
Number density (cm^{-3})	$< 10^{-2}$	10^2	1	10^2	10^4
Fractional ionization	1	1	0.1	10^{-3}	10^{-7}
Filling factor	$\geq 50\%$	$< 1\%$	$\leq 50\%$	3%	$< 1\%$
Mass fraction				~ 50%	~ 50%

Table 2: Approximate gas pressures in different regions of interstellar space

Region	Particle density (cm^{-3})	Temperature (K)	Pressure (dyne cm^{-2})
HII regions	$\geq 10^2$	8000	$\geq 1.1 \times 10^{-10}$
Diffuse atomic cloud	30	70	2.9×10^{-13}
Intercloud	0.3	6000	2.5×10^{-13}
Cool molecular cloud	10^3 - 10^4	20	2.8×10^{-12}
			2.8×10^{-11}
Coronal gas	$< 10^{-2}$	5×10^5	$< 6.9 \times 10^{-13}$

Why does the interstellar medium have this structure? In fact, it can be understood in terms of a "three phase model" in which an input of energy from supernovae explosions creates the very hot gas, and sufficiently frequent supernovae events create overlapping supernovae remnants that cool rather slowly. The model is obviously sensitive to the cooling function in the gas, and therefore directly to the abundance of heavy elements in the gas. It also depends on other heating sources, such as the cosmic rays, and the level of support given to the gas from magnetic fields. The general picture of an interstellar medium filled with hot, warm, and cool components is at least qualitatively similar to what is observed. Of course, other galaxies which may have different supernovae rates, different cooling functions because of different star formation histories, will have considerably different structures in their interstellar media.

2.   INTERSTELLAR DUST

In this section I review briefly the sources that we possess of information about interstellar dust, and summarize the present picture we have of its composition and properties. Excellent reviews have been given by Mathis (1990) and Whittet (1990).

The first indication of the presence of interstellar dust lay in the observation by William Herschel of regions in rich star fields which were apparently deficient in stars. At the time, it was not clear whether these "holes" in the sky were truly absences of stars or caused by some kind of obscuration. Gradually, it became clear that they were the result of obscuration which was not only localized but widespread, and efforts were made to describe the obscuring effect so that it could be discounted in the observation of stars and galaxies. These studies led to the definition of a standard interstellar extinction curve which shows how the extinction (i.e. absorption and scattering) varies with wavelength. The extinction was attributed to small particles of dust, and the extinction curve has been the guide for many models of dust. The extinction optical depth is relatively small in the infrared, rises steadily in the visual region, shows a peak in the ultraviolet near 220nm, then rises again from about 160nm into the far-ultraviolet. In fact, it is a curve that is simply fitted by many models, as the number of free parameters in any model is quite large. More instructive, perhaps, is the fact that the standard interstellar extinction curve is an average, and that along any particular line-of-sight significant variations can occur. Compared to the so-called standard interstellar extinction curve, extinction curves may show, for example, a weak or strong 220nm peak, and (independently) a weak or strong far-UV rise. It is evident that extinction measurements are insufficient to determine the nature of interstellar dust, and we need to involve new information sources.

A powerful indicator of the chemical nature of the dust is the set of elemental depletions along any particular line of sight. If one assumes that the interstellar gas is well mixed within a galaxy (and there are good reasons to believe so) then elemental abundances should be everywhere the same. In fact, in many regions the relative abundance of heavy atoms such as carbon, silicon, and iron with respect to hydrogen are low compared to a notional cosmic abundance. The missing atoms are assumed to be incorporated into dust. The depletions of elements relative to the cosmic abundance can be severe. For example, along many lines of sight more than 99% of the cosmic abundance of silicon is missing, while about one half of carbon is missing. The information contained in the depletion factors enables one to infer that the likely major constituents of interstellar dust are silicates and carbons.

To go further than this, one needs spectroscopy. Fortunately, this has been possible in the infrared from wavelengths of about 2 to 200 micrometres through the Infrared Space Observatory. Infrared spectroscopy for the Earth's surface is mostly blocked by the Earth's atmosphere, apart from a few windows. Figures 1, 2, and 3 show absorption by cold dust, emission by warm dust, and a crystalline silicate component emission.

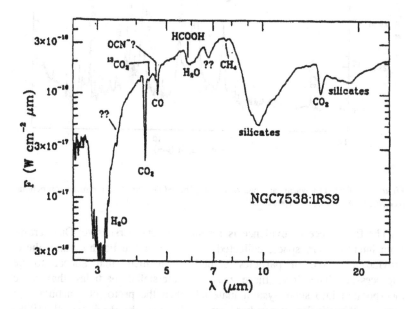

Figure 1: Absorption by cold ice-coated dust. The ISO spectrum of the embedded stellar object NGC7538:IRS9 (Whittet et al. 1996). The spectrum shows absorption features arising from various molecular species in the ice mantles on dust grains.

As well as these narrow band features, there is also a broad band feature, the so-called Extended Red Emission, peaking around 500 nanometres and about 100 nanometres wide. This has been attributed to luminescence in nanoparticles, particularly of silicon.

Of course, measurements of scattered light, and of linear polarization of starlight may also provide additional constraints on models of dust. Unfortunately, such measurements are generally difficult.

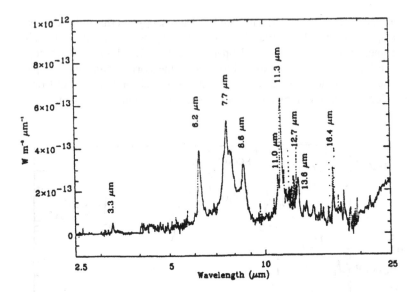

Figure 2: Emission by warm carbonaceous dust. The ISO spectrum of the reflection nebulae NGC7023 (Moutou et al., 1999).

The final piece of evidence is recent and very persuasive. Dust grains from interplanetary space, collected as they enter the Earth's atmosphere, provide a source of particles embedded in them that appear to be unprocessed. Therefore, these may be interstellar particles that were incorporated into solar system material when the proto-solar nebula was formed. Of particular interest has been the discovery by Bradley et al. (1999) of so-called GEMS - Glasses with Embedded Metals and Sulphides - found in the carbonaceous matrices of chondritic porous interplanetary dust. These particles have a size of about half a micrometre, they show infrared absorption features at wavelengths near 10 and 20 micrometres, and they contain nanometre-sized inclusions of Fe and FeS that are superparamagnetic. These characteristics are a catalogue of the properties that silicate grains are required to have in the interstellar medium to cause visual extinction and polarization. Thus, for the first time, it may be that we have the opportunity to study in the laboratory actual dust grains from the interstellar medium.

The nature of interstellar dust is important for solid-state astrochemistry. The dust provides surfaces on which heterogeneous reactions occur, and substrates on which ices can be deposited and processed. Our present knowledge of interstellar dust is becoming more precise and is summarized in Table 3. While some uncertainties remain, we do have enough guidance to consider and understand the details of the gas/dust interaction in interstellar space.

Table 3: Observational evidence for the composition of interstellar dust

Phenomenon	Carrier
Extinction	
• UV rise	small (~ nm) silicate and carbon grains
• 217.5 nm bump	carbons (stacked PAHs)
• Visible	GEMS
	crystalline silicates
	carbon grains/coatings
• IR	carbon grains; band gap
ERE	silicon or carbon nano-particles
UIBs	PAHs
	small (~nm) carbon (diamond) grains
IR absorptions	GEMS
	crystalline silicates
	ices (H_2O, CO, CO_2, CH_3OH...)
	saturated hydrocarbons

3. LOCATIONS OF SOLID-STATE ASTROCHEMISTRY: INTERSTELLAR

3.1 DIFFUSE INTERSTELLAR CLOUDS

Diffuse clouds are those through which starlight penetrates relatively easily, and typically have optical depths in the visual region, caused by dust extinction, of less than unity. Thus, chemical processes within them are dominated by photodissociation and photoionization by starlight. As a result, there are few molecules, although H_2 is a special case, see below. The gas typically has a density of about 100 H-nuclei cm^{-3}, though some of these may be in H_2 and some as free H, and a temperature around 100K. Molecules observed are usually simple diatomics such as CO, CH, CH^+, CN, OH, C_2, etc. with abundances relative to hydrogen of around 10^{-8} or less, though CO is a special case and can attain fractional abundances of around 10^{-6}. Recent observations have also indicated the presence of some simple polyatomics such as HCO^+, CH_2, HCN, etc. It seems likely that these regions may also contain large molecules, containing tens or even hundreds of atoms. Such large molecules have been proposed as the carriers of the so-called Diffuse Interstellar Bands - broad features (compared to atomic lines) that appear in the optical/infrared spectra of bright stars. Such large molecules cannot be formed by synthesis in interstellar space, and may be the result of degradation of interstellar carbon grains by shattering in shocks.

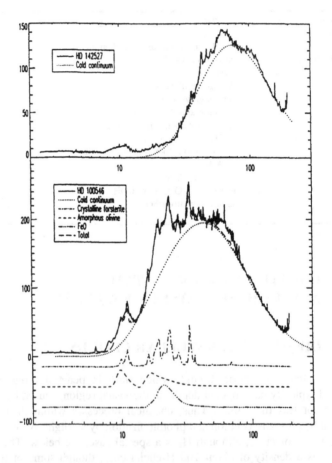

Figure 3: Emission from warm crystalline silicate dust. The ISO spectrum of warm silicate dust near the star HD100546. Wavelengths in micrometres are indicated on the x-axis, and intensity in Janskys on the y-axis (after Waters and Waelkens, 1999). The contribution to the emission by crystalline silicates is indicated.

There are three important roles for dust in diffuse clouds, affecting interstellar chemistry. Firstly, dust provides a modest extinction of starlight and a partial suppression of photodissociation and photoionization. Secondly, the dust - through the photoelectric effect - enables the energy of starlight to be coupled to the gas and provide the main heating source. Finally, dust grains enable heterogeneous chemistry to occur.

This chemical role of dust in diffuse clouds is indicated most clearly in the case of H_2. The rate of destruction of H_2 by photodissociation in optically thin regions (where the H_2 abundance is very low) can be calculated quite precisely. Assuming that chemical steady-state pertains, then this loss rate must be balanced by a formation rate. However, although gas phase processes to form H_2 can be proposed, such as

$$H + e \rightarrow H^- + h\nu$$

$$H^- + H \rightarrow H_2 + e,$$

these are found to be too slow, and the direct radiative association

$$H + H \rightarrow H_2 + h\nu$$

is very strongly forbidden and does not occur. By default, the suggestion that H_2 formation on dust grains has been adopted. The requirement of the observations is that most of the H atoms arriving at a grain surface must leave as part of a molecule. Modern studies of this process are discussed in the next chapter.

The process by which H_2 is destroyed is, however, a line excitation process, summarized

$$H_2 (X; v''= 0; J'' = 0, 1) + h\nu \rightarrow H_2(B, C; v', J') \rightarrow H_2(X; v''', J''') + h\nu'$$

The excitation occurs from ground electronic state X in the vibrational ground state (i.e. $v'' = 0$) and one of the two lowest rotational states (i.e. $J'' = 0, 1$) to excited electronic states B or C, from which relaxation rapidly occurs back to state X, in some new vibration and rotation states (represented by v''' and J'''). In about 10 or 20% of the relaxations, the final vibrational state is in the vibrational continuum, so that the molecule falls apart. In situations where the formation rate H_2 is high, the lines through which the excitation occurs may become saturated and high H_2 abundances build up. The formation of H_2 on dust, assuming a reasonable efficiency of reaction, is sufficient to produce regions that are optically thick in lines of H_2 so that the atomic/molecular balance in hydrogen is strongly in favour of H_2. This H_2 then feeds the chemistry of other species in ways that has been fully described elsewhere (e.g. van Dishoeck 1998).

3.2   MOLECULAR CLOUDS AND STAR-FORMING REGIONS

Molecular clouds are denser and more opaque than diffuse clouds. The higher density (typically $n_H \sim 10^4$ cm^{-3}) and larger opacity (typically, an optical depth in the visual of at least 5) mean that photochemistry is heavily suppressed, and that an ion-molecule chemistry initiated by cosmic ray ionization of hydrogen dominates. The temperatures of molecular clouds are generally low (~ 10 K) and so only exothermic barrier-free reactions are permitted. The products of this chemistry tend to emphasise a range of unsaturated hydrocarbons. The number of molecular species currently identified in molecular clouds is well over 100, and since many similar species implicated in the chemical networks may be unobservable, the total number of species must be at least several hundred. A current list can be found at the website http://www.cv.nrao.edu/~awootten/allmols.html

Table 4 summarizes the main processes that can occur in molecular clouds. The ion-molecule chemistry driven by cosmic ray ionization has a timescale that depends on the cosmic ray ionization rate, ζ s^{-1}. The timescale is calculated on the basis that enough ionization of H_2 to make H_3^+ via

$$H_2 \xrightarrow{\text{c.r.}} H_2^+ \xrightarrow{H_2} H_3^+$$

is required to make molecules through reactions such as

$$H_3^+ \xrightarrow{O} OH^+ \xrightarrow{H_2} H_2O^+ \xrightarrow{H_2} H_3O^+ \xrightarrow{e} H_2O, OH$$

The ionization caused by cosmic rays is also a heat source for the cloud. Unless this energy, and also energy released in dynamical events such as collapse, is removed, then the cloud temperature and pressure will rise and the cloud will dissipate. The molecules provide important coolants at low temperatures, though their rotational emission; CO and its isotopomers, OH and H_2O, are some important coolants.

Collapse under gravity is a process that may transform a molecular cloud into a star-forming region, though this is resisted by internal gas pressure (hence the importance of cooling) and by magnetic pressure (acting on ions and electrons) which couple though collisions with neutrals. The frictional coupling between ions and neutrals determines the effectiveness of the magnetic support; it is called ambipolar diffusion, and is determined through the chemistry by the level of the fractional ionization in the gas. Normally, $n(\text{ions})/n(H_2)$ is in the range of 10^{-7} to 10^{-8}. The molecules produced by the chemistry are therefore important in cooling and controlling the ionization, but can be removed from the gas by sticking to the surfaces of grains. There may also be desorption from grains occurring continuously by various non-thermal processes, or intermittently when, for example, a star nearby warms up the gas in its vicinity.

Table 4: Processes and Timescales in molecular clouds

Process	Mechanism	Timescale (y) at $n_H = 10^4 cm^{-3}$
Chemistry	cosmic ray ionization	$\sim 3 \times 10^5$
freeze-out	gas/grain collisions	$\sim 3 \times 10^5$
Cooling	radiative	$\sim 10^6$
Collapse	gravity	$\sim 10^6$
ambipolar diffusion	ion-neutral drift	$\sim 4 \times 10^5$

The important conclusion that one can draw from Table 4 is that as the density of the cloud increases the process of freeze-out on to dust becomes dominant. The uncertainty in sticking probabilities affects this conclusion, but at some density the dominance of the gas/grain interaction must occur. Therefore, solid-state astrochemistry becomes extremely important in the denser parts of molecular clouds, i.e. in star-forming regions. The consequences of this solid-state astrochemistry may be significant. Reaction paths in heterogeneous chemistry may be possible where equivalent gas phase paths are forbidden. The formation of ices on dust grains removes molecules from the gas and affects the balance of the elements as well as the cooling. The processing of ices by electromagnetic and particle fluxes creates new molecules which can be liberated into the gas phase directly, or stored for later injection.

Star-formation affects the molecular cloud in which it is occurring. The formation of low mass stars (similar to the Sun) is accompanied by jets and winds, even at the very earliest - protostellar - stages. These winds are powerful enough to disrupt the dense clump from which the star formed and to re-distribute the gas in molecular clouds to form new clumps from which new stars may form. This type of "stirring" was discussed originally by Norman and Silk (1980) and has been observed in some detail in various clouds, most notably by Goldsmith et al.(1986) for the cloud Barnard 5 (or B5). In this object, it is possible to identify proto-stars at various states of evolution. IRS1 is a protostar still embedded in its natal clump. Others are emerging, as their winds disrupt the clump, and one object, IRS4, is located in a cavity created by its wind. The details of this wind/gas interaction in the case of B5 IRS1 have been explored by Velusamy and Langer (1998) who showed that the opening angle of the wind increases with time - presumably as the wind erodes the confining interstellar disk (see Figure 4).

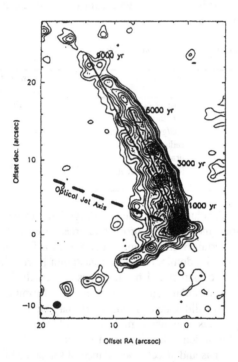

Figure 4: The outflow core from the proto-star IRS1 in the dark cloud Barnard 5 (B5). The contours are of emission intensity in CO 2-1. The extent and position of gas in the outflow are derived on the assumption that the outflow velocity is 6 km s^{-1}. The figure indicates that the opening angle of the core is widening (from Velusamy and Langer 1998).

The formation of massive stars (with masses around ten times larger than the solar mass) is accompanied by even more powerful winds and by intense radiation fields, perhaps a million times as intense as that of the Sun but peaking in the UV. These stars rapidly establish an ionized zone around themselves that is highly over-pressured compared to the cloud, so it expands rapidly driving the material from the cloud. This means that it is difficult to examine gas close to the star and to learn from that gas the history of the star formation process. However, there is a class of transient objects that do give some information - the so-called Hot Cores. These are small (≤ 0.1 pc) clumps associated with massive star formation. They are believed to be dense ($n_H \sim 10^7$ cm^{-3}) clumps of gas within which almost complete freeze-out and subsequent solid-state chemistry had occurred, before warming to ~ 200 K by the nearby star caused evaporation. The hot cores are unusually rich in large organic molecules and in small saturated molecules, and have a larger degree of deuterium fractionisation than is appropriate for their present temperature. The situation of hot cores close to the ultra-compact ionized regions set up by the new star ensures that these objects of unusual chemistry are transient. The detailed modelling (Millar 1993; Millar and Hatchell 1998) has been remarkably successful, and demonstrates that solid-state astrochemistry plays a key role in these objects.

4.   LOCATIONS OF SOLID-STATE ASTROCHEMISTRY: CIRCUMSTELLAR

4.1   COOL CIRCUMSTELLAR ENVELOPES

At late stages of evolution, stars of near-solar mass develop extended out-flowing envelopes that are chemically active. These envelopes may be oxygen-rich or carbon-rich. In the latter case, particularly, there is an extensive chemistry of organic and other molecules, and about 50 species have so far been identified. In oxygen-rich envelopes, the chemistry is more limited, and about 15 species, mostly inorganic, have been identified in them. These regions and their chemistry have been intensively studied and are now broadly understood (Millar 1998).

It is thought that the wind is generated by the star's inability to hold on to its atmosphere, so that it flows out to interstellar space with a speed observed to be uniform and about 10 km s^{-1}. The molecules present in the rather cool atmosphere (with temperatures around 2000 K) are simple stable species such as C_2, CH_4, C_2H_2 in the case of carbon-rich stars. As the material flows out into space, the ambient interstellar radiation penetrates the envelope and partially ionizes and dissociates these simple molecules. There is then an opportunity for new and more complex species to form from the interaction of the ions and the original atmospheric molecules. Detailed studies show that this model is a very good representation of the situation, and that excellent agreement between observations and theory can be obtained. However, the model description is almost entirely one of gas phase chemistry. Is there also a role for solid-state astrochemistry?

These envelopes are sources of dust, which is composed of carbonaceous or silicate materials. The processes of nucleation are rather poorly understood in either case, but it seems evident that chemistry, both gas-phase and heterogeneous, must be playing a part (Cherchneff 1998; Gail and

Sedlmayr 1998). Once dust is formed, it is carried with the outflow so that its shielding effect is diluted along with the gas density. It can, however, provide surfaces for heterogeneous chemistry. The appearance of CH_4 at about 125 stellar radii from one well-studied source (IRC + 10216) may be evidence for surface reactions (Nejad and Millar 1988; Keady and Ridgway 1993). The specific role of the dust in the chemistry of outflows has yet to be explored (Millar 1998).

4.2   PLANETARY NEBULAE

Eventually the cool outflowing envelope becomes detached from the star and moves steadily outwards into the interstellar medium, carrying with it the grains and molecules formed during the outflow phase. The star evolves to a new state: it contracts, and becomes much hotter, and develops an intense radiation field and a fast wind which impacts on the slow cool envelope in a complicated shock structure that gives rise to the diffuse optical emission by which planetary nebulae are recognised (Howe and Williams 1998). These nebulae are transient. After about 10,000 years, the shocked envelope drifts into interstellar space to mix with the interstellar gas, while the star cools to become a white dwarf.

Is there a role for solid-state astrochemistry in planetary nebulae? There is an opportunity for dust to be abruptly located in regions of high excitation, and for dust grains of different sizes to be accelerated differentially by shocks. Thus, Jura and Kroto (1990) have suggested that one may expect chemical products from the shattering and sputtering of the solid grains. It has been suggested that the large PAH molecule, chrysene $C_{18}H_{12}$, detected in the proto-planetary nebula CRL2688, may have arisen from the disruption of dust grains in shock conditions (Justtanont et al. 1996).

In the Helix nebula (NGC 7293) condensations of neutral gas are formed within the ionized region around the star, and have survived the passage of the fast shock. These objects contain dust and molecules, and are translucent to the stellar radiation. It seems evident that the dust in these globules contributes to their survival and to the chemistry that is maintained within them (Howe et al. 1994).

4.3   NOVAE AND SUPERNOVAE

The central star in a planetary nebula evolves into a white dwarf, a condensed star with strong gravity. If such a star is part of a binary system with another star which has an extended envelope, the white dwarf can drag material across from that envelope and compress it by gravity on the stellar surface sufficiently for it to ignite, in effect, as a hydrogen bomb. The result is a nova - the sudden brightening of a star by a factor of perhaps 10,000. From this emission peak, the nova declines in intensity back to its original state, usually over several years. During the early stages of the decline, some novae drop abruptly into a deep minimum for several months, and then recover. This decline is a signature of dust formation, and is accompanied by a rise in infrared emission from the warm dust. Evidently, in a period of only a few days, a complex chemistry can develop in the ejecta from the nova, in spite of the exceptionally harsh environment created by the nova's radiation field. This chemistry culminates in the formation of dust (Rawlings 1998). The relationship between the chemistry and the dust has been assumed to be that the chemistry provides molecules large enough to act as nucleation centres on which carbon grains may grow (Rawlings & Williams 1989). There are two separate epochs to be considered. Firstly, there is the phase of dust-free chemistry that forms simple molecules such as CO; from these intial species other molecules that can act as nucleation centres are formed. Secondly, there is the epoch of dust formation on these nucleation centres (e.g. Clayton and Wickramasinghe 1976).

Evidently, dust formation and chemistry are closely linked, at least in carbon-rich novae. No real understanding exists of the formation of silicates in oxygen-rich ejecta. Self-consistent chemistry and dust-formation models are now beginning to be explored (Pontefract and Rawlings 1999), but much work remains to be done.

Supernovae have some similarity to novae, in that a massive explosion occurs, generating conditions which may seem very harsh for the initial steps in chemistry (Liu 1998). Yet the infrared spectra of supernova SN1987A showed emission of CO in the 2-0 vibrational transition as early as 110 days after the explosion (Meikle et al. 1999) confirmed by the 1-0 transition from 157 days after the explosion. The emission from SiO 1-0 transition was observed from 160 days, and disappeared after 519 days (Bouchet and Danziger 1993), shortly before dust formed (530 days; Lucy et al. 1989; Danziger et al. 1991). Evidently, the SiO was being depleted into dust grains around this time, so a complex solid-state astrochemistry was taking place in the SN1987A ejecta within a year and a half of the explosion.

The chemistry of supernovae ejecta differs from that of interstellar clouds, principally because the ejecta are largely stratified, so that the shells containing hydrogen and helium are distinct from those containing oxygen and carbon. Thus, one needs to rely on a hydrogen-deficient chemistry. The ejecta are hot, dense, significantly ionized, and irradiated by fast electrons produced *in situ* by radioactive decay. Nevertheless, a complex chemistry can arise, the high densities overcoming the rather slow radiative associations that initiate the chemical network (Liu and Dalgaro 1994,

1995). The next steps to form dust, and the astrochemical role of the dust, remain to be fully explored.

4.4   COMETS

Comets are rich in dust and molecules, and are certainly sites where solid-state astrochemistry must be important.   The list of molecules identified in comet Hale-Bopp is very extensive, see Table 5. There is an excellent correspondence between molecules detected in interstellar ices and those in comets (Table 6) supporting the concept that comets have formed from largely unprocessed interstellar dust grains with icy mantles.

Table 5: Cometary parent molecules observed in comet Hale-Bopp, with their abundances

Molecule	Relative abundance	Molecule	Relative abundance
H_2O	100	HNC	0.03
CO	20	CH_3CN	0.02
CO_2	6	HC_3N	0.02
CH_4	~ 1	HNCO	0.1
C_2H_2	~ 0.5	NH_2CHO	0.01
C_2H_6	~ 0.5	H_2S	1.6
CH_3OH	2	SO	0.6
H_2CO	~ 1	SO_2	0.15
HCOOH	0.05	OCS	0.5
$HCOOCH_3$	0.05	CS_2	0.2
NH_3	0.6	S_2	0.005
HCN	0.2		

Abundances are given by number, relative to water. Table adapted from Crovisier (1998). S_2 was not observed in comet Hale-Bopp; the listed relative abundance refers to comet Hyakutake.

The role of solid-date astrochemistry in comets is concerned with the processing of the solid material under solar irradiation, and with the ejection of molecules from the solid-phase into the gas. The release of molecules from the solid may occur through sublimation, pyrolysis, and sputtering. Sublimation of small molecules should occur readily when grains are heated, and the products may be photodissociated to produce the radicals and molecules observed. Pyrolysis may occur in very small grains heated to a high temperature, and would be expected to give rise to small molecules such as CO and CN. Sputtering seems to be an unlikely process, given that the flux of energetic particles is low. The main effort in these areas is to understand the basic processes, and to apply them to particular comets in an attempt to account for the spatial distribution of molecules in the coma.

Table 6: Comparison of the compositions of interstellar and cometary ices

Species	Interstellar ices	Cometary ices	
H_2O	= 100	= 100	
CO	10-40	20	
CO_2	10	6	
CH_3OH	5	2	
H_2CO	2-6	~ 1	
HCOOH	3	0.05	
CH_4	1-2	~ 1	
other hydrocarbons	?	~ 1	$C_2H_2 + C_2H_6$
NH_3	< 10	0.6	
O_3	< 2	?	
XCN	< 0.5-10	0.37	nitriles + HNCO
OCS, XCS	0.2	0.7	$OCS + CS + H_2CS$
SO_2	?	0.15	
H_2	*ca.* 1	?	
N_2	?	?	
O_2	?	?	

All abundances are given by number, relative to water. Table adapted from Crovisier (1998).

5.   ACTIVE GALACTIC NUCLEI AND STARBURST GALAXIES

It was an interesting and surprising discovery that galaxies at very early epochs are rich in molecules and dust, and therefore potential sites of solid-state astrochemistry. Ohta et al. (1996) and Omont et al. (1996) reported the detection of CO and of dust continuum emission in the gravitationally lensed quasar BR1202-0725, at a redshift of $z = 4.69$. Since then, similar detections have been made for other galaxies, some even more distant. A redshift of 4.69 corresponds an age of the Universe of a few percent of its present value. Thus, chemistry in galaxies and consequent dust formation were established rather early in the history of the Universe.

The chemistry that took place at this early time would have significant differences from that in the Milky Way. The absolute elemental abundances would certainly be lower, and the relative abundances (especially of the crucial C:O ratio) may also differ from the galactic value. Even for nearby galaxies, the dust:gas ratio varies considerably, and this ratio at early times is likely to be quite different from the galactic value, and probably much smaller. Active and starburst galaxies are sources of intense electromagnetic and particle radiation which will have significant consequences on gas-phase chemistry, dust formation, and solid-state chemistry. All these effects may, for example, inhibit the growth of icy mantles as dust, and limit solid-state astrochemistry to heterogeneous reactions and grain disruption. Very little work in this area has yet been done.

A major uncertainty in trying to understand processes in galaxies at high redshift is in the calculation of the total mass of gas. Since H_2 is not observed directly, its abundance is estimated through observations of molecular lines and the abundances converted to H_2 by some assumed factor (Booth and Aalto 1998). The most commonly used molecule is CO but the conversion of the observed CO line strength into a column density of H_2 is fraught with difficulty. The conversion factor for the Milky Way Galaxy is uncertain, to some extent; it certainly cannot be expected to apply to very distant galaxies, but the Milky Way conversion factor is frequently used. This factor is certainly affected by the solid-state astrochemistry, through the contribution of heterogeneous chemistry in forming H_2, shielding of the gaseous molecules by dust from the radiation fields, and the freeze-out occurring in the gas/dust interaction. This is an area urgently requiring attention.

6.   CONCLUSION

The role of solid-state astrochemistry in interstellar clouds is well-established: it consists mainly of heterogeneous chemistry, and of the deposition, processing, and ejection of ices. These processes become dominant in regions of higher density where star-formation is occurring.

In circumstellar environments the role of solid-state astrochemistry has been less well explored. The grain formation processes have been studied, but more remains to be done in understanding nucleation and growth. The interaction of the dust formation and chemistry during this phase needs further study.

Most of the discussion in this chapter has been concerned with regions of the Milky Way Galaxy. Similar processes may be occurring in other galaxies, and there is evidence that chemistry and dust were abundant in galaxies even at very early times. It is clear that the physical parameters in such galaxies would be very different from those in the Milky Way, and it is important to understand how an efficient chemistry and dust formation can occur in such conditions.

REFERENCES

Booth, R.S. and Aalto, S. (1998), in *The Molecular Astrophysics of Stars and Galaxies*, Oxford Science Publications, Oxford, p.437.

Bouchet, P. and Danziger, I.J. (1993), *A&A*, **273**, 451.

Bradley, J.P., Keller, L.P., Snow, T.P., Hanner, M.S., Flynn, G.J., Gezo, J.C., Clemett, S.J., Brownlee, D.E. and Bowey, J.E. (1999), *Science*, **285**, 1716.

Clayton, D. and Wickramasinghe, N.C. (1976), *Astrophys. Spa. Sci.*, **42**, 463.

Crovisier, J. (1998), *Faraday Discussions*, **109**, 437.

Danziger, I.J., Lucy, L.B., Bouchet, P. and Gouiffes, C. (1991), in S.E. Woosley (ed.), *Supernovae*, Springer-Verlag, New York, p.69.

Goldsmith, P.F., Langer, W.D. and Wilson, R.W. (1986), *ApJ*, **303**, L11.

Howe, D.A. and Williams, D.A., in *The Molecular Molecular Astrophysics of Stars and Galaxies*, Oxford Science Publications, Oxford, p.347.

Jura, M. and Kroto, H.W. (1990), *ApJ*, **351**, 222.

Justtanont, K., Barlow, M.J., Skinner, C.J., Roche, P.F., Aitken, D.K. and Smith, C.H. (1996), *A&A*, **309**, 612.

Keady, J.J. and Ridgway, S.T. (1993), *ApJ*, **406** 199.

Kurtz, S. (2000), in Manning, V. et al. (eds), *Protostars and Planets IV*, University of Arizona Press, Tucson.

Liu, W. (1998), in *The Molecular Molecular Astrophysics of Stars and Galaxies*, Oxford Science Publications, Oxford, p.415.

Liu, W. and Dalgarno, A. (1994), *ApJ*, **428**, 769.

Liu, W. and Dalgarno, A. (1995), *ApJ*, **454**, 472.

Lucy, L.B., Danziger, I.J., Gouiffes, C. and Bouchet, P. 1989, in Tenorio-Tenagle, G., Moles, M. and Melnick, J. (eds.), *Proc. IAU Coll. 120, Structure and Dynamics of the Interstellar Medium*, Springer-Verlag, Berlin, p.164.

Mathis, J. (1990), *Ann. Rev. Astr. Astrophys.* , **28**, 37.

Meikle, W.P.S., Spyromilio, J., Allen, D.A., Varani, G.-F. and Cumming, R.J. (1993), *MNRAS*, **261**, 535.

Millar, T.J. (1993), in Millar, T.J. and Williams, D.A. (eds.), *Dust and Chemistry in Astronomy*, IoP Publishing, Bristol, p.249.

Millar, T.J. (1998), in *Molecular Astrophysics of Stars and Galaxies*, Oxford Science Publications, Oxford, p.331.

Millar, T.J. and Hatchell, J. (1998), *Faraday Discussions*, **109**, 15.

Moutou, C., Sellgren, K., Léger, A., Verstraete, L. and Le Coupanec, P. (1999), in d'Hendecourt, L., Joblin, C., Jones, A.P. (eds.), *Solid Interstellar Matter: the ISO Revolution*, Springer-Verlag, Berlin, p.89.

Nejad, L.A.M. and Millar, T.J. (1988), *MNRAS*, **230**, 79.

Norman, C. and Silk, J. (1980), *ApJ*, **238**, 138.

Ohta, K., Yamada, T., Nakanishi, K., Kohno, K., Akiyama, M. and Kawabe, R. (1996), *Nature*, **382**, 426.

Omont, A., Petitjean, P., Guilloteau, S., McMahon, R.C., Solomon, P.M. and Pontal, E. (1989), *Nature*, **382**, 428.

Pontefract, M. and Rawlings, J.M.C. (2000), in preparation.

Rawlings, J.M.C. (1998), in *The Molecular Astrophysics of Stars and Galaxies*, Oxford Science Publications, Oxford, p.393.

Rawlings, J.M.C. and Williams, D.A. (1989), *MNRAS*, **240**, 729.

van Dishoeck, E.F. (1998), in *The Molecular Molecular Astrophysics of Stars and Galaxies*, Oxford Science Publications, Oxford, p.53.

Velusamy, T. and Langer, W.D. (1998), *Nature*, **392**, 685.
Waters, L.B.F.M. and Waelkens, C., (1998), *Ann. Rev. Astron. Astrophys*, **36**, 219.
Whittet, D.C.B. (1990), in *Dust in the Galactic Environment*, IoP Publishing, Bristol.
Whittet, D.C.B., Schutte, W.A., Tielens, A.G.G.M., Boogert, A.C.A., de Graauw, T., Ehrenfreund, P., Gerakines, P.A., Helmich, F.P., Prusti, T. and van Dishoeck, E.F. (1996), *A&A*, **315**, L357.

IMPORTANT OPEN QUESTIONS IN ASTROCHEMISTRY: HOW CAN DUST HELP?

David A. Williams

Department of Physics & Astronomy, University College London, Gower Street, LONDON WC1E 6BT

Keywords: Dust - molecules - star forming regions

Abstract: Some of the most important open questions that are related with solid state astrochemistry are reviewed, ranging from the formation of molecular hydrogen to that of hydrides, to the formation of icy mantles on grains and their processing. Attention is paid to chemical processes occurring in star forming regions.

INTRODUCTION

In this article I shall divide the topic into two parts. Firstly, I shall discuss questions concerning fundamental astrochemical processes in which dust may play a role. Secondly, I shall consider questions of a more general nature in astronomy to which dust may contribute at least part of the solution. I shall conclude that dust grains play an important role in astronomy, but that there is still much to understand about the basic physical and chemical processes involving dust.

V. Pirronello et al. (eds.), Solid State Astrochemistry, 21–35.

1.   OPEN QUESTIONS IN ASTROCHEMISTRY

1.1   THE FORMATION OF MOLECULAR HYDROGEN

Molecular hydrogen is the most abundant molecule in the Universe. Within the Galaxy, perhaps half of the mass of interstellar gas is molecular hydrogen. While the destruction of H_2 by starlight is well-understood, its formation process remains one of considerable uncertainty.

The radiative association of two hydrogen atoms to form H_2 is very strongly forbidden indeed, and is not significant even over the very long timescales of astronomy. However, viable gas-phase reaction schemes do exist. One such scheme makes use of an electron

$$H + e^- \rightarrow H^- + h\nu$$

$$H^- + H \rightarrow H_2 + e$$

and there is a corresponding scheme using a proton

$$H + H^+ \rightarrow H_2^+ + h\nu$$

$$H_2^+ + H \rightarrow H_2 + H^+$$

These gas-phase schemes were probably important in providing H_2 in the early Universe, before the era of star formation and dust production began. The H_2 was the most important coolant in pre-galactic gas clouds, and played a part in encouraging the growth of inhomogeneities in the early Universe.

However, these gas-phase routes for H_2 formation are inadequate in the interstellar medium of the Galaxy, because the photodissociation of H_2 by starlight is rather rapid. In the unshielded interstellar radiation field in the Galaxy the mean lifetime of a lone H_2 molecule is only a few hundred years. This loss needs to be matched by an efficient formation process, and in the absence of sufficiently rapid gas-phase mechanisms attention turned to reactions on the surfaces of dust grains. The constraint of the observations, taken with the known population of dust grains causing the interstellar extinction, requires that nearly every H atom arriving at a grain surface in a diffuse cloud should leave as part of an H_2 molecule (see Duley and Williams 1984; Williams 1993).

The surface reaction has been examined in an important series of experiments by Pirronello, Vidali, and collaborators and this pioneering work is discussed in detail elsewhere in this volume (see Biham et al. 2002, this volume), and will not be discussed further here. This work has raised new questions about the microscopic behaviour of adsorbed atoms on surfaces at temperatures appropriate for interstellar dust. Other relevant experiments in France, Japan, and the UK are also under way (Williams et al. 2000).

Theoretical work on H_2 formation by surface chemistry is now of such sophistication and complexity that fairly reliable answers can be obtained. The most detailed calculation to date is that of Farebrother et al. (2000) which explores the interactions of an incident H-atom on another H-atom chemically bound to a C-atom in a graphite lattice. The interaction is assumed to be collinear. In this direct (or Eley-Rideal) interaction, the efficiency to form H_2 is found to be nearly unity, and the product molecule is vibrationally excited with many levels populated, with the population distribution peaking at level $v = 2$. A generalized version of this interaction, allowing for off-axis collisions, has also been completed (Meijer et al. 2002) and allows predictions of the rotational excitation of the product molecule.

Other kinds of reactions are also possible, in which both H atoms are adsorbed on the surface and (at least) one of them is mobile. This kind of interaction is called a Langmuir-Hinshelwood interaction. The nature of the surface plays a significant role in controlling the surface mobility of the H-atom. Mobility calculations have been completed by Farebrother and Clary (2001) for H atoms on a graphite lattice.

The conclusion is that definitive results for H_2 formation on dust are now being obtained. However, experiments will need to examine a range of materials and physical conditions, and the use of theoretical modelling to interpret the results is essential. As well as the efficiency of H_2 formation (comprising sticking, mobility, ejection and ejection of product), one needs to understand the energy budget and the excitation of H_2; all these remain open questions at present. However, we can look forward to rapid progress in the next few years.

1.2 HYDRIDE FORMATION ON SURFACES

Since hydrogen atoms apparently are combined efficiently into H_2 molecules on surfaces of interstellar dust grains, one may ask whether other atoms react similarly. Given the overwhelming abundance of hydrogen, then the most likely reaction is to form a hydride

$$X \rightarrow XH_n .$$

It is unclear at present, on the basis of chemistry, how efficient such processes may be under the physical conditions in the interstellar medium. However, there is some evidence from astronomy that such surface reactions play a role. The detection of the diatomic molecule NH in diffuse clouds along the lines of sight to several bright stars (Meyer and Roth 1991; Crawford and Williams 1997) has been interpreted as evidence of the surface hydrogenation of N-atoms (Wagenblast et al. 1993; Wagenblast & Williams 1993). Only a modest surface efficiency is required to account for the observed NH abundance; however, purely gas-phase routes fail to produce enough NH.

Similar ideas for hydrocarbon production in diffuse and translucent clouds have been explored by Viti et al. (2000) and a generalization to all species has also been carried out (Viti et al. 2001). Reactions of oxygen atoms to form water molecules on surfaces, with subsequent ejection of the H_2O can assist in the production of CO through the reactions

$$C^+ + H_2O \rightarrow HCO^+$$

$$HCO^+ + e^- \rightarrow H + CO$$

(Wagenblast and Williams 1993). Purely gas phase models of diffuse cloud chemistry fail in some cases to provide enough CO to match the observations.

Thus, it is plausible that hydrogenation of atoms at grain surfaces does occur, and there is observational evidence that tends to support this view. Little is known about the process, however. If hydrogenation occurs, does it go to complete saturation? At what stage, and how, is the hydrogenated molecule ejected? These are open questions, at present.

1.3   ONSET OF ICES

It is well known that interstellar H_2O ices are deposited on dust grains for lines of sight on which the visual extinction is above some critical value. In Taurus, the critical value is about 2.2 magnitudes whereas along other lines of sight the critical extinction for ice deposition to occur is greater, up to ten times larger. A similar behaviour is found for the deposition of CO ice, though here the critical values of extinction are larger than those for H_2O. What controls the onset of ice deposition? Why does the critical value of extinction vary significantly from one line of sight to another?

These are open questions, though several suggestions have been put forward (Williams, Hartquist and Whittet 1992). The simplest idea is that photodesorption by visible or ultraviolet light might be responsible for removing surface H_2O molecules before ice can form, up to some critical depth into the cloud. In fact, this seems unlikely because a variation by a factor 10 in the critical visual extinction would imply an enormous and unreasonable change in the incident flux. However, it is worth noting that photodesorption efficiencies have been rather poorly determined, and revised values are significantly larger than had been previously thought (see Section 2.5, below). It is also possible that infrared radiation may affect adsorbed lone H_2O molecules in a different way from H_2O molecules in bulk.

Other ideas for inhibiting H_2O-ice formation depend on surface chemistry. It seems possible that at some critical depth within a cloud, the abundance of some species that react on a surface is reduced sufficiently so that any side-effects of the reaction - such as ejection of a nearby H_2O-molecule - are suppressed. Duley and Williams (1993) considered the possibility that the reaction of atomic hydrogen to form H_2 might inadvertently remove any H_2O molecules in the vicinity. It is unclear, however, that sufficient energy is deposited into a surface to cause desorption of nearby molecules (Takahashi and Williams 2000).

1.4   ICE DEPOSITION

One of the findings of the SWAS mission (Melnick et al. 2000) is that H_2O is not an abundant gas phase molecule. This is consistent with theoretical studies which show that in low temperature environments the abundance of water in the gas phase remains low except for rather high density gas and for late times.

The H_2O-ice that is observed must be deposited within the age of the cloud; and for normal cold cloud conditions the H_2O abundance in the gas is too low to meet this constraint (Jones and Williams 1984). It has been shown that H_2O formed in the transient warm gas heated by shocks within dark clouds could be deposited on grains to form ice (Bergin et al. 1998) but O'neill and Williams (1999) have pointed out that for reasonable values of extinction the ion-molecule chemistry initiated by starlight in the gas destroys the H_2O before appreciable deposition can occur.

Hence, the inference is that water ice deposited on interstellar dust grains is formed by *in situ* hydrogenation of incident O-atoms. The studies have shown that if a reasonable fraction of oxygen is converted into ice, then the required timescale for ice deposition is compatible with the presumed age of the clouds. However, this process has not been studied in the laboratory, and the efficiency of hydrogenation and the possibility of desorption occurring at intermediate stages is unknown.

1.5   FREEZE-OUT

The freeze-out timescale varies inversely with density. As density increases, the freeze-out timescale becomes shorter than other relevant timescales of chemistry, collapse, cooling, or ambipolar diffusion (see Table 3 of Chapter 1). Hence, freeze-out of atoms and molecules on to dust will dominate all other processes in the highest density regions of interstellar space. These tend to occur in star-forming regions.

Even if one assumes that all species stick to grains with the same efficiency, selective effects occur in the chemistry. The most obvious of these is associated with hydrocarbon formation. Gaseous CO is dissociatively ionized by He^+, and normally the C^+ ions are quickly recycled into CO through reactions with OH or H_2O, or other molecules. However, if such molecules are reduced in abundance by freeze-out, then C^+ reacts preferentially in a rather slow process with the abundant H_2 and generates a hydrocarbon chemistry (Rawlings et al. 1992). How these effects occur in relation to density changes is controlled by the adopted sticking probability. It is usually assumed that the sticking probabilities are unity for all species other than hydrogen and helium. The interaction of ions with dust is treated somewhat differently, on the assumption that many of the grains are negatively charged so that the ion-grain interaction is enhanced by electrostatic effects (Umebayashi and Nakano 1980). The possible outcomes of the interaction are unknown, and require investigation. It is often assumed that ions will stick to charged grains, and that some selective effects in the chemistry may arise as a consequence (e.g. Ruffle et al. 1999).

Of course, it is unlikely that all neutral species stick with the same probability on dust grains. The physical and chemical nature of the surface and the electronic and kinetic state of the incident atom, radical, or molecule must influence the interaction considerably. The significance of such influences is, however, an open question at present. The information required must be obtained from experiments (Fraser and McCoustra 2001).

The question therefore arises: is freeze-out within molecular clouds a one-way process that is only reversed by intermittent star formation, or are there other mechanisms that operate continuously to keep at least some heavy molecules in the gas phase? There is now considerable observational evidence that depletion varies from point to point within a molecular cloud (Gibb and Little 1998); does this variation reflect local conditions within the cloud?

Several mechanisms that may drive desorptions from dust have been discussed, and all depend on the injection of energy locally into the surface, creating a transient hot spot from which molecules may desorb. The simplest is direct photodesorption by starlight (or other UV radiation). Although this process has largely been discounted on the assumption that the photodesorption efficiency has been considered to be low, more recent determinations of H_2O desorption (Westley et al. 1995) may call for a reassessment.

Other energy sources are cosmic rays (Léger et al. 1985) and chemical energy (Allen and Robinson 1975: Duley and Williams 1993). The proposals have been evaluated in the context of time-dependent chemistry in molecular clouds (Willacy et al. 1994).

A small energy input may lead to a modest rise in temperature which promotes the evaporation only of weakly bound molecules such as CO and N_2. Larger temperature fluctuations may be less discriminating so that any species in the ice may be desorbed. In either case, the gas-phase chemistry associated with the desorbed molecules is considerably prolonged, though at a lower level of abundance. For example, if desorption of CO and N_2 occurs, then reaction with He^+ provides a source of atomic C, O, and N, and ions from which new molecules will form and - ultimately - freeze-out. The prolongation can be for a significant fraction of the likely age of the cloud (see Figure 1, courtesy Dr S. Viti). Of course, the elemental abundances feeding the chemistry are somewhat different in this phase. For example, the C:O ratio for species in the gas phase is close to unity, since these atoms have mainly been derived from desorbed CO.

Therefore, it seems clear that desorption is occurring and that its consequences for interstellar chemistry are significant. However, the extent to which it occurs is unclear and the relative effectiveness by photodesorption, cosmic rays, and surface chemistry remains to be determined.

1.6   PROCESSING OF ICES

The balance of deposition and desorption determines the extent of ice mantles on the dust grains. Chemistry within the ice can be promoted by electromagnetic and fast particle radiation, and fairly large molecules can be formed in a solid solution of simple large species. This area is one of the most fully developed in solid state astrochemistry, and much of this work has been carried out in Leiden. The subject is described in detail in Ehrenfreund and Fraser (2002, this volume), and will not be discussed further here.

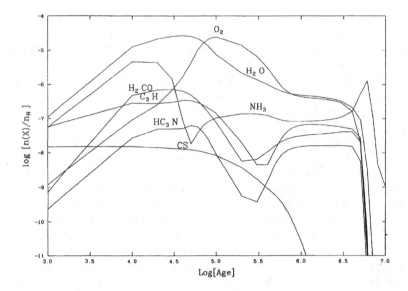

Figure 1. Computed fractional abundances within dense gas, as a function of time (figure courtesy of Dr S. Viti). The gas-phase chemistry is supplemented in this model by surface processes that inhibit partially the effects of freeze-out. These surface processes include the hydrogenation of atoms followed by immediate ejection of the products, and the immediate return to the gas-phase of any CO and N2 striking the grain. The consequence of this type of model is the prolongation of an active gas-phase chemistry up to about 6 My, whereas the freeze-out time is ~ 1 My.

2.   OPEN QUESTIONS IN ASTRONOMY

2.1   CHEMISTRY IN COLLAPSING CORES: THE INITIAL STAGES OF STAR FORMATION

Collapsing cores remain cool during much of the pre-stellar phase, and molecules are consequently the best tracers of the physical conditions in the gas (via emission in their rotational transitions) during this important phase of the star formation process. It is, therefore, essential to understand the interplay between chemistry and collapse in regions where density is increasing indefinitely. In principle, a characteristic signature for infall is a double-peaked line profile with the "blue" peak (i.e. the shorter wavelength peak) stronger than the "red", for an optically thick line. In fact, this signature is only detected in some cases. For example, Myers and Benson

(1983) showed that the line profile of NH_3 (assumed to be a good tracer of high density gas) in a core L1498 in which collapse was believed to be occurring was in fact a single-peaked profile corresponding in width to little more than the thermal temperature of 10 K. How could the evidence of collapse be hidden in the line profile of such a molecule?

Menten et al (1984) suggested that molecules in the densest and most rapidly moving gas are depleted by freezing out on to dust grains. Rawlings et al. (1992) explored the time-and-space-dependence of molecular abundances within a core collapsing according to the simple dynamical model of Shu (1977). The physical parameters were chosen to model L1498, and the results confirmed the plausibility of the suggestion of Menten et al. (1984) that NH_3 did not exhibit a broad or double-peaked profile because it was depleted in those regions where the infall velocities were greatest. The work of Rawlings et al. (1992) also demonstrated that other species should have broader profiles than NH_3, and that these species maintained (or even enhanced) their abundances in the deeper, denser, and faster-moving gas. This selective chemistry is directly attributable to the gas-dust interaction. As molecules such as H_2O are removed from the gas, ions that could react with abundant H_2O are now able to interact with other chemicals, and enhance the abundances of other species. Thus, even as molecules are being lost from the gas phase, its chemistry is - transitorily, of course - being modified because of the gas-dust interaction.

These special chemical effects arise because of the comparability of three important timescales within the dense core gas: the timescale for gas-phase ion-molecule chemistry driven by cosmic ray ionization; the timescale for freeze-out; and the collapse timescale. The first of these is simply the time for enough hydrogen to be ionized by cosmic rays to react with the main elements C, N, and O. This is independent of density, and is about 10^6 years. The second timescale depends on both density and the sticking coefficient, as discussed in Section 2.5, above, and is approximately $10^9/n_HS$) years, where $n_H(cm^{-3})$ is the total hydrogen number density and S is the sticking probability. The third timescale is determined by the mode of collapse, and for unrestricted free-fall is about $10^8/(n_H)^{1/2}$ years. It is, therefore, possible to imagine scenarios in which each of the three is the dominant process. The case of L1498 suggests that freeze-out occurs on a timescale shorter than that for dynamics. In fact, there is a good possibility that observational data can be used to help determine some of the important freeze-out parameters. Many molecules can be detected in collapsing cores, and several lines of each can be observed. Hence, there is a wealth of information available. Rawlings and Yates (2001) have shown how sensitive the line profiles are to the parameters involved.

Dust is important in the study of collapsing cores because thermal emission from dust can be used to constrain the density and temperature profile within the core. Of course, there are no chemical consequences to this thermal emission, but the information obtained from dust emission is a necessary starting point for detailed chemical modelling (Rawlings and Evans 2002).

2.2   MOLECULES ASSOCIATED WITH HERBIG-HARO OBJECTS

Herbig-Haro objects are small regions of diffuse emission often associated with star-forming regions. Their origin is in the interaction of stellar jets with ambient gas clouds in a shock that creates the emitting hot ($\sim 10^4$K) gas. Sometimes the objects are close to the young stellar object and the association is obvious. In other cases the emission may be quite far, ~ 1 parsec, from the star, and the association can be found by showing the alignment of several emission regions (Bachiller 1996).

In recent years, it has been found that small clumps of molecular emission (especially NH_3 and HCO^+) may be located close to the Herbig-Haro object. These clumps are quiescent, cool, and share the velocity of the ambient interstellar gas rather than of the jet or the H-H region (Rudolf and Welch 1988; Torrelles et al. 1992, 1993; Girart et al. 1994). It appears that these emission clumps are part of the ambient gas that is affected by the strong radiation field from the H-H shock that shines into gas that would otherwise be dark (Girart et al. 1994). This radiation helps to evaporate icy mantles on dust and to promote a photochemistry that generates transiently high abundances of certain molecules. Theoretical studies (Taylor and Williams 1996) have demonstrated the plausibility of this situation, and on this basis the stellar jet interaction is providing a new probe of the molecular cloud, in particular of its clumpiness and of the gas:dust interaction within those clumps.

More detailed models (Viti and Williams 1999a) have explored the extensive chemistry that arises once the photon source is switched on, and have predicted that detectable levels of abundance should be possible for many molecules. These predictions are largely confirmed by observations by Girart et al. (2001) of CH_3OH, H_2CO, CS, HCN, SO, as well as the molecules NH_3 and HCO^+ detected earlier.

In summary, the transient radiation source enables us to probe the structure and chemical content of molecular clouds in a new way. It is evident that the clouds are clumpy, and that within a clump significant amounts of material are frozen on to dust. The release of this material into the gas, and the pressure of some ionizing photons, creates a photochemistry which leads to the observed molecules. Evidently, the gas: dust interaction is helpful in elucidating the clumpy structure of molecular clouds. Such clumps may be transient or stable, or they may be in the initial stages of star formation. The origin and stability of the clumpiness is an open question in astronomy, but the gas:dust interaction is providing some insight.

2.3 ANOMALOUS MOLECULAR ABUNDANCES IN STAR-FORMING CORES

B335 is a star-forming core with an embedded young stellar object at its centre. It is approximately spherical, and shows evidence of infall in the characteristic line profiles showing double-peaked structure. One of the molecules showing this profile is HCO^+.

However, the HCO^+ abundance is large, about an order of magnitude larger than normally found within dense interstellar material. It is difficult to account for this high abundance on the basis of quasi-static chemical models, and its origin is an interesting open question in astrochemistry.

One possible solution concerns the dynamics of the region. The detected infall is probably accompanied by an outflow. If so, then the interaction between the infall and the outflow may be highly energetic, as outflow velocities are typically on the order of 100 km s^{-1}. When the outflow encounters inhomogeneities in the infall, the interaction should generate powerful shocks that stimulate a photochemistry similar to that ahead of an HH object. A theoretical study of such interactions has shown that an order of magnitude enhancement in HCO^+ over the level normally found in dense material, can be obtained (Rawlings et al. 2000). The interaction zones should be located at the outflow/infall interface, and in simple cases may have a conical morphology. Observations by Hogerheide et al. (1998) indicate that HCO^+ emission may be occurring with this morphology in the dense star-forming core L15270).

On this interpretation, HCO^+ should be regarded as a potential indicator of regions of high excitation, in which cold, dark, molecular cloud material with depleted gas phase abundances is abruptly exposed to radiation. This signal, however, depends on the role of grains as carriers of ice which is the seed of the photochemistry.

2.4 MOLECULES IN HOT CORES

Hot cores are distinct from the general interstellar molecular gas. They are dense, warm, compact regions found in association with young massive stars. They have a different chemical composition from that in quiescent molecular clouds, being rich in saturated molecules. Their existence poses two questions for the astrochemist: why do hot cores have a different chemistry from the ambient molecular cloud? What can we learn from that difference? These questions, at one time "open" questions in astrochemistry, have now largely been answered as an application of solid state astrochemistry.

An example of a hot core is shown in Figure 2 (Kurtz et al. 2000). The dotted contours indicate the 1.3 cm emission from ionized hydrogen around a massive star. Adjacent to this ultracompact HII region are contours of CH_3CN emission in the 6.5 rotational transition; these contours trace the hot core. Embedded in the hot core are OH masers, and their presence suggests that the core contains a high mass star. Maps in the lines of other molecules have also been made. The contours of CH_3OH emission overlap the CH_3CN emission, but extend beyond it (Pratap et al. 1999).

The generally accepted origin of hot cores is that they are parts of the collapsing molecular cloud that were not incorporated into the newly formed massive star. The density in these regions close to the star formation site was high and freeze-out was rapid. Therefore, nearly all species more massive than hydrogen and helium were frozen on to dust grains. When the star began to radiate, the whole region was suddenly warmed, and the molecules that had been in ices were now rapidly returned to the gas phase. Processing within the ice, as solid-state chemistry, may also have occurred. Therefore, all unsaturated species - common in quiescent interstellar clouds - were saturated by hydrogen addition in the ices. Deuteration was also favoured at the low temperatures of interstellar ice. The sudden evaporation of the ices therefore populates the region with molecules that are characteristic of a solid-state low-temperature astrochemistry. Eventually, reactions in the gas phase will try to restore the chemistry to its more normal interstellar form, so the anomalous chemistry of hot cores is transient, with a lifetime of some tens of thousands of years.

This picture of hot cores is well established by many observational and theoretical studies (e.g. Millar 1993; Walmsley and Schilke 1993; Millar and Hatchell 1998), and the particularly rich chemistry of hot cores is now recognised as a consequence of solid-state astrochemistry, followed by thermal evaporation of the products. These objects are therefore a direct motivation for the study of ice processing in the laboratory.

Hot cores are important from the point of view of astronomy, too. They provide a sample of material whose chemical composition reflects the physical conditions in the cloud prior to star formation. They also give some information about the earliest stages of the newly-formed star. For example, all young stars appear to have winds, but do these winds begin before or after the star begins to radiate? Such winds should introduce shocks into the neighbouring hot core, and these shocks may leave observable chemical signatures (Charnley and Kaufman 2000; Viti et al. 2001). How does a star begin to radiate? Does it attain "full power" instantaneously or over a period? One possible way to address this question is to ask whether "young" and "old" hot cores would appear the same, if the warming of the grains (and ice evaporation) took place over a period that was comparable with the duration of a hot core. Detailed studies (Viti and Williams 1999b) show that it may be possible to determine the "turn-on" time of a young star from changes in the chemistry of a neighbouring hot core. In fact, it may be that the differences in the CH_3CN and CH_3OH contours in G29.96-002 could be accounted for in this way (Macdonald et al. 2002).

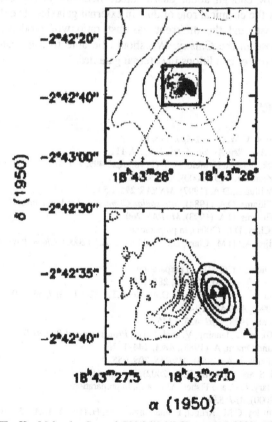

Figure 2. The Hot Molecular Core of G29.96-0.02 (Kurtz et al. 2000). The dotted contours are of 1.3 cm continuous emission from the ultracompact HII region around the young hot star. The continuous contours are of CH_3CN emission in the 6-5 rotational transition. These contours define the position of the hot core. Within the hot core are OH masers, indicate here by triangles. The presence of the masers suggests that there is a high mass star within this core.

3.   CONCLUSION

The interaction of gas and dust is important in many regions of the Galaxy. The most fundamental chemical role of dust is the formation of H_2, the seminal molecule for all gas-phase chemistry. In denser regions of interstellar space, its role in providing surfaces for ice deposition and the opportunity for solid-state chemistry becomes very important. Dust and ice when placed in energetic regions generate characteristic molecular tracers; for example, enhanced HCO^+ abundances may be a signature of violent mixing of cool and hot gas.

We now know enough about dust-related chemistry in the Galaxy to begin to consider the chemical role of dust in external galaxies. The different nature of the dust, and the different gas:dust ratio will produce effects distinct from those in the Galaxy. We should be able to use dust-related chemistry as a probe of conditions in external galaxies.

REFERENCES

Allen, M. and Robinson, G.W. (1975), *ApJ*, **195**, 81.
Bachiller, R. (1996), *Ann. Rev. Astron. Astrophys.*, **34**, 111.
Bergin, E.A., Melnick, G.J. and Neufeld, D.A. (1998), *ApJ*, **499**, 777.
Charnley, S.B. and Kaufman, M.J. (2000), *ApJ*, **529**, L111.
Crawford, I.A. and Williams, D.A. (1997), *MNRAS*, **291**, L53.
Duley, W.W. and Williams, D.A. (1984), *Interstellar Chemistry*, Academic Press, London.
Duley, W.W. and Williams, D.A. (1993), *MNRAS*, **260**, 37.
Farebrother, A. and Clary, D.C. (2001), in preparation.
Farebrother, A., Meijer, A.J.H.M., Clary, D.C. and Fisher, A.J. (2000), *Chem. Phys. Lett.*, **319**, 303.
Fraser, H. and McCoustra, M. (2001), in preparation.
Gibb, A.G. and Little, L.T. (1998), *MNRAS*, **285**, 299.
Girart, J.M., Estalella, R., Viti, S., Williams D.A., Ho, P.T.P. (2001), *ApJ*, **562**, L91.
Girart, J.M. et al. (1994), *ApJ*, **435**, L145.
Hogerheide, M.R. et al. (1998), *ApJ*, **435**, L145.
Kurtz, S. et al. (2000), in V. Manning, V. et al. (eds.), *Protostars and Planets IV*, p.299.
Léger, A., Jura, M. and Omont, A. (1985), *A&A*, **144**, 147.
Jones, A.P. and Williams, D.A. (1984), *MNRAS*, **209**, 955.
Macdonald, G., Viti, S. and Williams, D.A. (2002), in preparation.
Meijer, A.J.H.M., Clary, D.C. and Fisher, AJ (2002), in preparation.
Melnick, G. et al. (2000), *ApJ*, **539**, L77.
Menten, K.M., Walmsley, C.M., Krügel, E. and Ungerechts, H. (1984), *A&A*, **137**, 108.
Meyer, D.M. and Roth, K.C. (1991), *ApJ*, **376**, L49.
Millar, T.J. (1993), in Millar, T.J. and Williams, D.A. (eds.), *Dust and Chemistry in Astronomy*, Institute of Physics Publishing, Bristol, p.249.
Millar, T.J. and Hatchell, J. (1998), *Faraday Discussions*, **109**, 15.
Myers, P. and Benson, P.J. (1983), *ApJ*, **266**, 309.
O'neill, P.T. and Williams, D.A. (1999), *Ap. Space Sci.*, **266**, 539.

Pratap, P., Megeath, S.T. and Bergin, E.A. (1999), *ApJ,* **517**, 799.

Rawlings, J.M.C. and Evans, N. (2002), in preparation.

Rawlings, J.M.C., Hartquist, T.W., Menten, K.M. and Williams, D.A. (1992), *MNRAS,* **255**, 471.

Rawlings, J.M.C., Taylor, S.D. and Williams, D.A. (2000), *MNRAS,* **313**, 461.

Rawlings, J.M.C. and Yates, J. (2001), *MNRAS,* **326**, 1423.

Rudolf, S.A. and Welch, W.J. (1992), *ApJ,* **395**, 488.

Ruffle, D.P., Hartquist, T.W., Caselli, P. and Williams, D.A. (1999), *MNRAS,* **306**, 691.

Shu, F. (1977), *ApJ,* **214**, 488.

Takahashi, J. and Williams, D.A. (2000), *MNRAS,* **314**, 273.

Taylor, S.D. and Williams, D.A. (1996), *MNRAS,* **282**, 1343.

Torrelles, J.M., Gómez, J.F., Ho, P.T.P., Anglada, G., Rodriguez, L.F. and Canto, J. (1993), *ApJ,* **417**, 655.

Torrelles, J.M., Rodríguez, L.F., Cantó, J., Anglada, G., Gómez, J.F., Curiel, S. and Ho, P.T.P. (1992), *ApJ,* **396**, L95.

Umebayashi, T. and Nakano, T. (1980), *PASJ,* **32**, 401.

Viti, S., O'neill, P.T. and Williams, D.A. (2001), in preparation.

Viti, S. and Williams, D.A. (1999a), *MNRAS,* **310**, 517.

Viti, S. and Williams, D.A. (1999b), *MNRAS,* **255**, 471.

Viti, S., Williams, D.A. and O'neill, P.T. (2000), *A&A,* **354**, 1062.

Wagenblast, R. and Williams, D.A. (1993), in Millar, T.J. and Williams, D.A. (eds.), *Dust and Chemistry in Astronomy,* Institute of Physics Publishing, Bristol, p.171.

Wagenblast, R. and Williams, D.A., Millar, T.J. and Nejad, L.A.M. (1993), *MNRAS,* **260**, 420.

Walmsley, C.M. and Schilke, P. (1993), in Millar, T.J. and Williams, D.A. (eds.), *Dust and Chemistry in Astronomy,* Institute of Physics Publishing, Bristol, p.37.

Westley, M.S., Baragiola, R.A., Johnson, R.E. and Baratta, G.A. (1995), *Nature,* **373**, 405.

Willacy, K., Williams, D.A. and Duley, W.W. (1994), *MNRAS,* **267**, 949.

Williams, D.A. (1993), in Millar, T.J. and Williams, D.A. (eds.), *Dust and Chemistry in Astronomy,* Institute of Physics Publishing, Bristol, p.143.

Williams, D.A. et al. (2000), in Combes, F. and Pineau des Forêts, G. (eds.), H_2 *in Space,* Cambridge University Press, p.99.

Williams, D.A., Hartquist, T.W. and Whittet, D.C.B. (1992), *MNRAS,* **258**, 599.

IN DUST WE TRUST:
AN OVERVIEW OF OBSERVATIONS AND
THEORIES OF INTERSTELLAR DUST

Aigen Li
Princeton University Observatory, Peyton Hall, Princeton, NJ 08544, USA;
and Theoretical Astrophysics Program, University of Arizona, Tucson, AZ 85719, USA

J. Mayo Greenberg
The Raymond and Beverly Sackler Laboratory for Astrophysics, Sterrewacht Leiden,
Postbus 9513, 2300 RA Leiden, The Netherlands

Keywords: Interstellar dust - extinction - polarization - Comets

Abstract   The past century of interstellar dust has brought us from first ignoring it
to finding that it is an important component of the interstellar medium
and plays an important role in the evolution of galaxies, the formation
of stars and planetary systems, and possibly, the origins of life. Current
observational results in our galaxy provide a complex physical and chemical evolutionary picture of interstellar dust starting with the formation
of small refractory particles in stellar atmospheres to their modification
in diffuse and molecular clouds and ultimately to their contribution to
star forming regions. In this review, a brief history of the studies of interstellar dust is presented. Our current understanding of the physical
and chemical properties of interstellar dust are summarized, based on
observational evidences from interstellar extinction, absorption, scattering, polarization, emission (luminescence, infrared vibrational emission,
and microwave rotational emission), interstellar depletions, and theoretical modelling. Some unsolved outstanding problems are listed.

Introduction

It has been over 70 years since the existence of solid dust particles in
interstellar space was first convincingly shown by Trumpler (1930) based
on the discovery of color excesses. Interstellar dust has now become a

V. Pirronello et al. (eds.), Solid State Astrochemistry, 37–84.

subject of extensive study and one of the subjects in the forefront of astrophysics.

> "The role of dust is that of observer and of catalyst."
> — J. Mayo Greenberg [1963]

Historically, interstellar dust was regarded by astronomers as an annoying interstellar "fog" which prevented an accurate measurement of distances to stars. About 40 years ago, one of us (J.M.G.) wrote in the first volume of *Annual Review of Astronomy and Astrophysics* (Greenberg 1963):[1] "... Among the various performers of our Galaxy – the stars, the gas clouds, the cosmic rays – the grains seem to be the least dramatic. Their role is generally that of observer and of catalyst of events rather than prime mover. Why then are we so interested in these small particles whose total mass, by the most generous estimate, is only of the order of 1 per cent of that of the gas clouds? ... [It is because of] the three important activities of the grains: (a) the negative one of extinction [blocking the light from distant stars]; (b) the positive one of tracer of physical conditions [e.g. the Galactic magnetic fields and the gas temperature]; and (c) physical interactions with other components of the interstellar medium [the formation of molecules and stars]."

> "We now recognize, dust plays a role not only as a tracer of what goes
> on in space, a corrector for making different modifications in our idea
> of the morphology of galaxies, but also actively contributing to the
> chemical evolution of molecular clouds."
> — J. Mayo Greenberg [1996]

It is seen now that the role of interstellar dust was significantly underestimated 40 years ago. The advances of infrared (IR) astronomy, ultraviolet (UV) astronomy, laboratory astrophysics, and theoretical modelling over the past 40 years have had a tremendous impact on our understanding of the physical and chemical nature, origin and evolution of interstellar grains and their significance in the evolution of galaxies, the formation of stars and stellar systems (planets, asteroids, and comets), and the synthesis of complex organic molecules which possibly leads to the origins of life.

> "Dust is both a subject and an agent of the Galactic evolution."
> — J. Dorschner & Th. Henning [1995]

Instead of being a passive "observer", interstellar dust plays a vital role in the evolution of galaxies. Besides providing $\sim 30\%$ of the total Galactic luminosity via their IR emission, dust grains actively participate in the cycle of matter (gas and dust) from the interstellar medium (ISM) to stars and back from stars to the ISM: (1) solid grains condense in the cool atmospheres of evolved stars, Wolf-Rayet stars, planetary nebulae, and novae and supernovae ejecta, and are then ejected into the diffuse ISM; (2) in the diffuse ISM, interacting with hot (shocked) gas, stellar UV radiation, and cosmic rays, grains undergo destruction (sputtering by impacting gas atoms, vaporization and shattering by grain-grain collisions; see Tielens 1999 for a review); (3) in molecular clouds (formed through shock compression of diffuse gas, agglomeration of small clouds and condensation instabilities), grains are subject to growth through accretion of an ice mantle and coagulation; (4) cycling between the diffuse and molecular clouds, dust grains either form a carbonaceous organic refractory mantle as a result of UV processing of the ice mantle accreted on the silicate core in molecular clouds (Greenberg et al. 1972; Greenberg & Li 1999a), or perhaps grains re-condense in dense regions followed by rapid exchange of matter between diffuse gas and dense gas (Draine 1990); (5) collapse of dense molecular clouds leads to the birth of new stars. At the late stages of stellar evolution, gas and newly formed dust will eventually return to the ISM either through stellar winds or supernova explosions. It is clear that the life cycle of dust is associated with that of stars; stars are both a sink and a major source for the Galactic dust.

In addition to the fact that stars form out of interstellar dust and gas clouds, dust plays an important role in the process of star formation in molecular clouds: (1) IR emission from dust removes the gravitational energy of collapsing clouds, allowing star formation to take place; (2) dust grains provide shielding of molecular regions from starlight and thereby reduce the ionization levels and speed up the formation of protostellar cores (Ciolek 1995);[2] (3) IR emission from dust provides an effective probe for the star-formation processes (Shu, Adams, & Lizano 1989).

> "The importance of grains in various aspects of astrochemistry is evident: they shield molecular regions from dissociating interstellar radiation, catalyze formation of molecules, and remove molecules from the gas phase."
> — E.F. van Dishoeck, G.A. Blake, B.T. Draine, & J.I. Lunine [1993]

The interstellar chemistry problem concerns the chemical reactions between dust and atoms and molecules in space; dust plays an active role in those reactions (see van Dishoeck et al. 1993, van Dishoeck 1999 for reviews): (1) grain surfaces provide the site for the formation of molecular hydrogen (see Pirronello 2000 for a review), and probably other simple molecules through grain surface reactions and complex organic molecules through UV photoprocessing (e.g. see Greenberg et al. 2000; Allamandola 2002); (2) dust reduces the stellar UV radiation and protects molecules from photodissociation; (3) dust provides the major heating source for interstellar gas – photoelectrons ejected from grains; (4) dust grains are also involved in ion-molecule chemistry by affecting the electron/ion densities within the cloud.

Dust is one of the basic ingredients in comets. There is growing evidence from cometary observations that comets are a storage place for products of the chemical evolution which takes place in interstellar space. The complex chemistry and molecular evolution leading to what is now seen in comets may be the necessary precursor to life on the Earth and there is reason to believe from the evidence available that the oceans on the Earth were made of comets bringing interstellar ice to our young planet (Ehrenfreund & Charnley 2000).

In this review, we start in §1 with a historically oriented discussion of the discovery of interstellar extinction and dust, the development of early dust models as well as current modern models. Following up in §2 we present the present state-of-the-art understanding of interstellar dust observations and theories. In §3 we present a personal perspective of future dust studies.

1.   HISTORY OF DUST STUDIES

1.1.   INTERSTELLAR DUST: EARLY OBSERVATIONAL EVIDENCES

"Surely, there is a hole in the heavens!"
— Sir William Herschel [1785]

The Milky Way looks patchy with stars unevenly distributed: looking at the sky in the direction of Sagittarius it is clear that there are tremendously dark lanes especially in the region toward the Galactic Center. The subject of these dark patches and what makes them dark also have a very patchy history. The existence of dark regions in the Milky Way was first pointed out by Sir William Herschel in the late 18th century. At that time, these dark lanes – the dust clouds which obscure the light from the background stars, were considered as "holes in the heavens" (Herschel 1785).

"They are really Obscuring bodies!"
— Agnes Clerke [1903]

In August 1889, Edward Barnard started to take pictures and reported vast and wonderful cloud forms with their remarkable structure, lanes, holes and black gaps. At the beginning of the 20th century, astronomers started to realize that they "were really obscuring bodies" rather than holes devoid of stars (Barnard 1919). Agnes Clerke (1903) stated in an astrophysics text that "... The fact is a general one, that in all the forest of the universe there are glades and clearings. How they come to be thus diversified we cannot pretend to say; but we can see that the peculiarity is structural — that it is an outcome of the fundamental laws governing the distribution of cosmic matter. Hence the futility of trying to explain its origin, as a consequence, for instance, of the stoppage of light by the interposition of obscure bodies, or aggregations of bodies, invisibly thronging space."

Heber D. Curtis and Harlow Shapley[3] held a famous debate in 1920 (Shapley & Curtis 1921); among the points of contention was whether what is seen as the dark lanes in the Milky Way is caused by obscuring material. Curtis said the dark lanes observed in our Galaxy were obscuring material, while Shapley said he found no evidence of obscuring material in his observations of globular clusters. Later observers became aware that Shapley's argument was irrelevant because the globular clusters are out of the plane of the Galaxy. The obscuring dust was confined to the so called "plane of avoidance" which is the Galactic plane.

> "Stars are dimmed!"
>
> — Wilhelm Struve [1867]

The presence of interstellar extinction was pointed out as early as in 1847 by F.G. Wilhelm Struve. He found that the number of stars per unit volume seems to diminish in all directions receding from the Sun. This could be explained either if the Sun was at the center of a true stellar condensation, or if the effect was only an apparent one due to absorption (which may have been understood to include light scattering). He argued that there could be an visual extinction of about $1 \, \mathrm{mag \, kpc^{-1}}$ in interstellar space.

Jacobus C. Kapteyn (1904) had found a roughly spherical distribution of stars around the Sun. He assumed a constant stellar density and then used the observed density to arrive at a value for the extinction (absorption) of light $\approx 1.6 \, \mathrm{mag \, kpc^{-1}}$ (Kapteyn 1909), which differs little from current values ($\approx 1.8 \, \mathrm{mag \, kpc^{-1}}$ assuming a hydrogen density of $n_H = 1 \, \mathrm{cm^{-3}}$).[4]

In 1929 Schalén examined the question of stellar densities as a function of distance. He did a very detailed study of B and A stars, including those in Cygnus, Cepheus, Cassiopeia, and Auriga. He obtained rather different values of the absorption coefficient, particularly in Cygnus and Auriga where there are large dark patches. So obviously the absorption is more in some regions and less in others.

> "Cosmic dust particles produce the selective absorption."
>
> — Robert J. Trumpler [1930]

It was not until the work of Robert J. Trumpler in 1930 that the first evidence for interstellar reddening was found. Trumpler (1930) based this on his study of open clusters in which he compared the luminosities and distances of open clusters with the distances obtained by assuming that all their diameters were the same. By observing the luminosities and knowing the spectral distribution of stars he was able to find both absorption ($\approx 0.7\,\mathrm{mag\,kpc^{-1}}$) and selective absorption or color excess (between photographic and visual; $\approx 0.3\,\mathrm{mag\,kpc^{-1}}$) with increasing distance, and produce a reddening curve.[5] It was this work which led to the general establishment of the existence of interstellar dust.

The wavelength dependence of extinction in the optical was measured for the first time by Rudnick (1936) using the still-widely-used "pair-match" method. Further observations carried out by Hall (1937) and Stebbins, Huffer & Whitford (1939) pointed to an λ^{-1} reddening "law" (at that time limited to $1-3\,\mu\mathrm{m}^{-1}$): the reddening curve showed a rise inversely proportional to the wavelength λ.

In 1934 Paul W. Merrill reported the discovery of the 5780 Å, 5797 Å, 6284 Å, 6614 Å "unidentified interstellar lines". These widened absorption lines, now known as "Diffuse Interstellar Bands", still remain unidentified (see Krelowski 2002 for a review).

The existence of diffuse interstellar radiation, originally detected by van Rhijn (1921), was verified and attributed to small dust grains by Henyey & Greenstein (1941). Henyey & Greenstein (1941) found that interstellar particles are strongly forward scattering and have a high albedo.

> "Starlight is polarized!"
> — John S. Hall [1949]; W.A. Hiltner [1949]

At the end of 1940s, two investigators (Hall 1949; Hiltner 1949), inspired by a prediction of Chandrasekhar on intrinsic stellar polarization, independently discovered instead the general interstellar linear polarization. Magnetic fields were believed to confine cosmic rays and to play a role in the spiral structure of the Galaxy. The implication of the linear polarization was that the extinction was caused by non-spherical particles aligned by magnetic fields (Davis & Greenstein 1951). The wavelength dependent polarization curve was later shown to be well represented by the Serkowski law, an empirical formula (Serkowski 1973). In addition to the extinction curve which was later also extended to a wide wavelength range, the polarization law as well as the polarization to extinction ratio provide further insight into the physical and chemical nature of interstellar dust.

The circular polarization produced by interstellar birefringence (Martin 1972) was originally predicted by van de Hulst (1957). It was first detected along the lines of sight to the Crab Nebula by Martin, Illing, & Angel (1972) and to six early-type stars by Kemp & Wolstencroft (1972).

1.2.   INTERSTELLAR MATTER: THEORETICAL EVIDENCES

"There must be more matter than stars......"
— Jan H. Oort [1932]

In the 1930's Jan H. Oort took another approach to the problem by looking at the statistics of the motions of K giants perpendicular to the plane of the Galaxy, that is, at bulge objects. He used these to estimate the mass of material in the plane. He found that there had to be more material there than could be seen in stars. Oort (1932) estimated that the mass of the non-stellar material (dust and gas) is about 12×10^9 m$_\odot$. If this mass is distributed uniformly, the density of this non-stellar material is $\rho_{ism} \approx 6 \times 10^{-24}$ g cm^{-3} – this is the mass required to explain the observed motions.

The question then becomes what kind of material distributed with this density with what mass absorption coefficient could give rise to an extinction of about 1 mag kpc^{-1}, as observed. So what is required is that the scattering/extinction cross section of the material blocking the light per unit length is on the order of 1 mag kpc^{-1}.

1.3.   INTERSTELLAR DUST: EARLY MODELLING EFFORTS

1930s: Metallic grains
— C. Schalén; J.L. Greenstein

But what caused the interstellar reddening? Since hyperbolic meteors were thought to exist, first attempts were made to tie the interstellar dust to the meteors. Small metallic particles were among the materials initially proposed to be responsible for the interstellar reddening, based on an analogy with small meteors or micrometeorites supposedly fragmented into finer dust (Schalén 1936; Greenstein 1938).[6] Reasonably good fits to the λ^{-1} extinction law were obtained in terms of small metallic grains with sizes of the order of 0.01 μm.[7] It became evident later that meteors or micrometeorites are not of interstellar origin.

In 1948 Whitford published measurements of star colors versus spectral types over a wavelength range from about 3500Å (UV) to the near IR. The relation was not the expected straight line, but showed curvature at the near UV and IR regions. Things were beginning to make some physical sense from the point of view of small particle scattering.

1940s: Dirty ice grains
— J.H. Oort & H.C. van de Hulst

Based on the correlation between gas concentration and extinction, Lindblad (1935) argued that it seemed reasonable to grow particles in space since, as hypothesized by Sir Arthur Eddington, gaseous atoms and ions which hit a solid particle in space would freeze down upon it. Lindblad (1935) further put forward the hypothesis that interstellar dust could have formed by condensation (or more properly, accretion) of interstellar gas.

In the 1940's van de Hulst (1949) broke with tradition and published the results of making particles out of atoms that were known to exist in space: H, O, C, and N. He assumed these atoms combined on the surface to form frozen saturated molecules. The gas condensation scenario was further investigated by Oort & van de Hulst (1946) and led to what later became known as the "dirty ice" model.[8]

The dirty ice model of dust by van de Hulst was a logical followup of the then existing information about the interstellar medium and contained the major idea of surface chemistry leading to the ices H_2O, CH_4, NH_3. But it was not until the advent of IR astronomical techniques made it possible to observe silicate particles emitting at their characteristic $10\,\mu$m wavelength in the atmospheres of cool stars that we had the cores on which the matter could form. Interestingly, their presence was predicted on theoretical grounds by Kamijo (1963). As van de Hulst said, he chose to ignore the nucleation problem and just go ahead (where no one had gone before) with the assumption that "something" would provide the seeds for the mantle to grow on. By 1945 we had many of the theoretical basics to understand the sources of interstellar dust "ices" but it was not until about 1970 that the silicates were established. Without having a realistic dust model, van de Hulst developed the scattering tools to provide a good idea of dust properties.

1960s: Graphite grains
— F. Hoyle & N.C. Wickramasinghe

A "challenge" to the dirty ice model came out just after the discovery of interstellar polarization (Hall 1949; Hiltner 1949) since it seemed that the dirty ice model could not explain the rather high degree of polarization relative to extinction (see van de Hulst 1957; this was later shown not to be true by Greenberg et al. 1963a, b).

This led to the re-consideration of metallic grains and the consideration of graphite condensed in the atmospheres of carbon stars as a dust component (Cayrel & Schatzman 1954; Hoyle & Wickramasinghe 1962) because of their enormous potential for polarizing stellar radiation (as a result of its anisotropic optical properties of graphite). The graphite proposal seemed to be further supported by the detection of the 2175 Å hump extinction (Stecher & Donn 1965), although we know nowadays that the graphite model is not fully successful in explaining the 2175 Å hump and the ultimate identification is still not made (see §3.1.2).

Kamijo (1963) first proposed that SiO_2, condensed in the atmospheres of cool stars and blown out into the interstellar space, could provide condensation cores for the formation of "dirty ices". It was later shown by Gilman (1969) that grains around oxygen-rich cool giants are mainly silicates such as Al_2SiO_3 and Mg_2SiO_4. Silicates were first detected in emission in M stars (Woolf & Ney 1969; Knacke et al. 1969a), in the Trapezium region of the Orion Nebula (Stein & Gillett 1969), and in comet Bennett 1969i (Maas, Ney, & Woolf 1970); in absorption toward the Galactic Center (Hackwell, Gehrz, & Woolf 1970), and toward the Becklin-Neugebauer object and Kleinmann-Low Nebula (Gillett & Forrest 1973). Silicates are now known to be ubiquitous, seen in interstellar clouds, circumstellar disks around young stellar objects (YSOs), main-sequence stars and evolved stars, in HII regions, and in interplanetary and cometary dust (see Li & Draine 2001a for a review).

1.4. CONTEMPORARY INTERSTELLAR DUST MODELS

Since 1970s: Modern models
— J.M. Greenberg; J.S. Mathis; B.T. Draine; and their co-workers

The first attempt to find the 3.1 μm feature of H_2O was unsuccessful (Danielson, Woolf, & Gaustad 1965; Knacke, Cudaback, Gaustad 1969b). This was, at first, a total surprise to those who had accepted the dirty ice model. However, this gave the incentive to perform the early experiments on the UV photoprocessing of low temperature mixtures of volatile molecules simulating the "original" dirty ice grains (Greenberg et al. 1972; Greenberg 1973) to understand how and why the predicted H_2O was not clearly present. From such experiments was predicted a new component of interstellar dust in the form of complex organic molecules, as mantles on the silicates. This idea was further developed in the framework of the cyclic evolutionary silicate core-organic refractory mantle dust model (Greenberg 1982a; Greenberg & Li 1999a). Similar core-mantle models have also been proposed by others (Désert, Boulanger, & Puget 1990; Duley, Jones, & Williams 1989; Jones, Duley, & Williams 1990).

According to the cyclic evolutionary model, ices evolve chemically and physically in interstellar space, so do the organics. Where and how the interstellar dust is formed appears to involve a complex evolutionary picture. The rates of production of refractory components such as silicates in stars do not seem to be able to provide more than about 10% of what is observed in space because they are competing with destruction which is about 10 times faster by, generally, supernova shocks (Draine & Salpeter 1979a,b; Jones et al. 1994). At present the only way to account for the observed extinction amount is to resupply the dust by processes which occur in the interstellar medium itself. The organic mantles on the silicate particles must be created at a rate sufficient to balance their destruction. Furthermore, they provide a shield against destruction of the silicates. Without them the silicates would indeed be underabundant unless most of the grain mass was condensed in the ISM, as suggested by Draine (1990).

What is currently known about the organic dust component is based very largely on results of laboratory experiments which attempt to simulate interstellar processes. The organic refractories which are derived from the photoprocessing of ices contain a mixture of aliphatic and aromatic carbonaceous molecules (Greenberg et al. 2000). The laboratory analog suggests the presence of abundant prebiotic organic molecules in interstellar dust (Briggs et al. 1992).

The silicate core-organic mantle model is recently revisited by Li & Greenberg (1997) in terms of a trimodal size distribution consisting of (1) large core-mantle grains which account for the interstellar polarization, the visual/near-IR extinction, and the far-IR emission; (2) small carbonaceous grains of graphitic nature to produce the 2175 Å extinction hump; (3) polycyclic aromatic hydrocarbons (PAHs) to account for the far-UV extinction as well as the observed near- and mid-IR emission features at 3.3, 6.2, 7.7, 8.6, and 11.3 μm. This model is able to reproduce both the interstellar extinction and linear and circular polarization.

An alternative model was proposed by Mathis, Rumpl, & Nordsieck (1977) and thoroughly extended by Draine & Lee (1984). It consists of two separate dust components – bare silicate and graphite particles. Modifications to this model was later made by Sorrell (1990), Siebenmorgen & Krügel (1992), and Rowan-Robinson (1992) by adding new dust components (amorphous carbon, PAHs) and changing dust sizes.

Very recently, Draine and his co-workers (Li & Draine 2001b, 2002a; Weingartner & Draine 2001a) have extended the silicate/graphite grain model to explicitly include a PAH component as the small-size end of the carbonaceous grain population. The silicate/graphite-PAHs model provides an excellent quantitative agreement with the observations of IR emission as well as extinction from the diffuse ISM of the Milky Way Galaxy and the Small Magellanic Cloud.

Mathis & Whiffen (1989) have proposed that interstellar grains are composite collections of small silicates, vacuum ($\approx$ 80% in volume), and carbon of various kinds (amorphous carbon, hydrogenated amorphous carbon, organic refractories). However, the composite grains may be too cold and produce too flat a far-IR emissivity to explain the observational data (Draine 1994).[9] This is also true for the fractal grain model (Wright 1987).

In view of the recent thoughts that the reference abundance of the ISM (the abundances of heavy elements in both solid and gas phases) is subsolar (Snow & Witt 1995, 1996), Mathis (1996, 1998) updated the composite grain model envisioned as consisting of three components: (1) small silicate grains to produce the far-UV ($\lambda^{-1} > 6\,\mu m^{-1}$) extinction rise; (2) small graphitic grains to produce the 2175 Å extinction hump; (3) composite aggregates of small silicates, carbon, and vacuum ($\approx$ 45% in volume) to account for the visual/near-IR extinction. The new composite model is able to reproduce the interstellar extinction curve and the $10\,\mu m$ silicate absorption feature. But it produces too much far-IR emission in comparison with the observational data (Dwek 1997).

1.5.   SCATTERING OF LIGHT BY SMALL PARTICLES: AN ESSENTIAL TOOL FOR DUST STUDIES

Our knowledge about dust grains is mainly inferred from their interaction with starlight: a grain in the line of sight between a distant star and the observer reduces the starlight by a combination of scattering and absorption; the absorbed energy is then re-radiated in the IR. For a non-spherical grain, the light of distant stars is polarized as a result of differential extinction for different alignments of the electric vector of the radiation. Therefore, to model the observed interstellar extinction, scattering, absorption, polarization and IR emission properties, knowledge of the optical properties (extinction, absorption and scattering cross sections) of interstellar dust is essential. This requires knowledge of the optical constants of the interstellar dust materials [i.e., the complex index of refraction $m(\lambda) = m'(\lambda) - i\,m''(\lambda)$], and the dust sizes and shapes. Phrasing this differently, to infer the size, morphology, and chemical composition of interstellar dust, an important aspect of the modelling of interstellar grains involves the computation of extinction, absorption and scattering cross sections of particles comprised of candidate materials and in comparison with astronomical data.

During the recent years, dramatic progress has been made in measuring the complex refractive indices of cosmic dust analogues, e.g. by the

Jena Group (Henning & Mutschke 2000) and the Naples Group (Colangeli et al. 1999). Interstellar particles would in general be expected to have non-spherical, irregular shapes. However, our ability to compute scattering and absorption cross sections for nonspherical particles is extremely limited. So far, exact solutions of scattering problems exist only for bare or layered spherical grains ("Mie theory"; Mie 1908; Debye 1909), infinite cylinders (Lind & Greenberg 1966), and spheroids (Asano & Yamamoto 1975; Asano & Sato 1980; Voshchinnikov & Farafonov 1993). For grains with sizes much smaller than the wavelength of the incident radiation, the dipole approximation can be used to evaluate cross sections for bare or coated spheroidal grains (van de Hulst 1957; Gilra 1972; Draine & Lee 1984; Bohren & Huffman 1983). The "T-matrix" (transition matrix) method, originally developed by Barber & Yeh (1975) and recently substantially extended by Mishchenko, Travis, & Mackowski (1996), is able to treat axisymmetric (spheroidal or finite cylindrical) grains with sizes comparable to the wavelength. The discrete dipole approximation (DDA), originally developed by Purcell & Pennypacker (1973) and recently greatly improved by Draine (1988), is a powerful technique for irregular heterogeneous grains with sizes as large as several times the wavelength. The VIEF (volume integration of electric fields) method developed by Hage & Greenberg (1990), based on an integral representation of Maxwell's equations, is physically similar to the DDA method. The microwave analog methods originally developed by Greenberg, Pedersen & Pedersen (1960) provide an effective experimental approach to complex particles. This method is still proving to be powerful (Gustafson 1999; Gustafson et al. 1999) as the needs still outstrip the capacities of computers.

Although interstellar particles are obviously non-spherical as evidenced by the observed polarization of starlight, the assumption of spherical shapes (together with the Bruggeman or the Maxwell-Garnett effective medium theories for inhomogeneous grains; Bohren & Huffman 1983) is usually sufficient in modelling the interstellar absorption, scattering and IR emission. For IR polarization modelling, the dipole approximation for spheroidal grains is proven to be successful in many cases. The DDA method is highly recommended for studies of inhomogeneous (e.g. coated) grains and irregular grains such as cometary, interplanetary, and protoplanetary dust particles.

1.6.   COMETS

The origin of comets is closely linked to the solar system and they played an important role in cosmogony. Cometary nuclei have been created far away from the early Sun[10] and have been mostly kept intact since their formation. Although it is generally accepted that comets contain the most pristine material in the solar system, it is still a matter of considerable debate whether they are made of unmodified protosolar nebula interstellar dust or this material has been (completely or partially) evaporated before becoming a part of comets (see Mumma, Stern, & Weissman 1993; Crovisier 1999; Irvine & Bergin 2000 for reviews).

The internal structure and chemical composition of cometary nuclei have been a topic receiving much attention. Before 1950, the prevailing view was that the comet nucleus is composed of a coherent swarm of meteoroidal-type particles which are independent of each other (termed the "flying sand-bank" model). It was shown by Russell, Dugan, & Stewart (1926) that the "swarm" concept of the cometary nucleus was inadequate since the solar heating would vaporize icy bodies up to 30 cm in diameter.

Since 1950 a number of models for the comet nucleus have been proposed. For the most part, the icy-conglomerate model (also known as the "dirty snowball" model) of Whipple (1950) has been the standard which others have followed. According to Whipple (1950), the comet nucleus is a solid body consisting of a conglomerate of refractory dust grains and frozen gases (mostly of H_2O ice). More recently, the computer simulation of dust aggregates formed by random accumulation (Daniels & Hughes 1981) led Donn, Daniels, & Hughes (1985) to postulate the fractal model in which the comet nucleus is considered as a heterogeneous aggregate of ice and dust grains with substantial voids. Weissman (1986) proposed a primordial rubble-pile model as a modification of the basic icy-conglomerate model in which the cometary nucleus is envisaged as a loosely bound agglomeration of smaller fragments, weakly bonded by local melting at contact interfaces. Gombosi & Houpis (1986) suggested an icy-glue model. According to them, the comet nucleus is composed of rather large porous refractory boulders (tens of centimeters to hundreds of meters) "cemented" together with the icy-conglomerate type ice-dust grain mix ("Whipple glue").

Alternatively, Greenberg (1982b) proposed the interstellar dust model of comets in which the basic idea is that comets have formed directly through coagulation of interstellar dust (Greenberg 1982b; Greenberg 1998; Greenberg & Li 1999b). The morphological structure of comet nuclei is thus modelled as an aggregate of presolar interstellar dust grains whose mean size is of the order of one tenth micron. The representative individual presolar grain consists of a core of silicates mantled first by an organic refractory material and then by a mixture of water dominated ices in which are embedded thousands of very small carbonaceous particles/large molecules (Greenberg 1998; Greenberg & Li 1999b). Greenberg and his co-workers have further shown how the IR emission for several distinctly different types of comets bear a general resemblance to each other by reproducing the IR emission of various comets (Halley – a periodic comet [Greenberg & Hage 1990; Greenberg et al. 1996]; Borrelly – a Jupiter family short period comet [Li & Greenberg 1998a]; Hale-Bopp – a long period comet [Li & Greenberg 1998b]; and extra-solar comets in the β Pictoris disk [Li & Greenberg 1998c]) within the framework that all comets are made of aggregated interstellar dust.

2.   THE CURRENT STATE OF THE ART

The last 40 years have seen a revolution in the study of interstellar dust. This has been a four-fold process. First of all, the observational access to the UV and the IR brought into focus the fact that there had to be a very wide range of particle sizes and types to account for the blocking of the starlight. Secondly, the IR provided a probe of some of the chemical constituents of the dust. Thirdly, laboratory techniques were applied to the properties and evolution of possible grain materials. Fourthly, advances in numerical techniques and the speed and memory of computers have greatly enhanced our modelling capabilities.

Thanks to the successful performances of IUE (*International Ultra-violet Explorer*), IRAS (*Infrared Astronomical Satellite*), COBE (*Cosmic Background Explorer*), HST (*Hubble Space Telescope*), ISO (*Infrared Space Observatory*) as well as various ground-based UV, optical and IR instruments, we have witnessed an explosive accumulation of new observational information on interstellar extinction, polarization, scattering, IR continuum emission and spectral features, as well as elemental depletion. Rapid progress on laboratory experiments has also been made. Below we present an overview of our current knowledge of the dust which brings many related astrophysical problems to the fore.

2.1.   EXTINCTION (SCATTERING, ABSORPTION) AND POLARIZATION

2.1.1   Interstellar Extinction and Polarization Curves.

> "The extinction curve is a sharp discriminator of (dominant) size, but
> a very poor discriminator of composition."
> — H.C. van de Hulst [1989]

The most extensively studied dust property may be the interstellar extinction. The main characteristics of the wavelength dependence of interstellar extinction – "interstellar extinction curve" – are well established: a slow and then increasingly rapid rise from the IR to the visual, an approach to levelling off in the near UV, a broad absorption feature at about $\lambda^{-1} \approx 4.6\,\mu m^{-1}$ ($\lambda \approx 2175$ Å) and, after the drop-off, a "final" curving increase to as far as has been observed $\lambda^{-1} \approx 8\,\mu m^{-1}$. We note here that the access to the UV was only made possible when observations could be made from space, first by rockets and then by satellites (OAO2 and IUE).

The optical/UV extinction curves show considerable variations which are correlated with different regions.[11] Cardelli, Clayton, & Mathis (1989) found that the extinction curves over the wavelength range of $0.125\,\mu m \leq \lambda \leq 3.5\,\mu m$ can be fitted remarkably well by an analytical formula involving only one free parameter: $R_V \equiv A_V/E(B-V)$, the total-to-selective extinction ratio. Values of R_V as small as 2.1 (the high latitude translucent molecular cloud HD 210121; Larson, Whittet, & Hough 1996) and as large as 5.6 (the HD 36982 molecular cloud in the Orion nebula) have been observed in the Galactic regions. More extreme extinction curves are reported for gravitational lens galaxies: $R_V = 1.5$ for an elliptical galaxy at a lens redshift $z_l = 0.96$ and $R_V = 7.2$ for a spiral galaxy at $z_l = 0.68$ (Falco et al. 1999). The Galactic mean extinction curve is characterized by $R_V \approx 3.1$. The optical/UV extinction curve and the value of R_V depend on the environment: lower-density regions have a smaller R_V, a stronger 2175 Å hump and a steeper far-UV rise ($\lambda^{-1} > 4\,\mu m^{-1}$); denser regions have a larger R_V, a weaker 2175 Å hump and a flatter far-UV rise.

The near-IR extinction curve ($0.9\,\mu m \leq \lambda \leq 3.5\,\mu m$) can be fitted reasonably well by a power law $A(\lambda) \sim \lambda^{-1.7}$, showing little environmental variations. The extinction longward of $3\,\mu m$ is not as well determined as that of $\lambda \leq 3\,\mu m$. Lutz et al. (1996) derived the 2.5–$9\,\mu m$ extinction law toward the Galactic Center (GC) based on the ISO observation of the hydrogen recombination lines. They found that the GC extinction law has a higher extinction level in the 3–$8\,\mu m$ range than the standard Draine (1989a) IR extinction curve.[12]

Existing grain models for the diffuse interstellar medium are mainly based on an analysis of extinction (Mathis, Rumpl, & Nordsieck 1977; Greenberg 1978; Hong & Greenberg 1980; Draine & Lee 1984; Duley, Jones, & Williams 1989; Mathis & Whiffen 1989; Kim, Martin, & Hendry 1994; Mathis 1996; Li & Greenberg 1997; Zubko 1999a; Weingartner & Draine 2001a). All models are successful in reproducing the observed extinction curve from the near IR to the far-UV. The silicate core-organic refractory mantle model and the (modified) silicate/graphite model are also able to fit the HD 210121 extinction curve which has the lowest R_V value (Li & Greenberg 1998d; Larson et al. 2000; Clayton et al. 2000; Weingartner & Draine 2001a) and the Magellanic clouds extinction curves (Rodrigues et al. 1997; Zubko 1999b; Clayton et al. 2000; Weingartner & Draine 2001a). The fact that so many different materials with such a wide range of optical properties could be used to explain the observed extinction curve indicates that the interstellar extinction curve is quite insensitive to the exact dust composition. What the extinction curve does tell us is that interstellar grains span a size ranging from ~ 100 angstrom to submicron (Draine 1995).

The general shape of the polarization curve is also well established. It rises from the IR, has a maximum somewhere in the visual (generally) and then decreases toward the UV, implying that the aligned nonspherical grains are typically submicron in size, and the very small grain component responsible for the far-UV extinction rise is either spherical or not aligned.

The optical/UV polarization curve $P(\lambda)$ can also be fitted remarkably well by an empirical function known as the "Serkowski law", involving the very same free parameter R_V as the Cardelli et al. (1989) extinction functional form through λ_{max},[13] the wavelength where the maximum polarization P_{max} occurs: $P(\lambda)/P_{max} = \exp[-K \ln^2(\lambda/\lambda_{max})]$ (Serkowski 1973; Coyne, Gehrels, & Serkowski 1974; Wilking, Lebofsky, & Rieke 1982). The width parameter K is linearly correlated with λ_{max}: $K \approx 1.66 \lambda_{max} + 0.01$ (Whittet et al. 1992 and references therein).[14]

The near IR ($1.64 \,\mu m < \lambda < 5 \,\mu m$) polarization has been found to be higher than that extrapolated from the Serkowski law (Martin et al. 1992). Martin & Whittet (1990) and Martin et al. (1992) suggested a power law $P(\lambda) \propto \lambda^{-\beta}$ for the near IR ($\lambda > 1.64 \,\mu m$) where the power index β is independent of λ_{max} and in the range of 1.6 to 2.0, $\beta \simeq 1.8 \pm 0.2$.[15]

The observed interstellar polarization curve has also been extensively modelled in terms of various dust models by various workers. The silicate core-organic mantle model (Chlewicki & Greenberg 1990; Li & Greenberg 1997), the silicate-graphite model (only silicate grains are assumed to be efficiently aligned; Mathis 1986; Wolff, Clayton, & Meade 1993; Kim & Martin 1995, 1996), and the composite model (Mathis & Whiffen 1989) are all successful in reproducing the mean interstellar polarization curve ($\lambda_{max} = 0.55 \,\mu m$).

However, the processes leading to the observed grain alignment are still not well established. A number of alignment mechanisms have been proposed. The Davis-Greenstein paramagnetic dissipation mechanism (Davis & Greenstein 1951) together with other co-operative effects such as suprathermal rotation (Purcell 1979), superparamagnetic alignment (Jones & Spitzer 1967; Mathis 1986), radiative torques on irregular grains due to anisotropic starlight (Draine & Weingartner 1996, 1997) seems to be a plausible mechanism for dust in the diffuse ISM. In dense molecular clouds, non-magnetic alignment mechanisms such as streaming of grains through gas (Gold 1952; Purcell 1969; Lazarian 1994), through radiation (Harwit 1970), through ambipolar diffusion (Roberge 1996) have been studied.

The wavelength dependent albedo (the ratio of scattering cross section to extinction) measured from the diffuse Galactic light, reflecting the scattering properties of interstellar dust, provides another constraint on dust models. The silicate core-organic mantle model (Li & Greenberg 1997) and the silicate/graphite-PAHs model (Li & Draine 2001b) are shown to be in good agreement with the observationally determined albedos, whereas the albedos of the composite grain model (Mathis 1996) are too low (Dwek 1997).

Scatterings of X-rays by interstellar dust have also been observed as evidenced by "X-ray halos" formed around an X-ray point source by small-angle scattering. The intensity and radial profile of the halo depends on the composition, size and morphology and the spatial distribution of the scattering dust particles (see Smith & Dwek 1998 and references therein). A recent study of the X-ray halo around Nova Cygni 1992 by Witt, Smith, & Dwek (2001) pointed to the requirement of large interstellar grains, consistent with the recent Ulysses and Galileo detections of interstellar dust entering our solar system (Grün et al. 1994; Frisch et al. 1999; Landgraf et al. 2000).

2.1.2 Spectroscopic Extinction and Polarization Features.

It is the extinction (absorption) and emission spectral lines instead of the overall shape of the extinction curve provides the most diagnostic information on the dust composition.

1 The 2175 Å Extinction Hump

> "It is frustrating that almost 3 decades after its discovery, the identity of this (2175 Å) feature remains uncertain!"
> — B.T. Draine [1995]

The strongest spectroscopic extinction feature is the 2175 Å hump. Observations show that its strength and width vary with environment while its peak position is quite invariant. Its carrier remains unidentified 37 years after its first detection (Stecher 1965). Many candidate materials, including graphite (Stecher & Donn 1965), amorphous carbon (Bussoletti et al. 1987), graphitized (dehydrogenated) hydrogenated amorphous carbon (Hecht 1986; Goebel 1987; Sorrell 1990; Mennella et al. 1996; Blanco et al. 1999),[16] nano-sized hydrogenated amorphous carbon (Schnaiter et al. 1998), quenched carbonaceous composite (QCC; Sakata et al. 1995), coals (Papoular et al. 1995), PAHs (Joblin et al. 1992; Duley & Seahra 1998; Li & Draine 2001b), and OH^- ion in low-coordination sites on or within silicate grains (Duley, Jones & Williams 1989) have been proposed, while no single one is generally accepted (see Draine 1989b for a review).

Graphite was the earliest suggested and the widely adopted candidate in various dust models (Gilra 1972; Mathis et al. 1977; Hong & Greenberg 1980; Draine & Lee 1984; Dwek et al. 1997; Will & Aannestad 1999). However the hump peak position predicted from graphite particles is quite sensitive to the grain size, shape, and coatings (Gilra 1972; Greenberg & Chlewicki 1983; Draine 1988; Draine & Malhotra 1993) which is inconsistent with the observations. It was suggested that very small coated graphite particles ($a \leq 0.006\,\mu m$) could broaden the hump while keeping the hump peak constant (Mathis 1994).[17] However, this seems unlikely because the proposed particles are so small that temperature fluctuations will prevent them from acquiring a coating (Greenberg & Hong 1974a; Aannestad & Kenyon 1979). Furthermore, it was noted by Greenberg & Hong (1974b) that if the very small particles as well as the large particles accrete mantles the R_V value would *decrease* in molecular clouds, in contradiction with astronomical observations. Rouleau, Henning, & Stognienko (1997) proposed that the combined effects of shape, clustering, and fine-tuning of the optical properties of graphite could account for the hump width variability.

Another negative for graphite, is that the dust grains in circumstellar envelopes around carbon stars which are the major sources of the carbon component of interstellar dust are in amorphous form rather than graphitic (Jura 1986). It is difficult to understand how the original amorphous carbonaceous grains blown out from the star envelopes are processed to be highly anisotropic and evolve to the layer-lattice graphitic structures in interstellar space. Instead, it is more likely that the interstellar physical and chemical processes should make the carbonaceous grains even more highly disordered.

Recently, the PAH proposal is receiving increasing attention. Although a single PAH species often has some strong and narrow UV bands which are not observed (UV Atlas 1966), a cosmic mixture of many individual molecules, radicals, and ions, with a concentration of strong absorption features in the 2000–2400 Å region, may effectively produce the 2175 Å extinction feature (Li & Draine 2001b). This is supported by the correlation between the 2175 Å hump and the IRAS $12\,\mu$m emission (dominated by PAHs) found by Boulanger, Prévot, & Gry (1994) in the Chamaeleon cloud which suggests a common carrier. Arguments against the PAHs proposal also exist (see Li & Draine 2001b for references).

So far only two lines of sight toward HD 147933 and HD 197770 have a weak 2175 Å polarization feature detected (Clayton et al. 1992; Anderson et al. 1996; Wolff et al. 1997; Martin, Clayton, & Wolff 1999). Even for these sightlines, the degree of alignment and/or polarizing ability of the carrier should be very small (if both the hump excess polarization and the hump extinction are produced by the same carrier); for example, along the line of sight to HD 197770, the ratio of the excess polarization to the hump extinction is $P_{2175\,\text{Å}}/A_{2175\,\text{Å}} \simeq 0.002$ while the polarization to extinction ratio in the visual is $P_V/A_V \simeq 0.025$, thus $(P_{2175\,\text{Å}}/A_{2175\,\text{Å}})/(P_V/A_V)$ is only ~ 0.09. Therefore, it is reasonable to conclude that the 2175 Å carrier is either mainly spherical or poorly aligned.

The 2175 Å hump polarization was predicted by Draine (1988) for aligned non-spherical graphite grains. Wolff et al. (1993) and Martin et al. (1995) further show that the observed 2175 Å polarization feature toward HD 197770 can be well fitted with small aligned graphite disks.

Except for the detection of scattering in the 2175 Å hump in two reflection nebulae (Witt, Bohlin, & Stecher 1986), the 2175 Å hump is thought to be predominantly due to absorption, suggesting its carrier is sufficiently small to be in the Rayleigh limit.

2 The $9.7\,\mu$m and $18\,\mu$m (Silicate) Absorption Features

> "In my opinion, the only secure identification is that of the
> $9.7\,\mu$m and $18\,\mu$m IR features."
> — B.T. Draine [1995]

The strongest IR absorption features are the $9.7\,\mu$m and $18\,\mu$m bands. They are respectively ascribed to the Si-O stretch and O-Si-O bending modes in some form of silicate material, perhaps olivine $Mg_{2x}Fe_{2-2x}SiO_4$. The shape of the interstellar silicate feature is broad and featureless both of which suggest that the silicate is amorphous.[18] The originating source of the interstellar silicates is in the atmospheres of cool evolved stars of which the emission features often show a $9.7 \mu m$ feature consistent with amorphous silicates and sharper features arising from crystalline silicates (Waters et al. 1996).[19] How can interstellar crystalline silicate particles become amorphous? This is a puzzle which has not yet been completely solved. A possible solution may be that the energetic processes (e.g. shocks, grain-grain collision) operated on interstellar dust in the diffuse ISM have disordered the periodic lattice structures of crystalline silicates. Another possible solution may lie in the fact that, according to Draine (1990), only a small fraction of interstellar dust is the original stardust, i.e., most of the dust mass in the ISM was condensed in the ISM, rather than in stellar outflows. The re-condensation at low temperatures most likely leads to an amorphous rather than crystalline form.

First detected in the Becklin-Neugebauer (BN) object in the OMC-1 Orion dense molecular cloud (Dyck et al. 1973), the silicate polarization absorption feature in the $10 \mu m$ region is found to be very common in heavily obscured sources; some sources also have the $18 \mu m$ O-Si-O polarization feature detected (see Aitken 1996; Smith et al. 2000 for summaries). In most cases the silicate polarization features are featureless, indicating the amorphous nature of interstellar silicate material (see Aitken 1996) except AFGL 2591, a molecular cloud surrounding a young stellar object, has an additional narrow polarization feature at $11.2 \mu m$, generally attributed to annealed silicates (Aitken et al. 1988; Wright et al. 1999).

Reasonably good fits to the observed $10 \mu m$ Si-O polarization features can be obtained by elongated (bare or ice-coated) "astronomical silicate" grains (Draine & Lee 1984; Lee & Draine 1985; Hildebrand & Dragovan 1995; Smith et al. 2000). High resolution observations of the BN $10 \mu m$ and $18 \mu m$ polarization features provided a challenge to the "astronomical silicate" model since this model failed to reproduce two of the basic aspects of the observations: (1) the $10 \mu m$ feature was not broad enough and (2) the $18 \mu m$ feature was too low by a factor of two relative to the $10 \mu m$ peak (Aitken, Smith, & Roche 1989). Attempts by Henning & Stognienko (1993) in terms of porous grains composed of "astronomical silicates", carbon and vacuum were not successful. In contrast, it is shown by Greenberg & Li (1996) that an excellent match to the BN $10 \mu m$ and $18 \mu m$ polarization features in shape, width, and in relative strength can be obtained by the silicate core-organic mantle model using the experimental optical constants of silicate and organic refractory materials. It seems desirable to re-investigate the $18\,\mu m$ O-Si-O band strength of "astronomical silicates" (Draine & Lee 1984) by a combination of observational, experimental and modelling efforts.

In the mid-IR, interstellar grains of submicron size are in the Rayleigh limit. The scattering efficiency falls rapidly with increasing wavelength. Therefore the effects of scattering are expected to be negligible for the silicate extinction and polarization.

3   The $3.4\,\mu m$ (Aliphatic Hydrocarbon) Absorption Feature

"When the $3.4\,\mu m$ feature was detected in VI Cyg #12, the problem of why no H_2O was observed would appear to have been resolved."

— J. Mayo Greenberg [1999]

Another ubiquitous strong absorption band in the diffuse ISM is the $3.4\,\mu m$ feature. Since its first detection in the Galactic Center toward Sgr A W by Willner et al. (1979) and IRS 7 by Wickramasinghe & Allen (1980), it has now been widely seen in the Milky Way Galaxy and other galaxies (Butchart et al. 1986; Adamson, Whittet, & Duley 1990; Sandford et al. 1991; Pendleton et al. 1994; Wright et al. 1996; Imanishi & Dudley 2000; Imanishi 2000). Although it is generally accepted that this feature is due to the C-H stretching mode in saturated aliphatic hydrocarbons, the exact nature of this hydrocarbon material remains uncertain. Nearly two dozen different candidates have been proposed over the past 20 years (see Pendleton & Allamandola 2002 for a review). The organic refractory residue, synthesized from UV photoprocessing of interstellar ice mixtures, provides a perfect match, better than any other hydrocarbon analogs, to the observed $3.4\,\mu m$ band, including the $3.42\,\mu m$, $3.48\,\mu m$, and $3.51\,\mu m$ subfeatures (Greenberg et al. 1995).[20] But, at this moment, we are not at a position to rule out other dust sources as the interstellar $3.4\,\mu m$ feature carrier. This feature has also been detected in a carbon-rich protoplanetary nebula CRL 618 (Lequeux & Jourdain de Muizon 1990; Chiar et al. 1998) with close resemblance to the interstellar feature. However, after ejection into interstellar space, the survival of this dust in the diffuse ISM is questionable (see Draine 1990).

The $3.4\,\mu m$ feature consists of three subfeatures at $2955\,cm^{-1}$ (3.385

μm), 2925 cm^{-1} (3.420 μm), and 2870 cm^{-1} (3.485 μm) corresponding to the symmetric and asymmetric C-H stretches in CH$_3$ and
CH$_2$ groups in aliphatic hydrocarbons which must be interacting
with other chemical groups. The amount of carbonaceous material
responsible for the 3.4 μm feature is strongly dependent on the nature of the chemical groups attached to the aliphatic carbons. For
example, each carbonyl (C=O) group reduces its corresponding
C-H stretch strength by a factor of $\sim$ 10 (Wexler 1967). Furthermore, not every carbon is attached to a hydrogen as in saturated
compounds. Aromatic hydrocarbons do not even contribute to the
3.4 μm feature although they do absorb nearby at $\approx$ 3.28 μm. The
fact that the 3.4 μm absorption is not observed in molecular cloud
may possibly be attributed to dehydrogenation or oxidation (formation of carbonyl) of the organic refractory mantle by accretion
and photoprocessing in the dense molecular cloud medium (see
Greenberg & Li 1999a) – the former reducing the absolute number
of CH stretches, the latter reducing the CH stretch strength by a
factor of 10 (Wexler 1967) – the 3.4 μm feature would be reduced
per unit mass in molecular clouds.

Very recently, Gibb & Whittet (2002) reported the discovery of
a 6.0 μm feature in dense clouds attributed to the organic refractory. They found that its strength is correlated with the 4.62 μm
OCN$^-$ (XCN) feature which is considered to be a diagnostic of
energetic processing.

Attempts to measure the polarization of the 3.4 μm absorption
feature ($P_{\text{C-H}}^{\text{IRS7-obs}}$) was recently made by Adamson et al. (1999)
toward the Galactic Center source IRS 7. They found that this
feature was essentially unpolarized. Since no spectropolarimetric
observation of the 10 μm silicate absorption feature ($P_{\text{sil}}^{\text{IRS7}}$) has yet
been carried out for IRS 7, they estimated $P_{\text{sil}}^{\text{IRS7}}$ from the 10 μm
silicate optical depth $\tau_{\text{sil}}^{\text{IRS7}}$, assuming the IRS 7 silicate feature
is polarized to the same degree as the IRS 3 silicate feature; i.e.,
$P_{\text{sil}}^{\text{IRS7}}/\tau_{\text{sil}}^{\text{IRS7}} = P_{\text{sil}}^{\text{IRS3}}/\tau_{\text{sil}}^{\text{IRS3}}$ where $\tau_{\text{sil}}^{\text{IRS7}}$, $\tau_{\text{sil}}^{\text{IRS3}}$ and $P_{\text{sil}}^{\text{IRS3}}$ were
known. Assuming the IRS 7 aliphatic carbon (the 3.4 μm carrier)
is aligned to the same degree as the silicate dust, they expected the
3.4 μm polarization to be $P_{\text{C-H}}^{\text{IRS7-mod}} = P_{\text{sil}}^{\text{IRS7}}/\tau_{\text{sil}}^{\text{IRS7}} \times \tau_{\text{C-H}}^{\text{IRS7}}$. They
found $P_{\text{C-H}}^{\text{IRS7-obs}} \ll P_{\text{C-H}}^{\text{IRS7-mod}}$ (Adamson et al. 1999). Therefore,
they concluded that the aliphatic carbon dust is not in the form of a
mantle on the silicate dust as suggested by the core-mantle models
(Li & Greenberg 1997; Jones, Duley, & Williams 1990).[21] We note
that the two key assumptions on which their conclusion relies are

questionable: (1) $P_{\text{sil}}^{\text{IRS7}}/\tau_{\text{sil}}^{\text{IRS7}} = P_{\text{sil}}^{\text{IRS3}}/\tau_{\text{sil}}^{\text{IRS3}}$; (2) $P_{\text{C-H}}^{\text{IRS7}}/\tau_{\text{C-H}}^{\text{IRS7}} = P_{\text{sil}}^{\text{IRS7}}/\tau_{\text{sil}}^{\text{IRS7}}$ (see Li & Greenberg 2002 for details). We urgently need spectropolarimetric observations of IRS 7.

Hough et al. (1996) reported the detection of a weak $3.47\,\mu$m polarization feature in BN, attributed to carbonaceous materials with diamond-like structure, originally proposed by Allamandola et al. (1992) based on the $3.47\,\mu$m absorption spectra of protostars.[22]

4 The $3.3\,\mu$m and $6.2\,\mu$m (PAH) Absorption Features

Recently, two weak narrow absorption features at $3.3\,\mu$m and $6.2\,\mu$m were detected. The $3.3\,\mu$m feature has been seen in the Galactic Center source GCS 3 (Chiar et al. 2000) and in some heavily extincted molecular cloud sight lines (Sellgren et al. 1995; Brooke, Sellgren, & Geballe 1999). The $6.2\,\mu$m feature, about 10 times stronger, has been detected in several objects including both local sources and Galactic Center sources (Schutte et al. 1998; Chiar et al. 2000). They were attributed to aromatic hydrocarbons (Schutte et al. 1998; Chiar et al. 2000). The theoretical $3.3\,\mu$m and $6.2\,\mu$m absorption feature strengths (in terms of integrated optical depths) predicted from the astronomical PAH model are consistent with observations (Li & Draine 2001b). Note the 7.7, 8.6, and $11.3\,\mu$m PAH features are hidden by the much stronger $9.7\,\mu$m silicate feature, and therefore will be difficult to observe as absorption features.

The aromatic absorption bands allow one to place constraints on the PAH abundance if the PAH band strengths are known. But one should keep in mind that the strengths of these PAH absorption bands could vary with physical conditions due to changes in the PAH ionization fraction. In regions with increased PAH ionization fraction, the $6.2\,\mu$m absorption feature would be strengthened and the $3.3\,\mu$m feature would be weakened (see Li & Draine 2001b).

Although it was theoretically predicted that the PAH IR emission features can be linearly polarized (Léger 1988), no polarization has been detected yet (Sellgren, Rouan, & Léger 1988).

5 The Diffuse Interstellar Bands

In 1922, Heger observed two broad absorption features centering at 5780 Å and 5797 Å, conspicuously broader than atomic interstellar absorption lines. This marked the birth of a long standing astrophysical mystery – the diffuse interstellar bands (DIBs). But not until the work of Merrill (1934) were the interstellar nature of these absorption features established. So far, over 300 DIBs have been detected from the near IR to the near UV. Despite ~ 80 years' efforts, no definite identification (including the recent neutral/charged PAHs [Salama & Allamandola 1992], C_{60}^+ [Foing & Ehrenfreund 1994], and C_7^- [Tulej et al. 1998] proposals) of the carrier(s) of DIBs has been found. We refer the readers to the two extensive reviews of Krelowski (1999, 2002).

No polarization has been detected for the DIBs (Martin & Angel 1974, 1975; Fahlman & Walker 1975; Adamson & Whittet 1992, 1995; see Somerville 1996 for a review).

6 The Ice Absorption Features

The formation of an icy mantle through accretion of molecules on interstellar dust is expected to take place in dense clouds (e.g., the accretion timescale is only $\sim 10^5$ yrs for clouds of densities $n_H = 10^3 - 10^5 \, cm^{-3}$; Schutte 1996). The detection of various ice IR absorption features (e.g., H_2O [3.05, 6.0 μm], CO [4.67 μm], CO_2 [4.27, 15.2 μm], CH_3OH [3.54, 9.75 μm], NH_3 [2.97 μm], CH_4 [7.68 μm], H_2CO [5.81 μm], OCN^- [4.62 μm]; see Ehrenfreund & Schutte 2000 for a review) have demonstrated the presence of icy mantles in dark clouds (usually with an visual extinction > 3 magnitudes; see Whittet et al. 2001 and references therein). In comparison with that for a gas phase sample, the IR spectrum for the same sample in the solid phase is broadened, smoothed, and shifted in wavelength due to the interactions between the vibrating molecule with the surroundings, the suppression of molecular rotation in ices at low temperatures, and the irregular nature of the structure of (amorphous) solids (see Tielens & Allamandola 1987).

Note not only are the relative proportions of ice species variable in different regions but also the presence and absence of some species. In almost all cases, however, water is the dominant component. An important variability is in the layering of the various molecular components. Of particular note is the fact that the CO molecular spectrum is seen to indicate that it occurs sometimes embedded in the H_2O (a polar matrix) and sometimes not. This tells a story about how the mantles form. As was first noted by van de Hulst, the presence of surface reactions leads to the reduced species H_2O, CH_4, NH_3. Since we now know that CO is an abundant species as a gas phase molecule, we expect to find it accreted along with these reduced species — at least initially.

The two approaches to understanding how the grain mantles evolve are: (1) the laboratory studies of icy mixtures, their modification by UV photoprocessing and by heating; (2) theoretical studies combining gas phase chemistry with dust accretion and dust chemistry. In the laboratory one creates a cold surface (10K) on which various simple molecules are slowly deposited in various proportions. The processing of these mixtures by UV photons and by temperature variation is studied by IR spectroscopy. This analog of interstellar dust mantles is used to provide a data base for comparison with the observations (see Schutte 1999 for a review).

The $3.1\,\mu m$ ice polarization has been detected in various molecular cloud sources (see Aitken 1996 and references therein). The detection of the $4.67\,\mu m$ CO and $4.62\,\mu m$ OCN^- polarization was recently reported by Chrysostomou et al. (1996). The BN ice polarization feature was well fitted by ice-coated grains (Lee & Draine 1985), suggesting a core-mantle grain morphology. However, the AFGL 2591 molecular cloud shows no evidence for ice polarization (Dyck & Lonsdale 1980; Kobayashi et al. 1981) while having distinct ice extinction and silicate polarization (Aitken et al. 1988). Perhaps only the hot, partly annealed silicate grains close to the forming-star are aligned (say, by streaming of ambipolar diffusion) while the ice-coated cool grains in the outer envelope of the cloud are poorly aligned.

2.2.   DUST EMISSION

2.2.1   Dust Luminescence: The "Extended Red Emission".

> "The ERE has become an important observational aspect of
> interstellar grains that future models need to reproduce."
>
> — A.N. Witt [2000]

First detected in the Red Rectangle (Schmidt, Cohen, & Margon 1980), "extended red emission" (ERE) from interstellar dust consists of a broad, featureless emission band between $\sim$5400 Å and 9000 Å, peaking at $6100 \leq \lambda_p \leq 8200\,\text{Å}$, and with a width $600\,\text{Å} \leq \text{FWHM} \leq 1000\,\text{Å}$. The ERE has been seen in a wide variety of dusty environments: the diffuse ISM of our Galaxy, reflection nebulae, planetary nebulae, HII regions, and other galaxies (see Witt, Gordon, & Furton 1998 for a summary). The ERE is generally attributed to photoluminescence (PL) by some component of interstellar dust, powered by UV/visible photons.

The photon conversion efficiency of the diffuse ISM has been determined to be near $(10\pm3)\%$ (Gordon et al. 1998; Szomoru & Guhathakurta 1998) assuming that all UV/visible photons absorbed by interstellar grains are absorbed by the ERE carrier. The actual photoluminescence efficiency of the ERE carrier must exceed $\sim 10\%$, since the ERE carrier cannot be the only UV/visible photon absorber.

Various forms of carbonaceous materials – HAC (Duley 1985; Witt & Schild 1988), PAHs (d'Hendecourt et al. 1986), QCC (Sakata et al. 1992), C_{60} (Webster 1993), coal (Papoular et al. 1996), PAH clusters (Allamandola, private communication), carbon nanoparticles (Seahra & Duley 1999), and crystalline silicon nanoparticles (Witt et al. 1998; Ledoux et al. 1998) – have been proposed as carriers of ERE. However, most candidates appear to be unable to simultaneously match the observed ERE spectra and the required PL efficiency (see Witt et al. 1998 for details).

Although high photoluminescence efficiencies can be obtained by PAHs, the lack of spatial correlation between the ERE and the PAH IR emission bands in the compact HII region Sh 152 (Darbon et al. 2000), the Orion Nebula (Perrin & Sivan 1992), and the Red Rectangle (Kerr et al. 1999), and the detection of ERE in the Bubble Nebula where no PAH emission has been detected (Sivan & Perrin 1993) seem against PAHs as ERE carriers.

Seahra & Duley (1999) argued that small carbon clusters were able to meet both the ERE profile and the PL efficiency requirements. However, this hypothesis appears to be ruled out by non-detection in NGC 7023 of the $1\,\mu m$ ERE peak (Gordon et al. 2000) predicted by the carbon nanoparticle model.

Witt et al. (1998) and Ledoux et al. (1998) suggested crystalline silicon nanoparticles (SNPs) with 15Å – 50Å diameters as the carrier on the basis of experimental data showing that SNPs could provide a close match to the observed ERE spectra and satisfy the quantum efficiency requirement. Smith & Witt (2001) have further developed the SNP model for the ERE, concluding that the observed ERE in the diffuse ISM can be explained with $Si/H = 6$ ppm in SiO_2-coated SNPs with Si core radii $a \approx 17.5$ Å.

Li & Draine (2002b) calculated the thermal emission expected from such particles, both in a reflection nebula such as NGC 2023 and in the diffuse ISM. They found that Si/SiO_2 SNPs (both neutral and charged) would produce a strong emission feature at $20\,\mu m$. The observational upper limit on the $20\,\mu m$ feature in NGC 2023 imposes an upper limit of < 0.2ppm Si in Si/SiO_2 SNPs. The ERE emissivity of the diffuse ISM appears to require > 15 ppm ($\geq42\%$ of solar Si abundance) in Si/SiO_2

SNPs. In comparison with the predicted IR emission spectra, they found that the DIRBE (*Diffuse Infrared Background Experiment*) photometry appears to rule out such high abundances of free-flying SNPs in the diffuse ISM. Therefore they concluded that if the ERE is due to SNPs, they must be either in clusters or attached to larger grains. Future observations by SIRTF will be even more sensitive to the presence of free-flying SNPs.

2.2.2   Dust Temperatures and IR Emission.

The temperatures of interstellar dust particles depend on their optical properties and sizes (i.e., on the way they absorb and emit radiation) as well as on the interstellar radiation field (ISRF).[23] Most of the visible and UV radiation in galaxies from stars passes through clouds of particles and heats them. This heating leads to reradiation at much longer wavelengths extending to the millimeter. On the average, in spiral galaxies, $\sim 1/4 - 1/3$ of the total stellar radiation is converted into dust emission (Cox & Mezger 1989; Calzetti 2001). The converted radiation is a probe of the particles and the physical environments in which they find themselves.

There is a long history in the study of grain temperature (and emission) since Eddington's demonstration of a 3.2 K black body equilibrium dust temperature assuming a 10^4 K interstellar radiation field diluted by a factor of 10^{-14} (Eddington 1926). Van de Hulst (1949) was the first to provide a realistic dust model temperature, ~ 15 K for dielectric particles. A subsequent extensive investigation was made by Greenberg (1968, 1971) where the temperatures were calculated for various grain types in regions of various radiation fields. The first step to study the shape effects on dust temperatures was taken by Greenberg & Shah (1971). They found that the temperatures of non-spherical dielectric grains are generally lower than those of equivalent spheres, but insensitive to modest shape variations. Later efforts made by Chlewicki (1987) and Voshchinnikov, Semenov, & Henning (1999) essentially confirmed the results of Greenberg & Shah (1971).

The advent of the IRAS, COBE, and ISO space IR measurements provided powerful information regarding the far-IR emission of the large particles (the so-called "cold dust"). The "cold dust" problem has received much attention since it plays an important role in many astrophysical subjects; for example, the presence of "cold dust" would change the current concept on the morphology and physics of galaxies (Block 1996).

The presence of a population of ultrasmall grains was known long before the IR era. Forty-six years ago, Platt (1956) proposed that very small grains or large molecules with radii ≤ 10Å may be present in interstellar space. Donn (1968) further proposed that polycyclic aromatic hydrocarbon-like "Platt particles", may be responsible for the UV interstellar extinction.

These very small grains – consisting of tens to hundreds of atoms – are small enough that the time-averaged vibrational energy $\langle E \rangle$ is smaller than or comparable to the energy of the starlight photons which heat the grains. Stochastic heating by absorption of starlight therefore results in transient "temperature spikes", during which much of the energy deposited by the starlight photon is reradiated in the IR. The idea of transient heating of very small grains was first introduced by Greenberg (1968). Since then, there have been a number of studies on this topic (see Draine & Li 2001 and references therein).

Since the 1980s, an important new window on the "very small grain component" has been opened by IR observations. The near-IR continuum emission of reflection nebulae (Sellgren, Werner, & Dinerstein 1983) and the 12 and 25 μm "cirrus" emission detected by IRAS (Boulanger & Pérault 1988) explicitly indicated the presence of a very small interstellar dust component since large grains (with radii $\sim 0.1\mu$m) heated by diffuse starlight emit negligibly at such short wavelengths, whereas very small grains (with radii $\leq 0.01\mu$m) can be transiently heated to very high temperatures (≥ 1000 K depending on grain size, composition, and photon energy). Subsequent measurements by the DIRBE instrument on the COBE satellite confirmed this and detected additional broadband emission at 3.5 and 4.9 μm (Arendt et al. 1998).

More recently, spectrometers aboard the *Infrared Telescope in Space* (IRTS; Onaka et al. 1996; Tanaka et al. 1996) and ISO (Mattila et al. 1996) have shown that the diffuse ISM radiates strongly in emission features at 3.3, 6.2, 7.7, 8.6, and 11.3 μm.

> "PAHs, they are everywhere!"
> — L.J. Allamandola [1996]

These emission features, first seen in the spectrum of the planetary nebulae NGC 7027 and BD+30°3639 (Gillett, Forrest, & Merrill 1973), have been observed in a wide range of astronomical environments including planetary nebulae, protoplanetary nebulae, reflection nebulae, HII regions, circumstellar envelopes, and external galaxies (see Tielens et al. 1999 for a review for Galactic sources and Helou 2000 for extragalactic sources). Often referred to as "unidentified infrared" (UIR) bands, these emission features are now usually attributed to PAHs which are vibrationally excited upon absorption of a single UV/visible photon (Léger & Puget 1984; Allamandola, Tielens, & Barker 1985) although other carriers have also been proposed such as HAC (Duley & Williams 1981; Borghesi, Bussoletti, & Colangeli 1987; Jones, Duley, & Williams

1990), QCC (Sakata et al. 1990), coal (Papoular et al. 1993), fullerenes (Webster 1993), and interstellar nanodiamonds with sp^3 surface atoms reconstructed to sp^2 hybridization (Jones & d'Hendecourt 2000).

The emission mechanism proposed for the UIR bands – UV excitation of gas-phase PAHs followed by internal conversion and IR fluorescence[24] – is supported by laboratory measurements of the IR *emission* spectra of gas-phase PAH molecules (Cherchneff & Barker 1989; Brenner & Barker 1989; Kurtz 1992; Cook et al. 1998) and by theoretical investigations of the heating and cooling processes of PAHs in interstellar space (Allamandola, Tielens, & Barker 1989; Barker & Cherchneff 1989; d'Hendecourt et al. 1989; Draine & Li 2001a).

The near-IR $(1-5\,\mu m)$, mid-IR $(5-12\,\mu m)$ emission spectrum along with the far-IR $(>12\,\mu m)$ continuum emission of the diffuse Galactic medium yields further insights into the composition and physical nature of interstellar dust; in particular, the PAH emission features allow us to place constraints on the size distribution of the very small dust component.

Attempts to model the IR emission of interstellar dust have been made by various workers. Following the initial detection of 60 and $100\,\mu m$ cirrus emission (Low et al. 1984), Draine & Anderson (1985) calculated the IR emission from a graphite/silicate grain model with grains as small as 3Å and argued that the 60 and $100\,\mu m$ emission could be accounted for. When further processing of the IRAS data revealed stronger-than-expected 12 and $25\,\mu m$ emission from interstellar clouds (Boulanger, Baud, & van Albada 1985), Weiland et al. (1986) showed that this emission could be explained if very large numbers of 3–10Å grains were present. A step forward was taken by Désert, Boulanger, & Puget (1990), Siebenmorgen & Krügel (1992), Schutte, Tielens, & Allamandola (1993), and Dwek et al. (1997) by including PAHs as an essential grain component. Early studies were limited to the IRAS observation in four broad photometric bands, but Dwek et al. (1997) were able to use DIRBE and FIRAS data.

In recent years, there has been considerable progress in both experimental measurements and quantum chemical calculations of the optical properties of PAHs (Allamandola, Hudgins, & Sandford 1999; Langhoff 1996; and references therein). There is also an improved understanding of the heat capacities of dust candidate materials (Draine & Li 2001) and the stochastic heating of very small grains (Barker & Cherchneff 1989; d'Hendecourt et al. 1989; Draine & Li 2001), the interstellar dust size distributions (Weingartner & Draine 2001a), and the grain charging processes (Weingartner & Draine 2001b).

Li & Draine (2001b) have made use of these advances to model the full emission spectrum, from near-IR to submillimeter, of dust in the diffuse ISM. The model consists of a mixture of amorphous silicate grains and carbonaceous grains, each with a wide size distribution ranging from molecules containing tens of atoms to large grains $\geq 1\,\mu$m in diameter. The carbonaceous grains are assumed to have PAH-like properties at very small sizes, and graphitic properties for radii $a \geq 50\text{Å}$. On the basis of recent laboratory studies and guided by astronomical observations, they have constructed "astronomical" absorption cross sections for use in modelling neutral and ionized PAHs from the far UV to the far IR. Using realistic heat capacities (for calculating energy distribution functions for small grains undergoing "temperature spikes"), realistic optical properties, and a grain size distribution consistent with the observed interstellar extinction (Weingartner & Draine 2001a), Li & Draine (2001b) were able to reproduce the near-IR to submillimeter emission spectrum of the diffuse ISM, including the PAH emission features at 3.3, 6.2, 7.7, 8.6, and $11.3\,\mu$m.

The silicate/graphite-PAH model has been shown also applicable to the Small Magellanic Cloud (Li & Draine 2002a; Weingartner & Draine 2001a).

Li & Draine (2002c) have also modelled the excitation of PAH molecules in UV-poor regions. It was shown that the astronomical PAH model provides a satisfactory fit to the UIR spectrum of vdB 133, a reflection nebulae with the lowest ratio of UV to total radiation among reflection nebulae with detected UIR band emission (Uchida, Sellgren, & Werner 1998).

2.2.3   Microwave Emission: Spinning Dust Grains.

A number of physical processes including collisions with neutral atoms and ions, plasma drag, absorption and emission of photons can drive ultra-small grains to rapidly rotate (Draine & Lazarian 1998a, 1998b; Draine & Li 2002). The electric dipole emission from these spinning dust grains was shown to be able to account for the 10–100 GHz "anomalous" Galactic background component (See Draine & Lazarian 1998b and references therein).

2.3.   INTERSTELLAR DEPLETIONS

"Where have all these atoms gone?"

— J. Mayo Greenberg [1963]

Derivations of the relative abundances of the elements in our Galaxy are one of the principal needs for understanding the chemical evolution in interstellar space – and ultimately its memory in comets. A major factor in developing consistent dust models was the observation of the "depletion" in low density clouds (atoms locked up in grains are "depleted" from the gas phase) using the UV absorption line spectroscopy as a probe of the gas-phase abundances and assuming a reference abundance (abundances of atoms both in gas and in dust).

The deduced possible dust composition was initially only constrained to the extent that silicates alone could not be responsible for the interstellar extinction (Greenberg 1974). But in recent years, the problem of grain modelling has been exacerbated by the apparent decrease of the available condensible atoms (O, C, N, Si, Mg, Fe) by about 30% (Snow & Witt 1996) since the solar system was born.[25] This implies that the heavy elements are being consumed more than they are being created. However, if one goes back far enough in time, there were no condensible atoms because their initial production must follow the birth of stars. This brings us to the cosmological question of what do high-z galaxies look like and when and how was dust first found in them?

2.4. INTERSTELLAR DUST IN THE SOLAR SYSTEM

Interstellar grains have been found in primitive meteorites and in interplanetary dust particles based on the analysis of isotopic anomalies (see Kerridge 1999 and Bradley 1999 for recent reviews). Most presolar grains identified to date are carbonaceous: diamonds, SiC, and graphite (very small TiC, ZrC, and MoC grains have also been found as inclusions in SiC and graphite grains); also identified are oxides such as corundum (Al_2O_3) and silicon nitride (Si_3N_4). One should keep in mind that much of the less refractory dust incorporated into meteorites is lost during the chemical processing used to extract the refractory grains from meteorites. Therefore, the extracted presolar grains are compositionally not representative of the bulk interstellar dust;[26] for example, the procedures used to isolate interstellar grains in meteorites are designed to deliberately destroy silicate material which constitutes the bulk of the host meteorite (see Draine 1994).

The solar system is surrounded by the local interstellar cloud with a density $n_H \approx 0.3\,cm^{-3}$ and moving past the Sun with a velocity $\approx 26\,km\,s^{-1}$ (Lallement et al. 1994). Interstellar grains embedded in the local cloud with sufficiently low charge-to-mass ratios can penetrate the heliopause and enter the solar system on hyperbolic orbits (small, charged grains are deflected from the heliosphere; Linde & Gombosi 2000). The interplanetary spacecraft Ulysses and Galileo have detected over 600 grains flowing into the solar system and determined their speed, direction, and mass and therefore the mass flux (but not chemical composition since the grains were destroyed by the detection technique; Grün et al. 1993, 1994; Frisch et al. 1999; Landgraf et al. 2000). Including the mass of the large population of interstellar micrometeorites entering the Earth's atmosphere (Taylor et al. 1996; Baggaley 2000), Frisch et al. (1999) found that the total dust-to-gas mass ratio in the local interstellar cloud is about twice the canonical value determined from the interstellar extinction. They also found that there is a substantial amount of mass in large grains of $\sim 1\,\mu m$ in size which is difficult to reconcile with the interstellar extinction and interstellar elemental abundances.

2.5.   FROM INTERSTELLAR DUST TO COMETS

A major advance in our understanding of comets in the 20th century was made by the space probes Vega 1 and 2 and Giotto (see Nature, comet Halley issue, vol. 321, 1986). Until that time no one had ever seen a comet nucleus. The critical new discoveries were: (1) the low albedo (≈ 0.04) of comets; (2) the size distribution of the comet dust extending down to interstellar dust sizes ($10^{-15} - 10^{-18}$ g), (3) the organic fraction of comet dust (see Greenberg & Li 1999b and references therein). The current ground based observations of the volatile composition of comets implies a close connection with the ices of interstellar dust (see Crovisier 1999 and Irvine & Bergin 2000 for recent reviews).

Most of the current models of comet nuclei presume that to a major extent they are basically aggregates of the interstellar dust in its final evolved state in the collapsing molecular cloud which becomes the protosolar nebula. In addition to the chemical consequences of such a model, there is the prediction of a morphological structure in which the aggregate material consists of tenth micron basic units each of which contains (on average) a silicate core, a layer of complex organic material, and an outer layer of ices in which are embedded all the very small carbonaceous particles characterizing the interstellar UV hump and the far UV extinction. All these components have been observed in the comet comae in one way or another.[27] The implication is that space probes which can examine in detail the composition of comet nuclei will be able to provide us with hands-on data on most of the components of interstellar dust and will tell us what is the end product of chemical evolution in a collapsing protosolar molecular cloud. At this time many laboratories are preparing materials as a data base for comparison with what will be analyzed during the space missions.

3.   FUTURE

There are quite a few unsolved or partially solved problems related to interstellar dust which will be demanding close attention in the future (see below for a list). A number of new remote observational facilities which will be available early in the new millennium (Atacama Large Millimeter Array [MMA/LSA], Far Infrared and Submillimeter Telescope [FIRST], Next Generation Space Telescope [NGST], Space Infrared Telescope Facility [SIRTF], Stratospheric Observations for Infrared Astronomy [SOFIA], Submillimeter Wave Astronomy Satellite [SWAS]) will permit further tests of current dust models and promise new observational breakthroughs.

1 What is the source and nature of the Diffuse Interstellar Bands?

2 What is the carrier of the 2175 Å extinction hump?

3 What is the carrier of the "Unidentified Infrared Bands"? if it is PAHs, where are they formed? are they mainly from carbon star outflows or formed in situ by ion-molecule reactions (Herbst 1991) or from the organic refractories derived from photoprocessing of ice mixtures (Greenberg et al. 2000)?

4 What is the carrier of the Extended Red Emission?

5 What are all the sources and sinks (destruction) of interstellar dust? where are interstellar grains made? are they mainly made in the cold ISM (Draine 1990) or are the silicate cores mainly stardust (serving as "condensation seeds") while the organic mantles are formed in the ISM (Greenberg 1982a; Greenberg & Li 1999a)?

6 What are the exact composition and morphology of interstellar dust? are they separate bare silicate and graphite grains or silicate core-carbonaceous mantle grains or composite grains composed of small silicates, carbon and vacuum? if most of interstellar grain mass is condensed in the cold ISM, how can pure silicate and graphite grains form (see Draine 1995)?

7 What are the sizes of large dust grains ($> 0.25\,\mu$m)? how much can we learn from X-ray halos and from spacecraft in situ dust detections?

8 Why are crystalline silicates not seen in the ISM while they are present in stardust and cometary dust? how do cometary silicates become crystalized?

9 How do molecular hydrogen and other simple molecules form on grain surfaces? although considerable progress has been made in recently years in studies of the diffusion rates of adsorbed hydrogen atoms on the surfaces of variable dust materials, the recombination reactions, and the restoration of the new molecules to the gas phase (Pirronello et al. 1997; Pirronello et al. 1999; Manicò et al. 2001), the formation of molecular hydrogen is still not well understood (Herbst 2000; Pirronello 2000).

10 How do interstellar grains accrete and deplete mantles in dense molecular clouds? we need high spatial resolution observations of molecule distributions in the gas and in the solid as function of depth in the cloud – interiors of clouds as well as regions of low and high mass star formation.

11 How does dust evolve in protosolar regions? we need higher spatial resolution and sensitivity. Improvements in the theory of dust/grain chemistry, particularly in collapsing clouds leading to star formation as well as in quiescent molecular clouds.

12 Will the chemical and morphological analysis of comet nuclei and dust material reveal the true character of interstellar dust? will they provide further answers to the question of life's origin?

13 How can we resolve the evolution of interstellar matter leading to the material measured and analyzed in meteorites, in interplanetary dust particles?

14 What is the true atomic composition of the interstellar medium? how variable is it in time and space? are there global variation over distances of kiloparsecs?

15 When did dust first form in a galaxy? what are the composition and sizes of dust in extragalactic environments?

Acknowledgments

A. Li was deeply saddened by the passing away of Prof. J. Mayo Greenberg on November 29, 2001. As a pioneer in the fields of cosmic dust, comets, astrochemistry, astrobiology and light scattering, Mayo's passing was a great loss for the astrocommunity. Mayo had been scientifically active till his very last days. Just a few weeks before Mayo passed away, A. Li discussed future collaboration plans with him on dust in high-z galaxies. It was a great experience for A. Li to work with Mayo in Leiden. He will be remembered forever, as a great astrophysicist and as a great mentor. A. Li is also grateful to Profs. Bruce T. Draine and Ewine F. van Dishoeck for their continuous advice, encouragement and support. A. Li thanks Profs. Lou J. Allamandola, Bruce T. Draine, Jonathan Lunine, Valerio Pirronello for valuable discussions, comments and suggestions. Some of the materials of §2 (dust history) were taken from Greenberg & Shen (1999) and the "Introduction" Chapter of A. Li's PhD thesis (Leiden, 1998); of §3.2.1 (dust luminescence) from Li & Draine (2002a); and of §3.2.2 (dust IR emission) from Li & Draine (2001b). This work was supported in part by NASA grant NAG5-7030 and NSF grant AST-9988126 and by a grant from the Netherlands Organization for Space Research (SRON).

Notes

1. The contents in the square brackets were added by A. Li, for completeness, to summarize the then understanding as discussed in Greenberg (1963).

2. Molecular clouds are partially supported by magnetic fields; star formation occurs if ambipolar diffusion deprives cloud cores of magnetic support. Charged grains can couple to the magnetic field and increase the collisional drag on the neutrals, and thereby slow the rate of ambipolar diffusion within a cloud and increase the time needed to form a protostellar core (Ciolek 1995).

3. Henry Norris Russell (1922), Shapley's advisor at Princeton, believed that the existence of dark clouds accounted for the obscuration and argued that this obscuring matter had to be the form of fine dust. But Shapley did not follow his advice.

4. But Kapteyn did not take this seriously; for example, he assumed no extinction in his grand 1922 paper on the motion of stars in the Galaxy (Kapteyn 1922).

5. Trumpler's observations indicated reddening even where he saw no clouds. Dufay (1957) questioned whether interstellar space outside dark clouds and nebulae should be considered perfectly transparent.

6. Greenstein (1938) concluded that a dust size distribution of $dn/da \sim a^{-3.6}$ "seems to provide the best agreement of theory and observation (interstellar extinction)." Interesting enough, this power law distribution was very close to the $dn/da \sim a^{-3.5}$ distribution derived about 40 years later for the silicate/graphite dust model (Mathis, Rumpl, & Nordsieck 1977; Draine & Lee 1984).

7. Another possible influencing factor of proposing the metallic dust model might have been the fact that it was easier to compute the scattering by metallic particles using the Mie theory because to get a λ^{-1} law required smaller particles than if they were dielectric and computations for large particles were too tedious (van de Hulst 1986).

8. A. Li noted: the term "dirty ice" was invented by J.M. Greenberg as recalled by H.C. van de Hulst (1997).

9. Let $C_{abs}(\lambda) \propto \lambda^{-\beta}$ be the far-IR absorption cross section; T_d be the characteristic dust temperature; $j_\lambda \propto C_{abs}(\lambda) \times 4\pi B_\lambda(T_d) \propto \lambda^{-(4+\beta)}$ be the dust far-IR emissivity (where $B_\lambda[T_d]$ is the Planck function at wavelength λ and temperature T_d). While the observed emission spectrum between $100\,\mu m$ and $3000\,\mu m$ (Wright et al. 1991; Reach et al. 1995) is well represented by dust with $\beta = 1.7$, $T_d \approx 19.5\,K$ or $\beta = 2.0$, $T_d \approx 18.5\,K$ (Draine 1999), fluffy composite grains have $\beta \approx 1.60$ (Mathis & Whiffen 1989). A $\beta = 2.0$ emissivity law is naturally expected for solid compact dust: the far-IR absorption formula for spherical submicron-sized grains is $C_{abs}(\lambda)/V = 18\pi/\lambda \times \left\{ \epsilon_{im} / \left[(\epsilon_{re} + 2)^2 + \epsilon_{im}^2 \right] \right\}$ where V is the grain volume; ϵ_{re} and ϵ_{im} are respectively the real and imaginary part of the dielectric function. For dielectrics $\epsilon_{im} \propto \lambda^{-1}$ and $\epsilon_{re} \propto const$ ($\epsilon_{im} \ll \epsilon_{re}$) while for metals $\epsilon_{im} \propto \lambda$, $\epsilon_{re} \propto const$ ($\epsilon_{im} \gg \epsilon_{re}$) so that one gets the same asymptotic relation for both dielectrics and metals $C_{abs}(\lambda) \propto \lambda^{-2}$.

10. The current view is that Jupiter family comets (with small inclinations and orbital periods $P < 20$ yrs) formed in the trans-Neptune region now known as the "Kuiper Belt"; Halley-type comets (with relatively longer periods $20 < P < 200$ yrs and larger inclinations) as well as long-period comets (with all possible inclinations and orbital periods $200 < P < 10^7$ yrs) formed somewhere beyond the orbits of Jupiter and Saturn. See Li & Greenberg (1998a) and references therein.

11. Greenberg & Chlewicki (1983) found that the strength of the 2175 Å hump and the far-UV extinction can vary both independently and with respect to the visual extinction. This may imply that the 2175 Å hump and the far-UV extinction are produced by two different dust components.

12. The silicate core-organic refractory mantle model (Greenberg 1989; Li & Greenberg 1997) may provide a better fit to the 3–8 μm extinction curve than the silicate-graphite model (Mathis et al. 1977; Draine & Lee 1984) because the carbonaceous organic material is IR active at $\lambda > 5\,\mu m$ due to C = C, C = O, C − OH, C ≡ N, C − NH$_2$ stretches; CH, OH, and NH$_2$ deformations, and H wagging (Greenberg et al. 1995).

13. $R_V \approx (5.6 \pm 0.3)\lambda_{max}$ (λ_{max} is in micron; see Whittet 1992); i.e., the wavelength of maximum polarization λ_{max} shifts with R_V in the sense that it moves to longer wavelengths as R_V increases which is the effect of increasing the particle size.

14. The far-UV polarization observations only became available in recent years as a consequence of the Wisconsin Ultraviolet Photo-Polarimetry Experiment (WUPPE) (Clayton et al. 1992) and the UV polarimetry of the Hubble Space Telescope (Somerville et al. 1993, 1994; Clayton et al. 1995). It is found that (1) lines of sight with $\lambda_{max} \geq 0.54\,\mu m$ are consistent with the extrapolated Serkowski law; (2) lines of sight with $\lambda_{max} \leq 0.53\,\mu m$ show polarization in excess of the extrapolated Serkowski law; (3) two lines of sights show a polarization feature which seems to be associated with the 4.6 μm^{-1} extinction hump (Clayton et al. 1992; Anderson et al. 1996; Wolff et al. 1997; Martin, Clayton, & Wolff 1999).

15. Martin et al. (1999) proposed a more complicated formula – the "modified Serkowski law" – to represent the observed interstellar polarization from the near IR to the far UV.

16. Mennella et al. (1996) reported that a stable peak position can be obtained by subjecting small hydrogenated amorphous carbon grains to UV radiation. However, the laboratory produced humps are too wide and too weak with respect to the interstellar one. In a later paper (Mennella et al. 1998) they proposed that the 2175 Å carrier can be modelled as a linear combination of such materials exposed to different degrees of UV processing.

17. Hecht (1981) investigated the effect of coatings on graphite particles and concluded that small coatings could be present on spherical $a \approx 200$ Å particles and still cause the 2175 Å feature.

18. Very recently, Li & Draine (2001a) estimated that the abundances of $a < 1\,\mu m$ crystalline silicate grains in the diffuse ISM is $< 5\%$ of the solar Si abundance.

19. Crystalline silicates have also been seen in six comets (see Hanner 1999 for a summary), in dust disks around main-sequence stars (see Artymowicz 2000 for a summary), young stellar objects (see Waelkens, Malfait, & Waters 2000 for a summary), in interplanetary dust particles (IDPs) (Bradley et al. 1999), and probably also in the Orion Nebula (Cesarsky et al. 2000).

20. Pendleton & Allamandola (2002) questioned the robustness of the fit by the organic residue since the broad 5.5–10 μm band seen in the organic residue spectrum was not observed in astronomical spectra. We note that this band, largely attributed to the combined features of the C=O, C-OH, C≡N, C-NH$_2$, OH, and NH$_2$ stretches, bendings, and deformations, will become weaker if the organics are subject to further UV photoprocessing which will result in photodissociation and depletion of H, O, N elements. The organic residue samples presented in Greenberg et al. (1995) were processed at most to a degree resembling one cycle (from molecular clouds to diffuse clouds). According to the cyclic evolution model, interstellar grains will undergo ∼ 50 cycles before they are consumed by star formation or becomes a part of a comet (Greenberg & Li 1999a).

21. We note that the observed correlation between the 10 μm silicate and the 3.4 μm C-H hydrocarbon optical depths (Sandford, Pendleton, & Allamandola 1995) is consistent with the core-mantle scenario.

22. Hydrogenated nanodiamonds were identified in the circumstellar dust envelopes surrounding two Herbig Ae/Be stars HD 97048 and Elias 1 revealed by their 3.43 μm and 3.53 μm emission features (Guillois, Ledoux, & Reynaud 1999; van Kerckhoven, Tielens, & Waelkens 2002). Interstellar diamonds were first proposed as a dust component by Saslaw & Gaustad (1969) and first discovered in meteorites by Lewis, Anders, & Draine (1989). The fact that the 3.43 μm and 3.53 μm features are not observed in the ISM led Tielens et al. (1999) to infer an upper limit of ≤ 0.1 ppm for hydrogenated interstellar nanodiamonds. If interstellar diamonds are not hydrogenated, they could be much more abundant (van Kerckhoven et al. 2002). Analysis of the interstellar extinction observations show that up to 10% of the interstellar carbon can be locked up in diamond and escape detection (Lewis et al. 1989).

23. Originally, the ISRF was represented by a 10^4 K black-body radiation diluted by a factor of 10^{-14} (Eddington 1926) which is undoubtedly too crude but serves as a simple and adequate approximation for some purposes. Many attempts have been made to obtain a more reasonable determination of the ISRF either on the basis of direct measurements of the UV radiation from the sky or by calculating the radiation of hot stars using model atmospheres. Van Dishoeck (1994) has summarized the typical ISRF estimates (Habing 1968; Draine 1978; Gondhalekar et al. 1980; Mathis, Mezger, & Panagia 1983). As illustrated in Fig. 2 of van Dishoeck (1994), the various estimates agree within factors of two. The latest work on the local far-UV ISRF by Parravano, Hollenbach, & Mckee (2002) led to a value quite close to Draine (1978).

24. PAHs are actually excited by photons of a wide range of wavelengths (Li & Draine 2002c).

25. But see also Sofia & Meyer (2001), who argue that interstellar abundances are approximately solar. It is also possible that the solar system formed out of material with a higher metallicity than the average ISM at that time (there is evidence that stars with planets have higher metallicities than equal age stars without planets). So the metallicity of the ISM may not have declined since the solar system was formed (Draine, private communication).

26. We have already seen in Footnote-22 that diamond is not a major interstellar dust component. Whittet, Duley, & Martin (1990) found that the abundance of Si in SiC dust in the diffuse ISM is at most 5% of that in silicates.

27. PAH molecules, a significant constituent of interstellar dust (see §3.2.2), would also be present in comets if they indeed contain unprocessed interstellar matter. The presence of PAHs in comets has been suggested by the 3.28 μm emission feature detected in some comets (Bockelée-Morvan, Brooke, & Crovisier 1995). More specifically, a 3-ring PAH molecule – phenanthrene ($C_{14}H_{10}$) – has been proposed as the carrier for the 342–375 nm fluorescence bands seen in comet 1P/Halley (Moreels et al. 1994). Li & Draine (2002d) are currently working on this topic.

REFERENCES

Aannestad, P.A. and Kenyon, S.J. (1979), *ApJ*, **230**, 771.

Adamson, A.J. and Whittet, D.C.B. (1992), *ApJ*, **398**, L69.

Adamson, A.J. and Whittet, D.C.B. (1995), *ApJ*, **448**, L49.

Adamson, A.J., Whittet, D.C.B., Chrysostomou, A., Hough, J.H., Aitken, D.K., Wright, G.S. and Roche, P.F. (1999), *ApJ*, **512**, 224.

Adamson, A.J., Whittet, D.C.B. and Duley, W.W. (1990), *MNRAS*, **243**, 400.

Aitken, D.K. (1996), in Roberge, W.G. and Whittet, D.C.B. (eds.), *Polarimetry of the Interstellar Medium*, ASP Conf. Ser. 97, ASP, San Francisco, p.225.

Aitken, D.K., Roche, P.F., Smith, C.H., James, S.D. and Hough, J.H. (1988), *MNRAS*, **230**, 629.

Aitken, D.K., Smith, C.H. and Roche, P.F. (1989), *MNRAS*, **236**, 919.

Allamandola, L.J. (1996), in Greenberg, J.M. (ed.), *The Cosmic Dust Connection*, Kluwer, Dordrecht, p.81.

Allamandola, L.J. (2002), this volume.

Allamandola, L.J., Hudgins, D.M. and Sandford, S.A. (1999), *ApJ*, **511**, L115.
Allamandola, L.J., Sandford, S.A., Tielens, A.G.G.M. and Herbst, T.M. (1992), *ApJ*, **399**, 134.
Allamandola, L.J., Tielens, A.G.G.M. and Barker, J.R. (1985), *ApJ*, **290**, L25.
Allamandola, L.J., Tielens, A.G.G.M. and Barker, J.R. (1989), *ApJSS*, **71**, 733.
Anderson, C.M., Weitenbeck, A.J., Code, A.D. et al. (1996), *AJ*, **112**, 2726.
Arendt, R.G., Odegard, N., Weiland, J.L. et al. (1998), *ApJ*, **508**, 74.
Artymowicz, P. (2000), *Space Sci. Rev.*, **92**, 69.
Asano, S. and Sato, M. (1980), *Appl. Opt.*, **19**, 962.
Asano, S. and Yamamoto, G. (1975), *Appl. Opt.*, **14**, 29.
Baggaley, W.J. (2000), *J. Geophys. Res.*, **105**, 10353.
Barber, P.W. and Yeh, C. (1975), *Appl. Opt.*, **14**, 2864.
Barker, J.R. and Cherchneff, I. (1989), in Allamandola, L.J. and Tielens, A.G.G.M. (eds.), *Interstellar Dust*, IAU Symp. 135, Kluwer, Dordrecht, p.197.
Barnard, E.E. (1919), *ApJ*, **49**, 1.
Blanco, A., Cappello, D., Fonti, S., Ientile, A. and Orofino, V. (1998), *Adv. Space Res.*, **24**, 443.
Block, D.L., (1996), in Block, D.L. and Greenberg, J.M. (eds.), *New Extragalactic Perspectives in the New South Africa*, Kluwer, Dordrecht, p.1.
Bockelée-Morvan, D., Brooke, T.Y. and Crovisier, J. (1995), *Icarus*, **116**, 18.
Bohren, C.F. and Huffman, D.R. (1983), *Absorption and Scattering of Light by Small Particles*, Wiley, New York.
Borghesi, A., Bussoletti, E. and Colangeli, L. (1987), *ApJ*, **314**, 422.
Boulanger, F., Baud, B. and van Albada, G.D. (1985), *A&A*, **144**, L9.
Boulanger, F. and Pérault, M. (1988), *ApJ*, **330**, 964.
Boulanger, F., Prévot, M.L. and Gry, C. (1994), *A&A*, **284**, 956.
Bradley, J.P. (1999), in Greenberg, J.M. and Li, A. (eds.), *Formation and Evolution of Solids in Space*, Kluwer, Dordrecht, p.485.
Bradley, J.P., Keller, L.P., Snow, T.P. et al. (1999), *Science*, **285**, 1716.
Brenner, J.D. and Barker, J.R. (1992), *ApJ*, **388**, L39.
Briggs, R., Ertem, G., Ferris, J.P., Greenberg, J.M., McCain, P.J., Mendoza-Gómez, C.X. and Schutte, W.A. (1992), *Origin of Life and Evolution of the Biosphere*, **22**, 287.
Brooke, T.Y., Sellgren, K. and Geballe, T.R. (1999), *ApJ*, **517**, 883.
Bussoletti, E., Colangeli, L., Borghesi, A. and Orofino, V. (1987), *A&AS*, **70**, 257.
Butchart, I., McFadzean, A.D., Whittet, D.C.B., Geballe, T.R. and Greenberg, J.M. (1986), *A&A*, **154**, L5.
Calzetti, D. (2001), *PASP*, **113**, 1449.
Cardelli, J.A., Clayton, G.C. and Mathis, J.S. (1989), *ApJ*, **345**, 245.
Cayrel, R. and Schatzman, E. (1954), *Ann. d'Ap.*, **17**, 555.
Cesarsky, D., Jones, A.P., Lequeux, J. and Verstraete, L. (2000), *A&A*, **358**, 708.
Cherchneff, I. and Barker, J.R. (1989), *ApJ*, **341**, L21.
Chiar, J.E., Pendleton, Y.J., Geballe, T.R. and Tielens, A.G.G.M. (1998), *ApJ*, **507**, 281.
Chiar, J.E., Tielens, A.G.G.M., Whittet, D.C.B. et al. (2000), *ApJ*, **537**, 749.
Chlewicki, G. (1987), *A&A*, **181**, 127.
Chlewicki, G. and Greenberg, J.M. (1990), *ApJ*, **365**, 230.
Chrysostomou, A., Hough, J.H., Whittet, D.C.B., Aitken, D.K., Roche, P. F. and Lazarian, A. (1996), *ApJ*, **465**, L61.

Ciolek, G.E. (1995), in Ferrara, A., McKee, C.F., Heiles, C. and Shapiro, P.R. (eds.), *The Physics of the Interstellar Medium and Intergalactic Medium*, ASP Conf. Ser. 80, ASP, San Francisco, p.174.

Clayton, G.C., Anderson, C.M., Magalhaes, A.M. et al. (1992), *ApJ*, **385**, L53.

Clayton, G.C., Wolff, M.J., Allen, R.G. and Lupie, O. (1995), *ApJ*, **445**, 947.

Clayton, G.C., Wolff, M.J., Gordon, K.D. and Misselt, K.A. (2000), in Sitko, M.L., Sprague, A.L. and Lynch, D.K. (eds.), *Thermal Emission Spectroscopy and Analysis of Dust Disks and Regoliths*, ASP Conf. Ser. 196, ASP, San Francisco, p.41.

Clerke, A.M. (1903), *Problems in Astrophysics*, Black, London, p.567.

Colangeli, L., Mennella, V., Palumbo, P. and Rotundi, A. (1999), in Greenberg, J.M. and Li, A. (eds.), *Formation and Evolution of Solids in Space*, Kluwer, Dordrecht, p.203.

Cook, D.J., Schlemmer, S., Balucani, N. et al. (1998), *J. Phys. Chem.*, **102**, 1465.

Cox, P. and Mezger, P.G., (1989), *A&A Rev.*, 3.

Coyne, G.V., Gehrels, T. and Serkowski, K. (1974), *AJ*, **79**, 581.

Crovisier, J. (1999), in Greenberg, J.M. and Li, A. (eds.), *Formation and Evolution of Solids in Space*, Kluwer, Dordrecht, p.389.

Daniels, P.A. and Hughes, D.W. (1981), *MNRAS*, **195**, 1001.

Danielson, R.E., Woolf, N.J. and Gaustad, J.E. (1965), *ApJ*, **141**, 116.

Darbon, S., Zavagno, A., Perrin, J.-M., Savine, C., Ducci, V. and Sivan, J.-P. (2000), *A&A*, **364**, 723.

Davis, L. Jr. and Greenstein, J.L. (1951), *ApJ*, **114**, 206.

Debye, K.L. (1909), *Ann. Phys.*, NY **30**, 59.

Désert, F.X., Boulanger, F. and Puget, J.L. (1990), *A&A*, **237**, 215.

d'Hendecourt, L.B., Léger, A., Boissel, P. and Désert, F.X. (1989), in Allamandola, L.J. and Tielens, A.G.G.M. (eds.), *Interstellar Dust*, IAU Symp. 135, Kluwer, Dordrecht, p.207.

d'Hendecourt, L.B., Léger, A., Olofson, G. and Schmidt, W. (1986), *A&A*, **170**, 91.

Donn, B. (1968), *ApJ*, **152**, L129.

Donn, B., Daniels, P.A. and Hughes, D.W. (1985), *BAAS*, **17**, 520.

Dorschner, J. (1999), in Greenberg, J.M. and Li, A. (eds.), *Formation and Evolution of Solids in Space*, Kluwer, Dordrecht, p.229.

Dorschner, J. and Henning, Th. (1995), *A&A Rev.*, **6**, 271.

Draine, B.T. (1978), *ApJSS*, **36**, 595.

Draine, B.T. (1988), *ApJ*, **333**, 848.

Draine, B.T. (1989a), in Kaldeich, B.H. (ed.), *Infrared Spectroscopy in Astronomy*, Proc. 22nd Eslab Symp., ESA, Noordwijk, p.93.

Draine, B.T. (1989b), in Allamandola, L.J. and Tielens, A.G.G.M. (eds.), *Interstellar Dust*, IAU Symp. 135, Kluwer, Dordrecht, p.313.

Draine, B.T. (1990), in Blitz, L. (ed.), *The Evolution of the Interstellar Medium*, ASP Conf. Ser. 12, ASP, San Francisco, p.193.

Draine, B.T. (1994), in Cutri, M. and Latter, B. (ed.), *The First Symposium on the Infrared Cirrus and Diffuse Interstellar Medium*, ASP Conf. Ser. 58, ASP, San Francisco, p.227.

Draine, B.T. (1995), in Ferrara, A., McKee, C.F., Heiles, C. and Shapiro, P.R. (eds.), *The Physics of the Interstellar Medium and Intergalactic Medium*, ASP Conf. Ser. 80, ASP, San Francisco, p.133.

Draine, B.T. (1999), in Maiani, L., Melchiorri, F. and Vittorio, N. (eds.), *Proc. of the EC-TMR Conf. on 3K Cosmology*, AIP, Woodbury, p.283.

Draine, B.T. and Anderson, N. (1985), *ApJ*, **292**, 494.

Draine, B.T. and Lazarian, A. (1998a), *ApJ*, **494**, L19.

Draine, B.T. and Lazarian, A. (1998b), *ApJ*, **508**, 157.

Draine, B.T. and Lee, H.M. (1984), *ApJ*, **285**, 89.

Draine, B.T. and Li, A. (2001), *ApJ*, **551**, 807.

Draine, B.T. and Li, A. (2002), in preparation.

Draine, B.T. and Malhotra, S. (1993), *ApJ*, **414**, 632.

Draine, B.T. and Salpeter, E.E. (1979a), *ApJ*, **231**, 77.

Draine, B.T. and Salpeter, E.E. (1979b), *ApJ*, **231**, 438.

Draine, B.T. and Weingartner, J.C. (1996), *ApJ*, **470**, 551.

Draine, B.T. and Weingartner, J.C. (1997), *ApJ*, **480**, 633.

Dufay, J. (1957), *Galactic Nebulae and Interstellar Matter*, Philosophical Library, New York.

Duley, W.W. (1985), *MNRAS*, **215**, 259.

Duley, W.W., Jones, A.P. and Williams, D.A. (1989), *MNRAS*, **236**, 709.

Duley, W.W. and Seahra, S. (1998), *ApJ*, **507**, 874.

Duley, W.W. and Williams, D.A. (1981), *MNRAS*, **196**, 269.

Dwek, E. (1997), *ApJ*, **484**, 779.

Dwek, E., Arendt, R.G., Fixsen, D.J. et al. (1997), *ApJ*, **475**, 565.

Dyck, H.M., Capps, R.W., Forrest, W.J. and Gillett, F.C. (1973), *ApJ*, **183**, L99.

Dyck, H.M. and Londsale, C.J. (1981), in Wynne-Williams, C.G. and Cruickshank D.P. (eds.), *Infrared Astronomy*, IAU Symp. 96, Reidel, Dordrecht, p.223.

Eddington, A.S. (1926), *Diffuse Matter in Interstellar Space (Bakerian Lecture)*, Proc. Roy. Soc., **111A**, 424.

Ehrenfreund, P. and Charnley, S.B. (2000), *ARA&A*, **38**, 427.

Ehrenfreund, P. and Schutte, W.A. (2000), in Minh, Y.C. and van Dishoeck, E.F. (eds.), *Astrochemistry: From Molecular Clouds to Planetary Systems*, IAU Symp. 197, ASP, San Francisco, p.135.

Fahlman, G.G. and Walker, G.A.H. (1975), *ApJ*, **200**, 22.

Falco, E.E., Impey, C.D., Kochanek, C.S., Lehár, J., McLeod, B.A., Rix, H.-W., Keeton, C.R., Munoz, J.A. and Peng, C.Y. (1999), *ApJ*, **523**, 617.

Foing, B.H. and Ehrenfreund, P. (1994), *Nature*, **369**, 296.

Frisch, P.C., Dorschner, J., Geiss, J. et al. (1999), *ApJ*, **525**, 492.

Gibb, E.L. and Whittet, D.C.B. (2002), *ApJ*, **566**, L113.

Gillett, F.C. and Forrest, W.J. (1973), *ApJ*, **179**, 483.

Gillett, F.C., Forrest, W.J. and Merrill, K.M. (1973), *ApJ*, **184**, L93.

Gilman, R.C. (1969), *ApJ*, **155**, L185.

Gilra, D.P. (1972), in Code, A.D. (ed.), *The Scientific Results from the Orbiting Astronomical Observatory OAO-2*, NASA SP-310, 295.

Goebel, J.H. (1987), in Léger, A., d'Hendecourt, L. and Boccara, N. (eds.), *Polycyclic Aromatic Hydrocarbons and Astrophysics*, Reidel, Dordrecht, p.329.

Gold, T. (1952), *MNRAS*, **112**, 215.

Gombosi, T.I. and Houpis, H.L. (1986), *Nature*, **324**, 43.

Gondhalekar, P.M., Phillips, A.P. and Wilson, R. (1980), *A&A*, **85**, 272.

Gordon, K.D., Witt, A.N. and Friedmann, B.C. (1998), *ApJ*, **498**, 522.

Gordon, K.D., Witt, A.N., Rudy, R.J. et al. (2000), *ApJ*, **544**, 859.

Greenberg, J.M. (1963), *ARA&A*, **1**, 267.

Greenberg, J.M. (1968), in Middlehurst, B.M. and Aller, L.H. (eds.), *Stars and Stellar Systems*, Vol. VII, Univ. Chicago Press, Chicago, p.221.

Greenberg, J.M. (1971), *A&A*, **12**, 240.

Greenberg, J.M. (1973), in Gordon, M.A. and Snyder, L.E. (eds.), *Molecules in the Galactic Environment*, Wiley, New York, p.94.

Greenberg, J.M. (1974), *ApJ*, **189**, L81.

Greenberg, J.M. (1978), in McDonnell, J.A.M. (ed.), *Cosmic Dust*, Wiley, New York, p.187.

Greenberg, J.M. (1982a), in Beckman, J.E. and Phillips, J.P. (eds.), *Submillimeter Wave Astronomy*, Cambridge Univ. Press, New York, p.261.

Greenberg, J.M. (1982b), in Wilkening, L.L. (ed.), *Comets*, Univ. Arizona Press, Tucson, p.131.

Greenberg, J.M. (1989), in Allamandola, L.J. and Tielens, A.G.G.M. (eds.), *Interstellar Dust*, IAU Symp. 135, Kluwer, Dordrecht, p.345.

Greenberg, J.M. (1996), in Block, D.L. and Greenberg, J.M. (eds.), *New Extragalactic Perspectives in the New South Africa*, Kluwer, Dordrecht, p.609.

Greenberg, J.M. (1998), *A&A*, **330**, 335.

Greenberg, J.M. (1999), in Greenberg, J.M. and Li, A. (eds.), *Formation and Evolution of Solids in Space*, Kluwer, Dordrecht, p.53.

Greenberg, J.M. and Chlewicki, G. (1983), *ApJ*, **272**, 563.

Greenberg, J.M., Gillette, J.S., Munoz Caro, G.M., Mahajan, T.B., Zare, R.N., Li, A., Schutte, W.A., de Groot, M. and Mendoza-Gómez, C.X. (2000), *ApJ*, **531**, L71.

Greenberg, J.M. and Hage, J.I. (1990), *ApJ*, **361**, 260.

Greenberg, J.M. and Hong, S.S. (1974a), in Moorwood, A.F.M. (ed.), *HII Regions and the Galactic Center*, ESRSOSP-105, p.153.

Greenberg, J.M. and Hong, S.S. (1974b), in Kerr, F. and Simonson, S.C. (eds.), *Galactic and Radio Astronomy*, IAU Symp. 60, Reidel, Dordrecht, p.155.

Greenberg, J.M. and Li, A. (1996), *A&A*, **309**, 258.

Greenberg, J.M. and Li, A. (1998), *A&A*, **332**, 374.

Greenberg, J.M. and Li, A. (1999a), *Adv. Space Res.*, **24**, 497.

Greenberg, J.M. and Li, A. (1999b), *Space Sci. Rev.*, **90**, 149.

Greenberg, J.M., Li, A., Kozasa, T. and Yamamoto, T. (1996), in Gustafson, B.Å.S. and Hanner, M.S. (eds.), *Physics Chemistry and Dynamics of Interplanetary Dust*, ASP Conf. Ser. 104, ASP, San Francisco, p.497.

Greenberg, J.M., Li, A., Mendoza-Gómez, C.X., Schutte, W.A., Gerakines, P.A. and de Groot, M. (1995), *ApJ*, **455**, L177.

Greenberg, J.M., Libelo, L., Lind, A.C. and Wang, R.T. (1963b), *Int'l. Series of Monographs on Electromagnetic Waves*, Pergamon Press, 6, 81.

Greenberg, J.M., Lind, A.C., Wang, R.T. and Libelo, L. (1963a), *Proc. Inter-Disciplinary Conf. Electro. Scattering*, Pergamon Press, 123.

Greenberg, J.M., Pedersen, N.E. and Pedersen, J.C. (1961), *J. Appl. Phys.*, **32**, 233.

Greenberg, J.M. and Shah, G.A. (1971), *A&A*, **12**, 250.

Greenberg, J.M. and Shen, C. (1999), *Ap&SS*, **269**, 33.

Greenberg, J.M., Yencha, A.J., Corbett, J.W. and Frisch, H.L. (1972), *Ultraviolet Effects on the Chemical Composition and Optical Properties of Interstellar Grains*, Mém. Soc. Roy. Sci. Liège, 6e série, tome III, 425.

Greenstein, J.L. (1938), *Harvard Obs. Circ.*, No. 422.

Grün, E., Gustafson, B.Å.S., Mann, I., Baguhl, M., Morfill, G.E., Staubach, P., Taylor, A. and Zook, H.A. (1994), *A&A*, **286**, 915.

Grün, E., Zook, H.A., Baguhl, M. et al. (1993), *Nature*, **362**, 428.

Guillois, O., Ledoux, G. and Reynaud, C. (1999), *ApJ*, **521**, L133.

Gustafson, B.Å.S. (1999), in Greenberg, J.M. and Li, A. (eds.), *Formation and Evolution of Solids in Space*, Kluwer, Dordrecht, p.535.

Gustafson, B.Å.S., Kolokolova, L., Thomas-Osip, J.E., Waldemarsson, K.W.T., Loesel, J. and Xu, Y.-L. (1999), in Greenberg, J.M. and Li, A. (eds.), *Formation and Evolution of Solids in Space*, Kluwer, Dordrecht, p.549.

Habing, H.J. (1968), *Bull. Astron. Inst. Netherlands*, **19**, 421.

Hackwell, J.A., Gehrz, R.D. and Woolf, N.J. (1970), *Nature*, **227**, 822.

Hage, J.I. and Greenberg, J.M. (1990), *ApJ*, **361**, 251.

Hall, J.S. (1937), *ApJ*, **85**, 145.

Hall, J.S. (1949), *Science*, **109**, 166.

Hanner, M.S. (1999), *Space Sci. Rev.*, **90**, 99.

Harwit, M. (1970), *Nature*, **226**, 61.

Hecht, J.H. (1981), *ApJ*, **245**, 124.

Hecht, J.H. (1986), *ApJ*, **305**, 817.

Heger, M.L. (1922), *Lick Obs. Bull.*, **337**, 141.

Helou, G. (2000), in Casoli, F., Lequeux, J. and David, F. (eds.), *Infrared Space Astronomy Today and Tomorrow*, Springer, Berlin, p.337.

Henning, Th. and Mutschke, H. (2000), in Sitko, M.L., Sprague, A.L. and Lynch, D.K. (eds.), *Thermal Emission Spectroscopy and Analysis of Dust Disks and Regoliths*, ASP Conf. Ser. 196, ASP, San Francisco, p.253.

Henning, Th. and Stognienko, R. (1993), *A&A*, **280**, 609.

Henyey, L.G. and Greenstein, J.L. (1941), *ApJ*, **93**, 70.

Herbst, E. (1991), *ApJ*, **366**, 133.

Herbst, E. (2000), in Combes, F. and Pineau des Forêts, G. (eds.), *Molecular Hydrogen in Space*, Cambridge Univ. Press, Cambridge, p.85.

Herschel, W. (1785), *Phil. Trans.*, **75**, 213.

Hildebrand, R.H. and Dragovan, M. (1995), *ApJ*, **450**, 663.

Hiltner, W.A. (1949), *Science*, **109**, 165.

Hong, S.S. and Greenberg, J.M. (1980), *A&A*, **88**, 194.

Hough, J.H., Chrysostomou, A., Messinger, D.W., Whittet, D.C.B., Aitken, D.K. and Roche, P.F. (1996), *ApJ*, **461**, 902.

Hoyle, F. and Wickramasinghe, N.C. (1962), *MNRAS*, **124**, 417.

Imanishi, M. (2000), *MNRAS*, **319**, 331.

Imanishi, M. and Dudley, C.C. (2000), *ApJ*, **545**, 701.

Irvine, W.M. and Bergin, E.A. (2000), in Minh, Y.C. and van Dishoeck, E.F. (eds.), *Astrochemistry: From Molecular Clouds to Planetary Systems*, IAU Symp. 197, ASP, San Francisco, p.447.

Joblin, C., Léger, A. and Martin, P. (1992), *ApJ*, **393**, L79.

Jones, A.P. and d'Hendecourt, L. (2000), *A&A*, **355**, 1191.

Jones, A.P., Duley, W.W. and Williams, D.A. (1990), *QJRAS*, **31**, 567.

Jones, T.W. and Merrill, T.M. (1977), *ApJ*, **209**, 509.

Jones, R.V. and Spitzer, L. Jr. (1967), *ApJ*, **147**, 943.

Jones, A.P., Tielens, A.G.G.M., Hollenbach, D.J. and McKee, C.F. (1994), *ApJ*, **433**, 797.

Jura, M. (1986), *ApJ*, **303**, 327.

Kamijo, F. (1963), *PASJ*, **15**, 440.

Kapteyn, J.C. (1904), *ApJ*, **24**, 115.

Kapteyn, J.C. (1909), *ApJ*, **29**, 46.

Kapteyn, J.C. (1922), *ApJ*, **55**, 302.

Kemp, J.C., & Wolstencroft, R.D. 1972, ApJ, 176, L115
Kerr, T.H., Hurst, M.E., Miles, J.R., & Sarre, P.J. 1999, MNRAS, 303, 446
Kerridge, J.F. 1999, in Formation and Evolution of Solids in Space, ed. J.M. Greenberg & A. Li (Dordrecht: Kluwer), 447
Kim, S.H., & Martin, P.G. 1995, ApJ, 442, 172
Kim, S.H., & Martin, P.G. 1996, ApJ, 462, 296
Kim, S.H., Martin, P.G., & Hendry, P.D. 1994, ApJ, 422, 164
Knacke, R.F., Gaustad, J.E., Gillett, F.C., & Stein, W.A. 1969a, ApJ, 155, L189
Knacke, R.F., Cudaback, D.D., & Gaustad, J.E. 1969b, ApJ, 158, 151
Kobayashi, Y., Kawara, K., Sato, S., & Okuda, H. 1980, PASJ, 32, 295
Krelowski, J. 1999, in Formation and Evolution of Solids in Space, ed. J.M. Greenberg & A. Li (Dordrecht: Kluwer), 147
Krelowski, J. 2002, in Solid State Astrochemistry, ed. V. Pirronello, & J. Krelowski (Dordrecht: Kluwer), in press
Kurtz, J. 1992, A&A, 255, L1
Lallement, R., Bertin, P., Ferlet, R., Vidal-Madjar, A., & Bertaux, J.L. 1994, A&A, 286, 898
Landgraf, M., Baggaley, W.J., Grün, E., Krüger, H., & Linkert, G. 2000, J. Geophys. Res., 105, 10343
Langhoff, S.R. 1996, J. Phys. Chem., 100, 2819
Larson, K.A., Whittet, D.C.B., & Hough, J.H. 1996, ApJ, 472, 755
Larson, K.A., Wolff, M.J., Roberge, W.G., Whittet, D.C.B., & He, L. 2000, ApJ, 532, 1021
Lazarian, A. 1994, MNRAS, 268, 713
Ledoux, G., Ehbrecht, M., Guillois, O., et al. 1998, A&A, 333, L39
Lee, H.M., & Draine, B.T. 1985, ApJ, 290, 211
Léger, A. 1988, in Polarized Radiation of Circumstellar Origin, ed. G.V. Coyne, A.M. Magalhaes, A.F. Moffat, et al. (Tucson: Univ. Arizona Press), 769
Léger, A., & Puget, J.L. 1984, A&A, 137, L5
Lequeux, J., & Jourdain de Muizon, M. 1990, A&A, 240, L19
Lewis, R.S., Anders, E., & Draine, B.T. 1989, Nature, 339, 117
Li, A., & Draine, B.T. 2001a, ApJ, 550, L213
Li, A., & Draine, B.T. 2001b, ApJ, 554, 778
Li, A., & Draine, B.T. 2002a, ApJ, 576, 762
Li, A., & Draine, B.T. 2002b, ApJ, 564, 803
Li, A., & Draine, B.T. 2002c, ApJ, 572, 232
Li, A., & Draine, B.T. 2002d, ApJ, in preparation
Li, A., & Greenberg, J.M. 1997, A&A, 323, 566
Li, A., & Greenberg, J.M. 1998a, A&A, 338, 364
Li, A., & Greenberg, J.M. 1998b, ApJ, 498, L83
Li, A., & Greenberg, J.M. 1998c, A&A, 331, 291
Li, A., & Greenberg, J.M. 1998d, A&A, 339, 591
Li, A., & Greenberg, J.M. 2002, ApJ, 577, 789
Lind, A.C., & Greenberg, J.M. 1966, J. Appl. Phys., 37, 3195
Lindblad, B. 1935, Nature, 135, 133
Linde, T.J., & Gombosi, T.I. 2000, J. Geophys. Res., 105, 10411
Low, F.J., Young, E., Beintema, D., et al. 1984, ApJ, 278, L19
Lutz, D., Feuchtgruber, H., Genzel, R., et al. 1996, A&A, 315, L269
Maas, R.W., Ney, E.P., & Woolf, N.J. 1970, ApJ, 160, L101
Manicò, G., Raguní, G., Pirronello, V., Roser, J.E., & Vidali, G. 2001, ApJ, 548, 253

Martin, P.G. (1972), *MNRAS*, **159**, 179.

Martin, P.G., Adamson, A.J., Whittet, D.C.B., Hough, J.H., Bailey, J.A., Kim, S.-H., Sato, S., Tamura, M. and Yamashita, T. (1992), *ApJ*, **392**, 691.

Martin, P.G. and Angel, J.R.P. (1974), *ApJ*, **188**, 517.

Martin, P.G. and Angel, J.R.P. (1975), *ApJ*, **195**, 379.

Martin, P.G., Clayton, G.C. and Wolff, M.J. (1999), *ApJ*, **510**, 905.

Martin, P.G., Illing, R. and Angel, J.R.P. (1972), *MNRAS*, **159**, 191.

Martin, P., Somerville, W., McNally, D., Whittet, D.C.B., Allen, R., Walsh, J. and Wolff, M. (1995), in Tielens, A.G.G.M. and Snow, T.P. (eds.), *The Diffuse Interstellar Bands*, Kluwer, Dordrecht, p.271.

Martin, P.G. and Whittet, D.C.B. (1990), *ApJ*, **357**, 113.

Mathis, J.S. (1986), *ApJ*, **308**, 281.

Mathis, J.S. (1990), *ARA&A*, **28**, 37.

Mathis, J.S. (1994), *ApJ*, **422**, 176.

Mathis, J.S. (1996), *ApJ*, **472**, 643.

Mathis, J.S. (1998), *ApJ*, **497**, 824.

Mathis, J.S., Mezger, P.G. and Panagia, N. (1983), *A&A*, **128**, 212.

Mathis, J.S., Rumpl, W. and Nordsieck, K.H. (1977), *ApJ*, **217**, 425.

Mathis, J.S. and Whiffen, G. (1989), *ApJ*, **341**, 808.

Mattila, K., Lemke, D., Haikala, L.K. et al. (1996), *A&A*, **315**, L353.

Mennella, V., Colangeli, L., Bussoletti, E., Palumbo, P. and Rotundi, A. (1998), *ApJ*, **507**, L177.

Mennella, V., Colangeli, L., Palumbo, P., Schutte, W.A. and Bussoletti, E. (1996), *ApJ*, **464**, L191.

Merrill, P.W. (1934), *PASP*, **46**, 206.

Mie, G. (1908), *Ann. Phys.*, NY **25**, 377.

Mishchenko, M.I., Travis, L.D. and Mackowski, D.W. (1996), *J. Quant. Spectrosc. Radiat. Transfer*, **55**, 535.

Moreels, G., Clairemidi, J., Hermine, P., Brechignac, P. and Rousselott, P. (1994), *A&A*, **282**, 643.

Mumma, M.J., Stern, S.A. and Weissman, P.R. (1993), in Levy, E.H., Lunine, J.I. and Matthews, M.S. (eds.), *Planets and Protostars III*, Univ. Arizona Press, Tucson, p.1177.

Onaka, T., Yamamura, I., Tanabe, T. et al. (1996), *PASJ*, **48**, L59.

Oort, J.H. (1932), *Bull. Astron. Inst. Netherlands*, **6**, 249.

Oort, J.H. and van de Hulst, H.C. (1946), *Bull. Astron. Inst. Netherlands*, **10**, 187.

Papoular, R., Conard, J., Guillois, O., Nenner, I., Reynaud, C. and Rouzaud, J.-N. (1996), *A&A*, **315**, 222.

Papoular, R., Ellis, K., Guillois, O., Reynaud, C. and Nenner, I. (1993), *J. Chem. Soc. Farady Trans.*, **89**, 2289.

Papoular, R., Guillois, O., Nenner, I., Perrin, J.M., Reynaud, C. and Sivan, J.P. (1995), *Planet. Space Sci.*, **63**, 1287.

Parravano, A., Hollenbach, D.J. and McKee, C.F. (2002), *ApJ*, in press.

Pendleton, Y.J. and Allamandola, L.J. (2002), *ApJSS*, **138**, 75.

Pendleton, Y.J., Sandford, S.A., Allamandola, L.J., Tielens, A.G.G.M. and Sellgren, K. (1994), *ApJ*, **437**, 683.

Perrin, J.-M. and Sivan, J.-P. (1992), *A&A*, **255**, 271.

Pirronello, V., Biham, O., Liu, C., Shen, L. and Vidali, G. (1997), *ApJ*, **483**, L131.

Pirronello, V., Biham, O., Manicò, G., Roser, J., Vidali, G. (2000), in Combes, F. and Pineau des Forêts, G. (eds.), *Molecular Hydrogen in Space*, Cambridge Univ. Press, Cambridge, p.71.

Pirronello, V., Liu, C., Roser, J.E. and Vidali, G. (1999), *A&A*, **344**, 681.

Platt, J.R. (1956), *ApJ*, **123**, 486.

Purcell, E.M. (1969), *Physica*, **41**, 100.

Purcell, E.M. (1976), *ApJ*, **206**, 685.

Purcell, E.M. (1979), *ApJ*, **231**, 404.

Purcell, E.M. and Pennypacker, C.R. (1973), *ApJ*, **186**, 705.

Reach, W.T., Dwek, E., Fixsen, D.J. et al. (1995), *ApJ*, **451**, 188.

Roberge, W.G. (1996), in Roberge, W.G. and Whittet, D.C.B. (eds.), *Polarimetry of the Interstellar Medium*, ASP Conf. Ser. 97, ASP, San Francisco, p.401.

Rodrigues, C.V., Magalhães, A.M., Coyne, G.V. and Piirola, V. (1997), *ApJ*, **485**, 618.

Rouleau, F., Henning, Th. and Stognienko, R. (1997), *A&A*, **322**, 633.

Rowan-Robinson, M. (1992), *MNRAS*, **258**, 787.

Rudnick, J. (1936), *ApJ*, **83**, 394.

Russell, H.N. (1922), *Drak Nebulae*, Proc. Nat. Acad. Sci., **8**, 115.

Russell, H.N., Dugan, R.S. and Stewart, J.Q. (1926), *Astronomy*, Ginn and Co., New York, Vol.1 Chap.13.

Sakata, A., Wada, S., Narisawa, T., Asano, Y., Iijima, Y., Onaka, T. and Tokunaga, A.T. (1992), *ApJ*, **393**, L83.

Sakata, A., Wada, S., Onaka, T. and Tokunaga, A.T. (1990), *ApJ*, **353**, 543.

Sakata, A., Wada, S., Tokunaga, A.T. and Narisawa, T. (1995), *Planet. Space Sci.*, **63**, 1223.

Salama, F. and Allamandola, L.J. (1992), *ApJ*, **395**, 301.

Sandford, S.A., Allamandola, L.J., Tielens, A.G.G.M., Sellgren, K., Tapia, M. and Pendleton, Y.J. (1991), *ApJ*, **371**, 607.

Sandford, S.A., Pendleton, Y.J. and Allamandola, L.J. (1995), *ApJ*, **440**, 697.

Saslaw, W.C. and Gaustad, J.E. (1969), *Nature*, **221**, 160.

Schalén, C. (1929), *Astron. Nachr.*, **236**, 249.

Schalén, C. (1936), *Medd. Uppsala Astron. Obs.*, No.64.

Schmidt, G.D., Cohen, M. and Margon, B. (1980), *ApJ*, **239**, L133.

Schnaiter, M., Mutschke, H., Dorschner, J., Henning, Th. and Salama, F. (1998), *ApJ*, **498**, 486.

Schutte, W.A. (1996), in Greenberg, J.M. (ed.), *The Cosmic Dust Connection*, Kluwer, Dordrecht, p.1.

Schutte, W.A. (1999), in Greenberg, J.M. and Li, A. (eds.), *Formation and Evolution of Solids in Space*, Kluwer, Dordrecht, p.177.

Schutte, W.A., Tielens, A.G.G.M. and Allamandola, L.J. (1993), *ApJ*, **415**, 397.

Schutte, W.A., van der Hucht, K.A., Whittet, D.C.B. et al. (1998), *A&A*, **337**, 261.

Seahra, S.S. and Duley, W.W. (1999), *ApJ*, **520**, 719.

Sellgren, K., Brooke, T.Y., Smith, R.G. and Geballe, T.R. (1995), *ApJ*, **449**, L69.

Sellgren, K., Rouan, D. and Léger, A. (1988), *A&A*, **196**, 252.

Sellgren, K., Werner, M.W. and Allamandola, L.J. (1996), *ApJS*, **102**, 369.

Serkowski, K. (1973), in Greenberg, J.M. and van de Hulst, H.C. (eds.), *Interstellar Dust and Related Topics*, IAU Symp. 52, Reidel, Dordrecht, p.145.

Shapley, H. and Curtis, H.D. (1921), *Bull. Nat. Res. Coun.*, **2**, 217.

Shu, F.H., Adams, F.C. and Lizano, S. (1989), in Bonetti, A., Greenberg, J.M. and Aiello, S. (eds.), *Evolution of Interstellar Dust and Related Topics*, North Holland, Amsterdam, p.213.

Siebenmorgen, R. and Krügel, E. (1992), *A&A*, **259**, 614.

Sivan, J.-P. and Perrin, J.-M. (1993), *ApJ*, **404**, 258.

Smith, C.H., Wright, C.M., Aitken, D.K., Roche, P.F. and Hough, J.H. (2000), *MN-RAS*, **312**, 327.

Smith, R.K. and Dwek, E. (1998), *ApJ*, **550**, L201.

Smith, T.L. and Witt, A.N. (2002), *ApJ*, **565**, 304.

Snow, T.P. and Witt, A.N. (1995), *Science*, **270**, 1455.

Snow, T.P. and Witt, A.N. (1996), *ApJ*, **468**, L65.

Sofia, U.J. and Meyer, D.M. (2001), *ApJ*, **554**, L221.

Somerville, W. (1996), in Roberge, W.G. and Whittet, D.C.B. (eds.), *Polarimetry of the Interstellar Medium*, ASP Conf. Ser. 97, ASP, San Francisco, p.143.

Somerville, W., Allen, R.G., Carnochan, D.J. et al. (1994), *ApJ*, **427**, L47.

Sorrell, W.H. (1990), *MNRAS*, **243**, 570.

Stebbins, J., Huffer, C.M. and Whitford, A.E. (1939), *ApJ*, **90**, 209.

Stecher, T.P. (1965), *ApJ*, **142**, 1683.

Stecher, T.P. and Donn, B. (1965), *ApJ*, **142**, 1681.

Stein, W.A. and Gillett, F.C. (1969), *ApJ*, **155**, L197.

Struve, F.G.W. (1847), Etudes d'Astronomie Stellaire.

Szomoru, A. and Guhathakurta, P. (1998), *ApJ*, **494**, L93.

Tanaka, M., Matsumoto, T., Murakami, H. et al. (1996), *PASJ*, **48**, L53.

Taylor, A., Baggaley, W.J. and Steel, D.I. (1996), *Nature*, **380**, 323.

Tielens, A.G.G.M. (1999), in Greenberg, J.M. and Li, A. (eds.), *Formation and Evolution of Solids in Space*, Kluwer, Dordrecht, p.331.

Tielens, A.G.G.M. and Allamandola, L.J. (1987), in Morfill, G.E. and Scholer, M. (eds.), *Physical Processes in Interstellar Clouds*, Reidel, Dordrecht, p.333.

Tielens, A.G.G.M., Hony, S., van Kerckhoven, C. and Peeters, E. (1999), in Cox, P. and Kessler, M.F. (eds.), *The Universe as Seen by ISO*, ESA-SP427, p.579.

Trumpler, R.J. (1930), *PASP*, **42**, 214.

Tulej, M., Kirkwood, D.A., Pachkov, M. and Maier, J.P. (1998), *ApJ*, **506**, L69.

Uchida, K.I., Sellgren, K. and Werner, M.W. (1998), *ApJ*, **493**, L109.

UV Atlas (1966), *UV Atlas of Organic Compounds*, Plenum Press, New York.

van de Hulst, H.C. (1949), *The Solid Particles in Interstellar Space*, Rech. Astron. Obs. Utrecht, 11, part 2.

van de Hulst, H.C. (1957), *Light Scattering by Small Particles*, Wiley, New York.

van de Hulst, H.C. (1986), in Israel, F.P. (ed.), *Light on Dark Matter*, Reidel, Dordrecht, p.161.

van de Hulst, H.C. (1989), in Bonetti, A., Greenberg, J.M. and Aiello, S. (eds.), *Evolution of Interstellar Dust and Related Topics*, North Holland, Amsterdam, p.1.

van de Hulst, H.C. (1997), in van Dishoeck, E.F. (ed.), *Molecules in Astrophysics: Probes and Processes*, IAU Symp. 178, Kluwer, Dordrecht, p.13.

van Dishoeck, E.F. (1994), in Cutri, M. and Latter, B. (ed.), *The First Symposium on the Infrared Cirrus and Diffuse Interstellar Medium*, ASP Conf. Ser. 58, ASP, San Francisco, p.319.

van Dishoeck, E.F. (1999), in Greenberg, J.M. and Li, A. (eds.), *Formation and Evolution of Solids in Space*, Kluwer, Dordrecht, p.91.

van Dishoeck, E.F., Blake, G.A., Draine, B.T. and Lunine, J.I. (1993), in in Levy, E.H., Lunine, J.I. and Matthews, M.S. (eds.), *Planets and Protostars III*, Univ. Arizona Press, Tucson, p.163.

van Kerckhoven, C., Tielens, A.G.G.M. and Waelkens, C. (2002), *A&A*, **384**, 568.

van Rhijn (1921), *Pub. Astron. Lab. Groningen*, 31.

Verstraete, L., Pech, C., Moutou, C. et al. (2001), *A&A*, **372**, 981.

Voshchinnikov, N.V. and Farafonov, V.G. (1993), *Ap&SS*, **204**, 19.

Voshchinnikov, N.V., Semenov, D.A. and Henning, Th. (1999), *A&A*, **349**, L25.

Waelkens, C., Malfait, K. and Waters, L.B.F.M. (2000), in Minh, Y.C. and van Dishoeck, E.F. (eds.), *Astrochemistry: From Molecular Clouds to Planetary Systems*, IAU Symp. 197, ASP, San Francisco, p.435.

Waters, L.B.F.M., Molster, F.J., de Jong, T. et al. (1996), *A&A*, **315**, L36.

Webster, A. (1993), *MNRAS*, **264**, L1.

Weiland, J.L., Blitz, L., Dwek, E. et al. (1986), *ApJ*, **306**, L101.

Weingartner, J.C. and Draine, B.T. (2001a), *ApJ*, **548**, 296.

Weingartner, J.C. and Draine, B.T. (2001b), *ApJSS*, **134**, 263.

Weissman, P.R. (1986), *Nature*, **320**, 242.

Wexler, A.S. (1967), *Appl. Spectrosc. Rev.*, **1**, 29.

Whipple, F.L. (1950), *ApJ*, **111**, 375.

Whitford, A.E. (1948), *ApJ*, **107**, 102.

Whittet, D.C.B. (1992), *Dust in the Galactic Environment*, Inst. Phys. Publ., New York.

Whittet, D.C.B., Duley, W.W. and Martin, P.G. (1990), *MNRAS*, **244**, 427.

Whittet, D.C.B., Gerakines, P.A., Hough, J.H. and Shenoy, S.S. (2001), *ApJ*, **547**, 872.

Whittet, D.C.B., Martin, P.G., Hough, J.H., Rouse, M.F., Bailey, J.A. and Axon, D.J. (1992), *ApJ*, **386**, 562.

Wickramasinghe, D.T. and Allen, D.A. (1980), *Nature*, **287**, 518.

Wilking, B.A., Lebofsky, M.J. and Rieke, G.H. (1982), *AJ*, **87**, 695.

Will, L.M. and Aannestad, P.A. (1999), *ApJ*, **526**, 242.

Willner, S.P., Russell, R.W., Puetter, R.C., Soifer, B.T. and Harvey, P.N. (1979), *ApJ*, **229**, L65.

Witt, A.N. (2000), in Minh, Y.C. and van Dishoeck, E.F. (eds.), *Astrochemistry: From Molecular Clouds to Planetary Systems*, IAU Symp. 197, ASP, San Francisco, p.317.

Witt, A.N., Bohlin, R.C. and Stecher, T.P. (1986), *ApJ*, **305**, L23.

Witt, A.N., Gordon, K.D. and Furton, D.G. (1998), *ApJ*, **501**, L111.

Witt, A.N. and Schild, R. (1988), *ApJ*, **325**, 837.

Witt, A.N., Smith, R.K. and Dwek, E. (2001), *ApJ*, **550**, L201.

Wolff, M.J., Clayton, G.C., Kim, S.H., Martin, P.G. and Anderson, C.M. (1997), *ApJ*, **478**, 395.

Wolff, M.J., Clayton, G.C. and Meade, M.R. (1993), *ApJ*, **403**, 722.

Woolf, N.J. and Ney, E.P. (1969), *ApJ*, **155**, L181.

Wright, C.M. (1999), in Greenberg, J.M. and Li, A. (eds.), *Formation and Evolution of Solids in Space*, Kluwer, Dordrecht, p.77.

Wright, E.L. (1987), *ApJ*, **320**, 818.

Wright, E.L., Mather, J.C., Bennett, C.L. et al. (1991), *ApJ*, **381**, 200.

Wright, G., Geballe, T., Bridger, A. and Pendleton, Y.P. (1996), in Block, D.L. and Greenberg, J.M. (eds.), *New Extragalactic Perspectives in the New South Africa*, Kluwer, Dordrecht, p.143.

Zubko, V.G. (1999a), in Greenberg, J.M. and Li, A. (eds.), *Formation and Evolution of Solids in Space*, Kluwer, Dordrecht, p.85.
Zubko, V.G. (1999b), *ApJ*, **513**, L29.

COSMIC SILICATES - A REVIEW

Thomas Henning

Astrophysical Institute and University Observatory (AIU), Friedrich Schiller University, Schillergäßchen 2-3, D-07745 Jena, Germany

Keywords: Cosmic Dust – Silicates – Infrared Bands – Laboratory Astrophysics

Abstract   The paper reviews our knowledge about cosmic silicates obtained from a combination of recent infrared observations and dedicated laboratory experiments. The formation and properties of crystalline silicates and their mineralogy will be discussed.

Introduction

Silicate grains make up a major fraction of cosmic dust. They contribute to the visual and near-infrared extinction curve, cause interstellar polarization by dichroic extinction, produce spectral features at mid-infrared wavelengths, and emit thermal radiation at far-infrared and millimetre wavelengths. These particles also provide the surface for chemical reactions and the accretion of atoms and molecules.

During the last years, we reached a lot of progress in our understanding of cosmic silicates by a combination of high-quality astronomical data, mainly provided by the spectrometers aboard the Infrared Space Observatory ISO, dedicated laboratory experiments on analogue materials and a thorough theoretical modeling of the dust lifecyle. This review will mainly summarize the new information we obtained about cosmic silicates. For a summary of earlier results and more background information about the structure of cosmic silicates and their characterization, I refer the reader to the reviews by Dorschner & Henning (1986), Nuth et al. (2000), and Henning & Mutschke (2000). In addition, I want to mention our series of papers about interstellar silicate mineralogy (see, e.g., Jäger et al. 1998, Fabian et al. 2001) and the reviews by Waters & Molster (1999) and Waters et al. (2000) about ISO results.

V. Pirronello et al. (eds.), Solid State Astrochemistry, 85–103.

1.   COSMIC SILICATES IN DIFFERENT ENVIRONMENTS

Spectroscopic evidence for silicates has been found in such different environments as Seyfert galaxies, the galactic centre, disks and envelopes around young stellar objects and oxygen-rich late-type stars, planetary nebulae, novae, and finally in comets and interplanetary dust grains. The widely observed broad features at about 10 and 18 μm are generally attributed to Si–O stretching and Si–O–Si bending modes in silicates of amorphous state. The features are observed in absorption, self-absorption, and emission depending on the optical depth of the configuration (Henning 1983).

The elemental components of silicates, i.e. the silicon, oxygen, magnesium, and iron atoms, are heavily depleted in the interstellar gas. The depletion of oxygen and silicon is roughly in agreement with these atoms being bound in SiO_4 tetrahedra (Snow & Witt 1996). Materials which are commonly considered as candidates for cosmic silicates are olivines and pyroxenes (Ossenkopf et al. 1992). Olivines are solid solutions of forsterite (Mg_2SiO_4) and fayalite (Fe_2SiO_4), whereas pyroxenes are solid solutions of ferrosilite ($FeSiO_3$) and enstatite ($MgSiO_3$). The olivines are neso-silicates (island silicates) because they consist of isolated SiO_4 tetrahedra which are linked by divalent cations. The pyroxenes form a large group of silicates belonging to the inosilicates (chain silicates). The SiO_4 tetrahedra form chains, i.e. each tetrahedron shares two oxygen atoms with its neighbours.

SiO_4 tetrahedra, connected by oxygen bridges, form the basic building blocks of silicates. The incorporation of metal oxides into an SiO_2 framework leads to non-bridging oxygen atoms - a situation present in amorphous silicates. Amorphous olivines and pyroxenes are characterized by a mixture of monomers, chains, and three-dimensional network structures (Mysen et al. 1982). This implies a short-range order determined by the stoichiometry, but a lack of long-range order.

The investigation of gas phase abundances in diffuse halo clouds in the Milky Way point to an (Mg+Fe)/Si ratio of 3.26±0.64 or 3.91±0.64 in dust grains if the solar or B-star reference abundances are used, respectively. This ratio, which is larger than what we can expect for silicates, implies that there must be grains composed of oxides and perhaps pure iron (Sembach & Savage 1996). Na, Ca, and Al atoms might be also constituents of the silicate grains.

Spitzer & Fitzpatrick (1993, 1995) found a very interesting abundance pattern along different lines of sight in the disk and halo. In regions with low or non-existent Mg depletion, they found that the number ratio of Fe to Si atoms in the grains is nearly 2:1. In regions with higher depletion, a ratio of $(Fe+Mg)/Si$ of 2:1 has been found. These observations suggest a structure where the grain cores consist of Fe_2SiO_4 and the grain mantle of Mg_2SiO_4. The mantles are being destroyed in clouds with low Mg depletion.

First evidence for the presence of crystalline silicates in the envelopes of very young massive stars, comets, and β Pictoris-type objects came from the observation of an feature at 11.2 μm (see Jäger et al. 1998 for references). This feature is typical of crystalline olivine. Observations with the spectrometers aboard ISO revealed the presence of a wealth of "crystalline" silicate features in the wavelength range between 20 and 70 μm due to metal-oxygen vibrations (see, e.g., Waters et al. 2000). These features are of great diagnostic value and allow a detailed characterization of the mineralogy of crystalline silicates. From the positions of the features, we have strong indications that Mg-rich olivines (see Fig. 1) and pyroxenes are the major components of crystalline silicates.

Figure 1 Three-dimensional model of olivine together with a terrestrial olivine rock. The black spheres represent Si atoms, the white balls are the metal cations (Mg and/or Fe) and the grey spheres are the oxygen atoms.

The very narrow features observed point to the presence of very pure silicate crystals in circumstellar space and will allow a better characterization of the formation routes of silicates. In addition, the wavelength location of their peak positions may be used as a dust thermometer and offers the possibility for the determination of the Mg/Fe ratio. Here we should note that the presence of these features usually does not imply that the crystalline silicates are the major component of the silicate dust observed (Molster 2000).

1.1.   CIRCUMSTELLAR ENVELOPES AROUND EVOLVED STARS

Ground-based infrared spectra and IRAS data of oxygen-rich circum-stellar envelopes, especially in the 10 μm range, implied the presence of amorphous silicates (see, e.g., Ossenkopf et al. 1992). The IRAS low-resolution spectrograph data in the wavelength range between 8 and 23 μm showed a rather wide diversity in the shape of the emission profiles (see, e.g., Hron et al. 1997). A recent extended ground-based study of infrared spectra of oxygen-rich evolved stars confirms the diversity of the dust features (Speck et al. 2000). Ground-based data also revealed that a large fraction of variable stars with 9.7 μm emission features shows spectral profile changes as a function of pulsation cycle (Monnier et al. 1998). A significant sharpening of the profile near maximum light suggests that the optical properties change and that heating leads to cleaner silicate dust.

ISO observations demonstrated the presence of crystalline silicates in the envelopes of AGB and post-AGB stars (Waters et al. 1996, Molster 2000). Their presence can be inferred from narrow features on top of the structureless dust continuum. The features divide in several complexes with a set of components which vary independently (Molster 2000). Many of the features are typical of Mg-rich olivines and pyroxenes. Variations in band positions and bandwidths are probably related to different temperatures and Mg/Fe ratios, the degree of crystallization, and size/shape effects among the particles. A significant fraction of the features cannot be identified with either olivines or pyroxenes and point to the presence of other materials in the envelopes.

The strongest bands of crystalline silicates have been found in objects with a disk-like geometry of the envelope. The temperature of crystalline silicates is often below 150 K. Therefore, they do not show strong features at wavelengths shortward of 20 μm.

An interesting ISO observation is the detection of crystalline silicate components in the Red Rectangle, which was known to be associated with an carbon-rich outflow (Waters et al. 1998a). The cystalline sili-cates are very probably part of the dusty circumstellar disk associated with the object. A possible precursor to this type of objects is the J-type carbon star IRAS 09425−6040 which is characterized by one of the highest fractions of crystalline silicates in the oxygen-rich circumstellar disk component (Molster et al. 1999, 2001). The spectrum of this object is shown together with the spectra of comet Hale-Bopp (Crovisier et al. 1997) and the binary AC Her in Fig. 2. The carriers of the features of AC Her are a mixture of amorphous and crystalline silicates and are very similar to those of comet Hale-Bopp.

Evidence for crystalline silicates has also been found in the ISO spectra of planetary nebulae (Waters et al. 1998b, Cohen et al. 1999) and OH/IR stars (Sylvester et al. 1999, Demyk et al. 2000).

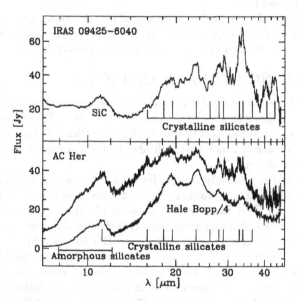

Figure 2 ISO-SWS grating scan of the carbon-rich giant IRAS 09425−6040 (upper panel) and of the post-AGB star AC Her (lower panel). Note the strong crystalline silicate bands in IRAS 09425−6040. The ISO-SWS spectra of comet Hale-Bopp (Crovisier et al. 1997) and the binary AC Her are shown for comparison. At short wavelengths ($\lambda < 14$ μm), the spectrum of IRAS 09425−6040 is characterized by carbon molecular bands (not shown here) and the SiC dust feature at 11 μm. After Molster et al. (1999).

1.2.   DIFFUSE INTERSTELLAR MEDIUM

The silicates in HII regions and the diffuse interstellar medium (ISM) seem to be essentially amorphous. Spectra of the diffuse ISM along the line of sight towards the galactic centre show a broad aborption band peaking at about 9.7 μm (Chiar et al. 2000). From the contribution to the spectra, crystalline silicates cannot be more than 2% of the total amount of silicates. In the emission spectra of the Orion nebula and the

HII region M17-SW, evidence of a small fraction of crystalline silicates, probably of interstellar origin, has been found (Cox et al. 1999, Cesarsky et al. 2000).

Draine & Lee (1984) introduced the term "astronomical silicates" for this type of amorphous silicate dust and compiled a set of optical constants for this material. The paper by Dorschner et al. (1995) provides a comprehensive compilation of optical data for amorphous silicates measured in the laboratory.

The amorphous silicates are likely produced by the irradiation of cosmic rays in the diffuse interstellar medium. However, the destruction and reformation of silicates may also play a role (Jones et al. 1994).

1.3.   YOUNG STELLAR OBJECTS

Mid-infrared observations of deeply dust-embedded luminous young stellar objects (Willner et al. 1982, Demyk et al. 1999), T Tauri stars (Cohen & Witteborn 1985), and Herbig Ae/Be stars (Berrilli et al. 1987) revealed the presence of amorphous silicates in the circumstellar environment of these objects.

ISO spectra of the relatively bright Herbig Ae/Be stars show a large variety of spectral features, ranging from spectra dominated by silicates to those with PAH features (Waelkens et al. 1996, Meeus et al. 2001). Strong silicate features have been detected in the disk sources HD 163 296 (van den Ancker et al. 2000) and HD 100 546 (Malfait et al. 1998). Most of the features in these two sources can be identified with crystalline forsterite. Fig. 3 shows the amazingly perfect match of the HD 100 546 "crystalline" features with those of comet Hale-Bopp. An interesting example for a very young B9 star (age about 10^5 years), which shows evidence for Mg-rich silicates in its spectrum, is the source HD 179 218 (Malfait 1999).

1.4.   INTERPLANETARY DUST PARTICLES AND COMETS

Interplanetary dust particles (IDPs) of asteroidal and cometary origin, collected in the stratosphere, provide presently the only opportunity for the direct investigation of cosmic silicates. Based on infrared spectroscopy of 26 IDPs, Sandford & Walker (1985) found three different silicate materials, i.e. olivines, pyroxenes, and layer-lattice silicates such as montmorillonite. An extremely interesting component of IDPs is the so-called GEMS (glass with embedded metals and sulfides; Bradley 1994). The GEMS consists of glassy Mg-rich silicates with inclusions of FeNi and Fe-rich sulfides. These particles have many properties thought

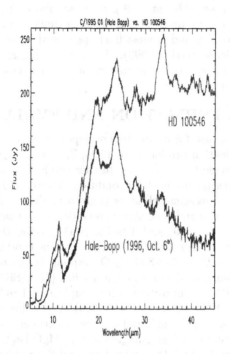

Figure 3 Comparison of the ISO spectrum of HD 100 546 with the spectrum of comet Hale-Bopp

to be typical of interstellar dust grains. However, the final isotopic proof of their interstellar (stellar) orgin is still lacking. Bradley et al. (1999) succeeded in obtaining mid-infrared spectra of the silicate component of IDPs. In the 10-μm range these spectra are very similar to the characteristic infrared spectra of the diffuse interstellar medium (see, e.g., Martin 1995).

Cometary dust spectra show a large diversity pointing to different size distributions and chemical composition of the grains (Hanner 1999). Evidence of crystalline silicates has been found for comet P/Halley from KAO observations in the wavelength range from 16 to 30 μm (Herter et al. 1987). The spectacular ISO observations of comet C/1995 O1 (Hale-Bopp) by Crovisier et al. (1997; see Fig. 3) clearly proved the presence of crystalline Mg-rich olivines and pyroxenes in this comet, building a bridge to the spectra of circumstellar material around Herbig Ae/Be stars. Ground-based data obtained by Wooden et al. (1999), Harker (1999), and Wooden & Harker (2000) support this view. The

11.2 μm peak and a shoulder at 11.9 μm are attributed to crystalline olivine. The broad maximum at 9.8–10.0 μm is similar to that seen in many interstellar sources and is most likely produced by amorphous or glassy silicate particles (Harker 1999). The ratio of the amorphous to the crystalline component is rather uncertain and depends very much on the assumed dust parameters.

2.   GRAIN FORMATION AND EVOLUTION

Fresh silicate stardust forms in the envelopes of oxygen-rich stars. The first likely nucleation products are TiO_2, ZrO_2 and Al_xO_y clusters which probably provide the surface for further condensation processes. Such a condensation sequence both for conditions in protoplanetary disks and envelopes around oxygen-rich stars is given in Fig. 4. Apart from the presence of forsterite and enstatite, observational evidence for spinel ($MgAl_2O_4$, Posch et al. 1999) and diopside ($CaMgSi_2O_6$, Demyk et al. 2000) has been found. An interesting result of recent condensation experiments, starting with Fe-Mg-SiO-H_2-O_2 vapour, is the finding that this process yields only solids with magnesiosilica ($MgO \cdot SiO_2$) and ferrosilica (Fe-oxide·SiO_2) compositions plus simple metal oxides (MgO, SiO_2, FeO or Fe_2O_3) (Rietmeijer et al. 1999). No solids with mixed Mg-Fe-O composition have been found in these experiments nor is there evidence for the formation of ferromagnesiosilica, $MgO \cdot Fe_yO_x \cdot SiO_2$.

The temperatures at which the grains form and the residence time in the condensation zone are critical parameters determining the fraction of crystalline silicates (see Sect. 4.3 for the relevant experiments). There is some evidence that the fraction of cystalline silicates increases with mass loss rate (Sogawa & Kozasa 1999). However, it is difficult or even impossible to prove with ISO data that objects with low mass loss rates do not produce crystalline silicates (Kemper et al. 2001).

In the diffuse interstellar medium, the grain size distribution is modified by shocks leading to sputtering, vaporization, shattering, and disaggregation (Jones et al. 1994, 1996). A redistribution occurs from large grains (radius $a \geq 100$ nm) into small grains ($a \leq 50$ nm). The lifetime for silicate grains against destruction is about 4×10^8 yr. This seems to be in contradiction to the observational fact that silicate grains are widely present in the diffuse interstellar medium and that we do not have any observational evidence of the presence of very small silicate particles. Such particles should show 10-μm emission features because of their elevated temperatures. A solution to the problem may be provided by the coagulation of grains in dense molecular cloud cores leading to fluffy aggregates (Ossenkopf & Henning 1994). Such porous grains may sur-

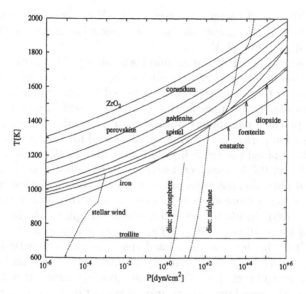

Figure 4 Condensation trajectories for the condensation process on pre-existing nuclei. The accretion disk model is for a steady-state Keplerian disk with a central stellar mass of 1 $M_\odot$, an accretion rate of 10^{-7} $M_\odot$/yr and an α value of 3×10^{-3}. The wind model stands for a steady-state wind with a stellar luminosity of 10^4 $L_\odot$ and a mass loss rate of 10^{-5} $M_\odot$/yr.

vive shocks better and seem also to be required for the explanation of the observed silicate absorption and polarization features (Henning & Stognienko 1993, Mathis 1998).

Based on observed cloud and intercloud depletion values of Si, Fe, and Mg, Tielens (1998) derived even shorter destruction timescales for Si-bearing dust (6×10^7 yr). The lifetime of Fe-bearing dust (5×10^8 yr) is compatible with a refractory reservoir for this material. Tielens concluded that a major fraction (about 30%) of elemental silicon is locked up in a relatively volatile dust component. This component is very likely a mantle grown on existing dust grains. Such a process could also be of importance for the amorphization of silicates and the formation of silicon nanoparticles embedded in such a mantle.

Apart from disruption, destruction, growth and coagulation processes, silicate grains in the diffuse interstellar medium are very probably modified by energetic cosmic-ray particles. The dust grains are subjected to a steady cosmic-ray energy density exceeding a value of 1.8 eV/cm^3 (Webber 1998). In astrophysics we have virtually to do with all kinds of ions comprising H$^+$, He$^+$, (CNO)$^+$, and (other ions)$^+$ with a ratio of

1: 10^{-1}:10^{-3}:10^{-4} and energies between a few eV to GeV. Interactions between the solid particles and ions may be expected in an energy range of a few eV up to a few MeV.

For the understanding of the chemical and structural processing of interstellar dust, it is necessary to study the transition of silicates from the crystalline to the amorphous state triggered by ion irradiation. In early experiments, Krätschmer & Huffman (1979) used intense ion irradiation to produce amorphous silicates. An amorphization study of enstatite has been performed by Wang et al. (1998) for 1.5 MeV Xenon ions. They determined the amorphization treshold to be 0.20 dpa at 298 K and 0.14 dpa at 100 K. However, the question remained open if corresponding amorphization reactions can be observed also for the low-mass ions which are much more abundant in cosmic rays than heavy nuclei. Demyk et al. (2001) irradiated crystalline olivine samples with He^+ ions at energies of 4 and 10 keV and found amorphization to occur.

In recent investigations, we irradiated enstatite with energetic ions. The samples consisted of nanometre- up to micrometre-sized enstatite particles with irregular shapes. The particles were dispersed on a TEM (Transmission Electron Microscopy) grid supported by a Lacey carbon film which was directly used for the irradiation experiments. The conversion process was controlled by TEM images in combination with electron diffraction. A series of particles with different sizes have been selected and electron diffraction patterns before and after the ion irradion have been used for monitoring the conversion of the structural state.

He^+, C^+, and Ar^+ ions with different energies have been used for the investigation of the crystalline-to-amorphous transition. The target temperature (70-300 K) and the irradiation doses ($10^{14} - 10^{18}$ ions/cm^2) have been varied in a wide range. The experiments with low-mass and high-mass ions in energy ranges between 50 keV and 3 MeV have demonstrated that the nuclear energy loss represents the crucial factor for the amorphization process. In contrast, electronic energy deposition seems to accelerate the healing processes of the induced disorder.

The displacements per atom, the ratio between the nuclear and electronic energy loss as well as the penetration depths were derived by SRIM (Stopping and Range of Ions in Matter)-2000.39 calculations. Our experiments have demonstrated that only He^+ ions with low energies (50 keV) are able to trigger the amorphization process. Nearly the same ratio between nuclear and electronic energy loss can be realized with 400 keV Ar^+ ions. The critical amorphization dose for 400 keV Ar^+ ions was determined to be about 3 10^{14} ions/cm^2 (Jäger et al. 2001).

The opposite process, the transition from the amorphous to the crystalline state, observed in circumstellar shells and protoplanetary disks, can also be triggered by a combination of ion irradiation and annealing (Williams et al. 1985, Grun et al. 1997). The process seems to require a high ratio of electronic to nuclear energy deposition. Defects generated by the ion beam can act as nucleation seeds resulting in a decrease of the crystallization temperature. First experiments with 400 and 150 keV He$^+$ and 400 keV Ar$^+$ ions in combination with annealing at 823 K have shown that the crystallization process of amorphous MgSiO$_3$ could not be initiated so far. Further studies with high-energy He$^+$ or other low-mass ions are required.

3.   DUST ANALOGS IN THE LABORATORY

3.1.   PRODUCTION METHODS

Based on the observational constraints discussed in the previous sections, we may try to produce Si-bearing materials in the laboratory, which have spectroscopic properties similar to the observed astronomical features. However, the goal of these experiments is not only to obtain optical data, but also to find out how these data are related to fundamental structural parameters and how silicates are formed from the gas phase, which annealing and irradiation processes can occur, and which solid-solid transformations are of importance. This usually requires a detailed characterization of the materials by various analytic techniques such as Raman and infrared spectroscopy (local and global ordering), high-resolution electron microscopy (global ordering, detailed structure), X-ray diffraction analysis (local ordering), and qualitative and quantitive analysis by EDX (Energy Dispersive X-Ray Spectroscopy).

In Table 1 we summarize the main production techniques applied in the synthesis of cosmic silicate dust analogues. These techniques cover nearly all of the synthesis methods usually used in chemical physics.

3.2.   OPTICAL PROPERTIES

The various steps necessary and the pitfalls to be avoided in deriving optical properties for silicate grains have been summarized in Henning & Mutschke (2000). Table 2 presents a compilation of the various measurements of optical constants/dielectric functions for olivines and pyroxenes. We will not discuss all these data, but will concentrate on a few problems relevant to astrophysics.

1 Absorption at near-infrared wavelengths - The dirty silicates

In order to explain the temperature of circumstellar dust in oxygen-rich envelopes around evolved stars, one has to assume that the silicate grains are much more absorptive at near-infrared wave-

Table 1 Techniques for the synthesis of silicate dust analogues.

Material	Technique	Key references
Silicates (cryst.)	Crystal pulling Annealing of amorphous silicates	Takai et al. (1984) Burlitch et al. (1991)
Silicates (amorph.)	Glass melting (with quenching)	Jäger et al. (1994) Dorschner et al. (1995)
	Condensation from the gas phase (i) Laser ablation (ii) Thermal evaporation (iii) Reactive ion sputtering (iv) DC arc plasma (v) Silane flame Sol-gel reactions Ion bombardment of crystals	Stephens et al. (1995) Nuth & Donn (1982, 1983) Day (1979, 1981) Tanabe et al. (1986) Hallenbeck et al. (1998) Day (1974, 1976) Krätschmer & Huffman (1979)

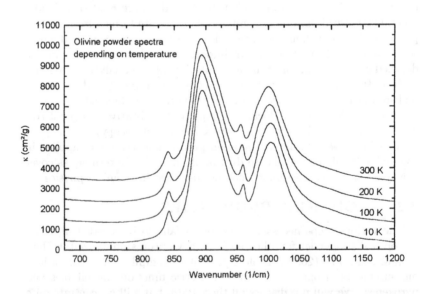

Figure 5 Mass absorption coefficient of olivine powder dispersed in KBr/PE (diameter < 1 µm) at 300, 200, 100, and 10K. For clarity, the spectra have been vertically shifted.

lengths than most of the terrestrial minerals or glasses. Jones & Merrill (1976) coined the term "dirty silicates" for materials with high absorption in this wavelength range (high value of the imaginary part k of the refractive index; $k=0.1$). Two possibilities exist to produce such dirty silicates: (a) an admixture of finely dispersed impurities of iron or iron oxides to the silicates (Ossenkopf et al. 1992) or (b) the presence of Fe^{2+} cations in the silicates leading to crystal-field transitions of the 3d electrons in the iron cations (Dorschner et al. 1995). In the latter case a systematic increase of the NIR absorptivity with iron content has been found in pyroxene glasses.

2 Anisotropy of optical behaviour

Many crystals, including olivine crystals, show an anisotropic optical behaviour. The absorption efficiency of such particles will depend on the alignment of the crystallographic axes. After the orientational averaging, this anisotropy can be removed for an ensemble of particles (Michel et al. 1999). The overall extinction efficiency is calculated for each set of optical constants and averaged:

$$Q_{ext} = 1/3(Q_{ext}(x) + Q_{ext}(y) + Q_{ext}(z)).$$

This approach is valid for ellipsoidal particles, but may fail for particles of other shape. If there are preferred orientations during crystal growth, the above averaging will also fail.

3 Position of the absorption bands

The wavelength positions of the infrared bands in silicates change with composition and temperature as well as the size and shape of the particles (Jäger et al. 1998, Henning & Mutschke 1997, Fabian et al. 2001). Jäger et al. (1998) found that all peaks of the olivine series (different Mg/Fe ratio) shifted towards longer wavelengths with growing Fe-content. The same trend is visible in the great majority of the bands of the pyroxene series. The mass percentage of FeO is closely related to the wavenumber shift.

Henning & Mutschke (1997, see for earlier references and physical basis), Mennella et al. (1998), Bowey et al. (2000), and Fabian et al. (2001) investigated how the bands shift at cryogenic temperatures. With decreasing temperature, the position of the features shifts to larger wavenumbers (see Fig. 5) and the bands become sharper.

4 Far-infrared behaviour - Crystalline vs. amorphous silicates

The far-infrared (FIR) and submillimeter optical properties of the grains are an important factor in determining the temperature distribution of the grains. In addition, the knowledge of the opacities is required if dust masses should be derived from optically thin submillimetre and millimetre continuum emission (Henning et al. 1995). The wavelength dependence of the FIR extinction is often parametrized by a power law with $Q_{ext} \propto \lambda^{-\beta}$. Here, we should note that the exponent β may also be wavelength-dependent.

The size, shape, chemical composition and the physical structure of the grains as well as their temperature are all factors which influence the FIR optical behaviour. The transition from amorphous to crystalline materials is a structural transformation which can strongly influence the FIR absorptivity. In contrast to small crystalline or metallic particles, where a λ^{-2} behaviour is expected, amorphous layered materials should show an emissivity index close to 1. Fabian et al. (2000) found a significant FIR absorptivity drop below wavenumbers of 250 cm^{-1} for MgSiO$_3$ and Mg$_2$SiO$_4$ smoke annealed at T=1000 K for 30 h.

3.3.   ANNEALING EXPERIMENTS

The formation of crystalline silicates in the dusty outflows of evolved stars probably occurs through the annealing of amorphous low-pressure condensates (Gail & Sedlmayr 1998, Sogawa & Kozasa 1999). Cystallization is a very complicated process, usually including nucleation and crystal growth. In the case of silicates, it is characterized by the formation of an ordered arrangement of the silicate tetrahedra by thermal diffusion processes. In addition, in systems that tend to decompose, chemical fractionation takes place.

Diffusion in solid materials occurs via lattice defects that are thermally activated. This process can be stimulated by irradiation and OH enrichment of the silicates. Irradiation by energetic cosmic-ray particles can increase the defect concentration. OH anions incorporated in silicates reduce the viscosity and promote crystallization. The temperature and time required to transform silicates from an amorphous to an crystalline state depends on the chemical structure and the presence of defects.

Astrophysically interesting experiments have been performed by Hallenbeck et al. (1998, see also Hallenbeck et al. 2000), Brucato et al. (1999), Fabian et al. (2000), and Thompson & Tang (2001). Fabian et al. studied the annealing behaviour of pure magnesium silicates and amorphous silica particles. They found that annealing magnesium silicate

Table 2 Compilation of optical data for olivines and pyroxenes.

Material	Wavelength Range	Remarks	Key references
Crystalline	Olivines		
Forsterite	8-67 μm	Reflectivity/Oscillator fits	Servoin & Piriou 1973
Forsterite	Infrared	Oscillator fits	Iishi 1978
Olivine	3-250 μm	Reflectivity/Dielectr. funct.	Steyer 1974
Olivine	0.03-250 μm	Optical constants (k)	Huffman 1975
Olivine	0.4-200 μm	Reflectivity/Dielectr. funct.	Fabian et al. 2001
Fayalite	7.7-200 μm	Reflectivity/Dielectr. funct.	Hofmeister 1997
Fayalite	0.4-200 μm	Reflectivity/Dielectr. funct.	Fabian et al. 2001
Amorphous	Olivines		
Olivine	7-33 μm	Optical constants	Day 1979
Olivine	7-300 μm	Optical constants	Day 1981
Olivine	0.04-62 μm	Optical constants	Scott & Duley 1996
Olivine	0.2-500 μm	Optical constants	Dorschner et al. 1995
Crystalline	Pyroxenes		
Enstatite	0.25-500 μm	Reflectivity/Dielectr. funct.	Jäger et al. 1998
Bronzite	7-500 μm	Optical constants/Low temp.	Henning & Mutschke 1997
Amorphous	Pyroxenes		
Pyroxene	0.04-62 μm	Optical constants	Scott & Duley 1996
Pyroxene	6-40 μm	Optical constants	Dorschner et al. 1988
Pyroxene	7-500 μm	Optical constants/Low temp.	Henning & Mutschke 1997
Pyroxene	7-300 μm	Optical constants	Day 1981
Pyroxene	0.2-500 μm	Optical constants	Jäger et al. 1994 Dorschner et al. 1995

smokes at 1000 K finally leads to a product which contains crystalline forsterite, trydimite (a crystalline modification of SiO_2) and amorphous silica (a-SiO_2) according to the initial Mg/Si-ratio of the smoke. Cystallization occured within a few hours for the Mg_2SiO_4 smoke (see Fig. 6) and within one day for the $MgSiO_3$ smoke. The annealing performance of silicates is certainly influenced by the presence of iron, aluminium, and less abundant elements (see, e.g., Karner et al. 1996). Experiments with smoke containing iron and as a minor component aluminium and produced by laser ablation were performed by Brucato et al. (1999). They found that their magnesium-iron smoke evolved into pyroxene. In any case, the formation of iron-rich crystalline silicates cannot be regarded independently of the gas phase composition, especially the abundance of oxygen. In an oxygen-dominated atmosphere, iron will quickly be oxydized. Therefore, annealing will lead to Mg-rich silicates and iron oxides. In contrast, annealing under oxygen-poor conditions would yield Mg-silicates, silica, and metallic iron.

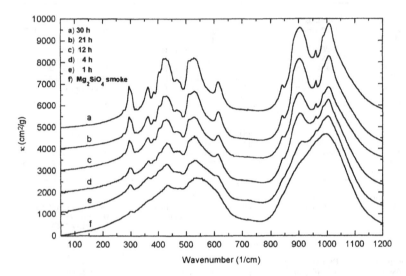

Figure 6   Evolution of the mid-infrared spectrum of Mg_2SiO_4 smoke after annealing at 1000 K up to 30 h. The annealing times are indicated. For clarity, the spectra are vertically shifted:a, b, c, d, e by +5000, +4000, +3000, +2000, +1000 cm^2/g

4.   SUMMARY

Thanks to the ISO mission and a detailed study of cosmic dust analogues in the laboratory, we reached a new quality of our understandig of cosmic silicates. For the first time, astromineralogy is no longer a dream, but becomes reality. The spectroscopic properties of the particles can be used to characterize their formation routes and lifecycle. From the observational material it became clear that we have to deal with a large diversity of silicates.

The ISO spectrometers could only obtain spectra of relatively bright objects. With the SIRT mission, we will obtain spectra of a large number of fainter sources, including the disks around T Tauri stars and Vega-type objects.

Acknowledgments

The author thanks the members of the Jena Laboratory Astrophysics Group for their support in writing this paper and the DFG for funding the experimental research. Drs. J. Gürtler and C. Jäger carefully read the paper. In addition, I thank D. Fabian, H.-P. Gail, F. Molster, R. Waters, and C. Waelkens for providing figures.

REFERENCES

Berrilli, F., Lorenzetti, D., Saraceno, P. and Strafella, F. (1987), *MNRAS*, **228**, 833.

Bowey, J.E., Lee, C., Tucker, C., Hofmeister, A.M. and Ade, P.A.R., in *ISO Beyond the Peaks*, ESA SP-456, 339.

Bradley, J.P. (1994), *Science*, **265**, 925.

Bradley, J.P. et al. (1999), *Science*, **285**, 1716.

Brucato, J.R., Colangeli, L., Mennella, V., Palumbo, P. and Bussoletti, E. (1999), *A&A*, **348**, 1012.

Burlitch, J.M., Beeman, M.L., Riley, B. and Kohlstedt, D.L. (1991), *Chem. Mater.*, **3**, 692.

Cesarsky, D., Jones, A.P., Lequeux, J. and Verstraete, L. (2000), *A&A*, **358**, 708.

Chiar, J.E., Tielens, A.G.G.M., Whittet, D.C.B., Schutte, W.A., Boogert, A.C.A., Lutz, D., van Dishoeck, E.F. and Bernstein, M.P. (2000), *ApJ*, **537**, 749.

Cohen, M. et al. (1999), *ApJ*, **513**, L135.

Cohen, M. and Witteborn, F.C. (1985), *ApJ*, **294**, 345.

Cox, P. and Roelfsema, P.R. (1999), in d'Hendecourt, L., Joblin, C. and Jones, A. (eds.), *Solid Interstellar Matter: The ISO Revolution*, EDP Sciences and Springer, Berlin, 151.

Crovisier, J., Leech, K., Bockelée-Morvan, D., Brooke, T.Y., Hanner, M.S., Altieri, B., Keller, H.U. and Lellouch, E. (1997), *Science*, **275**, 1904.

Day, K.L. (1974), *ApJ*, **192**, L15.

Day, K.L. (1976), *ApJ*, **210**, 614.

Day, K.L. (1979), *ApJ*, **234**, 158.

Day, K.L. (1981), *ApJ*, **246**, 110.

Demyk, K., Dartois, E., Wiesemeyer, H., Jones, A.P. and d'Hendecourt, L. (2000), *A&A*, **364**, 170.

Demyk, K. et al. (2001), *A&A*, **368**, L38.

Demyk, K., Jones, A.P., Dartois, E., Cox, P. and d'Hendecourt, L. (1999), *A&A*, **349**, 267.

Dorschner, J., Begemann, B., Henning, Th., Jäger, C. and Mutschke, H. (1995), *A&A*, **300**, 503.

Dorschner, J. and Henning, Th. (1986), *Astrophys. Space Sci.*, **128**, 47.

Draine, B.T. and Lee, H.M. (1984), *ApJ*, **285**, 89.

Fabian, D. et al. (2001), *A&A*, **378**, 228.

Fabian, D., Jäger, C., Henning, Th., Dorschner, J. and Mutschke, H. (2000), *A&A*, **364**, 282.

Gail, H.-P. and Sedlmayr, E. (1998), in Hartquist, T.W. and Williams, D.A. (eds.), *The Molecular Astrophysics of Stars and Galaxies*, Clarendon Press, Oxford, p.285.

Grun, J., Manka, C.K., Hoffman, C.A., Meyer, R.J., Glembocki, O.J., Kaplan, R., Quadri, S.B., Skelton, E.F., Donnelly, D. and Covington, B. (1997), *Phys. Rev. Letters*, **78**, 1584.

Hallenbeck, S.L., Nuth, J.A. and Daukantas, P.L. (1998), *Icarus*, **131**, 198.

Hallenbeck, S.L., Nuth, J.A. and Nelson, R.N. (2000), *ApJ*, **535**, 247.

Hanner, M.S. (1999), *Space Science Rev.*, **90**, 99.

Harker, D. (1999), Silicate Mineralogy of C/1995 O1 (Hale-Bopp) and its Implications to the Study of Pre-Main Sequence Stars and the Origins of the Solar System. *AAS*, **195.5803**.

Henning, Th. (1983), *Astrophys. Space Sci.*, **97**, 405.

Henning, Th., Michel, B. and Stognienko, R. (1995), *Planet. Space Sci.*, **43**, 1333-1343.

Henning, Th. and Mutschke, H. (1997), *A&A*, **327**, 743.

Henning, Th. and Mutschke, H. (2000), in M.L. Sitko, A.L. Sprague, D.V. Lynch (eds.): *Thermal Emission Spectroscopy and Analysis of Dust, Disks, and Regoliths*, ASP Conf. Ser., **196**, 253.

Henning, Th. and Stognienko, R. (1993), *A&A*, **280**, 609.

Herter, T., Humberto Campins and Gull, G.E. (1987), *A&A*, **187**, 629.

Hofmeister, A.M. (1997), *Phys. Chem. Miner.*, **24**, 535.

Hron, J., Aringer, B. and Kerschbaum, F. (1997), *A&A*, **322**, 280.

Huffman, D.R. (1975), *Astrophys. Space Sci.*, **34**, 175.

Iishi, K. (1978), *American Mineralogist*, **63**, 1198.

Jäger, C. et al. (2001), in preparation.

Jäger, C., Molster, F.J., Dorschner, J., Henning, Th., Mutschke, H. and Waters, L.B.F.M. (1998), *A&A*, **339**, 904.

Jäger, C., Mutschke, H., Begemann, B., Dorschner, J. and Henning, Th. (1994), *A&A*, **292**, 641.

Jones, A.P., Tielens, A.G.G.M. and Hollenbach, D.J. (1996), *ApJ*, **469**, 740.

Jones, A.P., Tielens, A.G.G.M., Hollenbach, D.J. and McKee, C.F. (1994), *ApJ*, **433**, 797.

Jones, T.W. and Merrill, K.W. (1976), *ApJ*, **209**, 509.

Karner, J.M., Rietmeijer, F.J.M. and Janeczek, J. (1996), *Meteoritics & Planetary Science*, **31**, A69.

Kemper, F., Waters, L.B.F.M., de Koter, A. and Tielens, A.G.G.M. (2001), *A&A*, **369**, 132.

Krätschmer, W. and Huffman, D.R. (1979), *Astrophys. Space Sci.*, **61**, 195.

Malfait, K. (1999), PhD Thesis, University of Leuven.

Malfait, K., Waelkens, C., Waters, L.B.F.M., Vandenbussche, B., Huygen, E. and de Graauw, M.S. (1998), *A&A*, **332**, L25.

Martin, P.G. (1995), *ApJ*, **445**, L63.

Mathis, J.S. (1998), *ApJ*, **497**, 824.

Meeus, G., Waters, L.B.F.M., Bouwman, J., van den Ancker, M.E., Waelkens, C. and Malfait, K. (2001), *A&A*, **365**, 476.

Mennella, V., Brucato, J.R., Colangeli, L., Palumbo, P., Rotundi, A. and Bussoletti (1998), *ApJ*, **496**, 1058.

Michel, B., Henning, Th., Kreibig, U. and Jäger, C. (1999), *Carbon*, **37**, 391.

Molster, F.J. (2000), PhD Thesis, Univ. of Amsterdam.

Molster, F.J. et al. (1999), *Nature*, **401**, 563.

Molster, F.J., Yamamura, I., Waters, L.B.F.M., Nyman, L.-A., Käufl, H.-U., de Jong, T. and Loup, C. (2001), *A&A*, **366**, 923.

Monnier, J.D., Geballe, T.R. and Danchi, W.C. (1998), *ApJ*, **502**, 833.

Mysen, B.O., Virgo, D. and Seifert, F.A. (1982), *Reviews of Geophysics and Space Physics*, **20**, 353.

Nuth, J.A. and Donn, B. (1982), *J. Chem. Phys.*, **77**, 2639.

Nuth, J.A. and Donn, B. (1983), *J. Chem. Phys.*, **78**, 1618.

Nuth, J.A., Hallenbeck, S.L. and Rietmeijer, F.J.M. (2000), *J. of Geophys. Res.*, **105**, 10387.

Ossenkopf, V. and Henning, Th. (1994), *A&A*, **291**, 943.

Ossenkopf, V., Henning, Th. and Mathis, J.S. (1992), *A&A*, **261**, 567.

Posch, T., Kerschbaum, F., Mutschke, H., Fabian, D., Dorschner, J. and Hron, J. (1999), *A&A*, **352**, 609.

Rietmeijer, F., Nuth III, J.A., Karner, J.M. (1999), *ApJ*, **527**, 395.

Sandford, S.A. and Walker, R.M. (1985), *ApJ*, **291**, 838.

Scott, A. and Duley, W. (1996), *ApJSS*, **105**, 401.

Sembach, K.R. and Savage, B.D. (1996), *ApJ*, **457**, 211.

Servoin, J.L. and Piriou, B. (1973), *Phys. Stat. Sol.*, **B 55**, 677.

Snow, T. and Witt, A. (1996), *ApJ*, **468**, L65.

Sogawa, H. and Kozasa, T. (1999), *ApJ*, **516**, L33.

Speck, A.K., Barlow, M.J., Sylvester, R.J. and Hofmeister, A.M. (2000), *A&A Suppl. Ser.*, **146**, 437.

Spitzer, L. Jr. and Fitzpatrick, E.L. (1993), *ApJ*, **409**, 299.

Spitzer, L. Jr. and Fitzpatrick, E.L. (1995), *ApJ*, **445**, 196.

Stephens, J.R., Blanco, A., Bussoletti, E., Colangeli, L., Fonti, S., Mennella, V. and Orofino, V. (1995), *Planet. and Space Sci.*, **43**, 1241.

Steyer, T.R. (1974), PhD Thesis, Univ. of Arizona, Tucson.

Sylvester, R.J. et al. (1999), *A&A*, **352**, 587.

Takai, H., Hosoya, S. and Ozima, M. (1984), in Sunagawa, I. (ed.), *Materials Science of the Earth's Interior*, Terrapub, Tokyo, p.107.

Tanabe, T., Kamijo, F., Onaka, T., Sakata, A. and Wada, S. (1986), *Astrophys. Space Sci.*, **119**, 147.

Thompson, S.P. and Tang, C.C. (2001), *A&A*, **368**, 721.

Tielens, A. G. G. M. (1998), *ApJ*, **499**, 267.

van den Ancker, M.E., Bouwman, J. and Wesselius, P.R. (2000), *A&A*, **357**, 325.

Waelkens, C. et al. (1996), *A&A*, **315**, L245.

Wang, S.X., Wang, L.M., Ewing, R.C. and Doremus, R.H. (1998), *J. Non-Cryst. Solids*, **238**, 198.

Waters, L.B.F.M. et al. (1996), *A&A*, **315**, L361.

Waters, L.B.F.M. et al. (1998a), *Nature*, **391**, 868.

Waters, L.B.F.M. et al. (1998b), *A&A*, **331**, L61.

Waters, L.B.F.M. et al. (2000), in Sitko, M.L., Sprague, A.L., Lynch, D.V. (eds.), *Thermal Emission Spectroscopy and Analysis of Dust, Disks, and Regoliths*, ASP Conf. Ser., **196**, p.3.

Waters, L.B.F.M. and Molster, F.G. (1999), in Le Bertre, T., Lebre, A., Waelkens, C. (eds.), *Asymptotic Giant Branch Stars*, IAU Symp., **191**, p.209.

Webber W.R. (1998), *ApJ*, **506**, 329.

Williams J.S., Elliman R. G., Brown W.L. and Seidel, T.E. (1985), *Phys. Rev. Letter*, **55**, 1482.

Willner, S.P. et al. (1982), *ApJ*, **253**, 53.

Wooden, D. and Harker, D. (2000), in Sitko, M.L., Sprague, A.L., Lynch, D.V. (eds.), *Thermal Emission Spectroscopy and Analysis of Dust, Disks, and Regoliths*, ASP Conf. Ser., **196**, p.99.

Wooden, D., Harker, D., Woodward, C.E., Butner, H., Koike, C., Witteborn, F.C. and McMurty, C.W. (1999), *ApJ*, **517**, 1034.

INTERSTELLAR CHEMISTRY
IN THE GAS AND ON THE SURFACES
OF DUST PARTICLES

Eric Herbst

Departments of Physics and Astronomy, The Ohio State University, 174 W. 18th Ave., Columbus, OH 43210-1106, USA

Keywords: molecules – interstellar clouds – gas-phase chemistry – dust particles – surface chemistry – diffusion – models – spectroscopy – kinetics – rate processes

Abstract   After an introduction to molecular observations pertaining to interstellar clouds, we discuss current ideas on the chemical processes occurring in the gas phase and on the surfaces of dust particles in clouds. Various pathways to the synthesis of polyatomic molecules are then illuminated. We summarize detailed gas-phase and gas-grain models of the chemistry in the cores of dense interstellar clouds and compare their results with observations. Extensions of modelling efforts to star formation regions are mentioned. We close with a discussion of current difficulties and likely directions of future activity.

Introduction

The cool component of the interstellar medium is concentrated into interstellar clouds, which can be many parsecs in extent and contain up to a million solar masses of matter in the form of gas and dust particles (Spitzer 1978; Hollenbach & Thronson 1987). The gas phase, which contains most of the matter, is mainly neutral hydrogen in atomic or molecular form at densities from 10^1–10^4 cm^{-3} with temperatures in the range 10–100 K. Atomic hydrogen is the dominant species in the more diffuse clouds, while molecular hydrogen is dominant in the denser regions. Dense clouds are the sites of star formation; in a still only partially understood process, small portions of the clouds collapse and eventually heat up to the point that nuclear fusion onsets to counteract

V. Pirronello et al. (eds.), Solid State Astrochemistry, 105–146.

© 2003 *Kluwer Academic Publishers. Printed in the Netherlands.*

the gravitational force. The process of star formation exhibits many manifestations depending on the mass of the collapsing object (Bergin 2000).

The composition of the gas and, to a lesser extent, dust phases of interstellar clouds can be determined from spectroscopic studies in various wavelength regions. Background stars can be used as lamps to study the optical absorption spectra of diffuse cloud gas which, although principally atomic, does contain detectable concentrations of diatomic species such as H_2, CO, CN, CH, OH, and CH^+ (Van Dishoeck & Black 1988) and a few polyatomic species such as C_3 (Maier et al. 2001), CCH, HCO^+, and C_3H_2 (Lucas & Liszt 2000). The detections of the last three species were made with a millimeter-wave lamp. The most abundant ion in diffuse clouds is singly-ionized atomic carbon, which is produced by ultra-violet radiation and possesses a fractional abundance with respect to the overall gas density of $\sim 10^{-4}$. Depletion of heavy elements in the gas from their abundances in stellar atmospheres yields clues to the elemental composition of the dust particles. Although stars have differing amounts of each element, the cosmic (number) average, as represented by nearby stars, is well defined (Snow & Witt 1996). Table 1 shows four sets of elemental abundances normalized with respect to hydrogen (Leung et al. 1984). The "high metal" abundances pertain to diffuse clouds, while the "low metal" abundances, often used for dense clouds, are more uncertain. Depletions of heavy elements from the solar and cosmic values can be seen.

Table 1   Elemental Abundances With Respect to Hydrogen

Element	Solar	Cosmic	High Metal	Low metal
He	6.3(-2)	1.0(-1)	1.40(-1)	1.40(-1)
O	7.4(-4)	4.6(-4)	1.76(-4)	1.76(-4)
C	4,0(-4)	2.1(-4)	7.30(-5)	7.30(-5)
N	9.3(-5)	6.6(-5)	2.14(-5)	2.14(-5)
S	2.6(-5)	1.2(-5)	8.00(-6)	8.00(-8)
Si	3.5(-5)	1.9(-5)	8.00(-7)	8.00(-9)
Na	2.1(-6)	—	2.00(-7)	2.00(-9)
Mg	3.8(-5)	2.5(-5)	7.00(-7)	7.00(-9)
Fe	3.2(-5)	2.7(-5)	3.00(-7)	3.00(-9)
P	2.8(-7)	—	3.00(-9)	3.00(-9)
Cl	1.1(-7)	—	4.00(-9)	4.00(-9)

In addition to the assigned optical absorption lines, there are many unassigned features ranging from narrow to broad in width known as the diffuse interstellar bands, or DIBs for short. Although debate has raged for many years on whether the DIBs are caused by the gas or the dust particles, current belief centers on exotic, large gas-phase molecules as their carriers (Tielens & Snow 1995).

Because visible and ultra-violet radiation do not penetrate dense interstellar clouds, they must be studied at longer wavelengths where the extinction, caused mainly by dust particles, lessens (Spitzer 1978). Despite the low temperatures, the gas emits in the millimeter-wave region of the spectrum due mainly to collisional excitation. The emission is in the form of many individual lines produced as molecules relax from higher to lower rotational levels. The lines can be matched with those studied by microwave spectroscopists in the laboratory for many years (Winnewisser et al. 1992; Herbst 1999; Wall et al. 1999). The atmosphere of the earth is transparent through the millimeter-wave; at submillimeter wavelengths there are still windows of relative transparency allowing some ground-based observations although these must be supplemented by aircraft and satellite-based studies (Melnick et al. 2000). The term "radio astronomy" is still applied to these studies, although most molecular rotational transitions lie at wavelengths shorter than the non-astronomical definition of "radio."

In the last decade, infra-red astronomy has become important in the study of dense interstellar clouds. Since the atmosphere of the earth is not particularly transparent in this wavelength region, mountaintop and especially satellite observations are needed. Particularly important has been the European-based satellite named ISO (Ehrenfreund & Schutte 2000). Unlike the situation in the millimeter, clouds are not typically warm enough to emit in the infrared except in localized hot regions. The study of cold gas and dust must be accomplished in absorption by utilizing two types of lamps: background stars and infra-red continuum sources within the clouds such as young stellar and near-stellar type objects. The detected spectral features are caused by vibrational transitions involving both gas-phase and condensed-phase molecules. The gas-phase molecules exhibit sharp lines with rotational substructure, while the condensed-phase species show much broader features due mainly to inhomogeneous environments. Infrared absorption spectra are particularly critical in the detection of non-polar gaseous molecules - only one of which, H_2, can be detected by quadrupolar rotational transitions. In recent years, an exciting infrared identification has been that of the most basic polyatomic species - H_3^+ (Geballe & Oka 1996). Infrared emission spectra from hot gas-phase molecules can be detected from a variety of sources. For example, "photon-dominated regions (PDRs)", in which radiation from newly-formed bright stars is pervasive, exhibit the well-known broad UIR (unidentified infrared) features, now assigned (Allamandola et al. 1989) to a collection of polycyclic aromatic hydrocarbons (PAHs), closely akin to automobile soot on earth.

Because low density conditions present in the interstellar medium often lead to non-thermal situations, population inversions are occasionally obtained. Interstellar clouds are even the sites of masers, which are located in regions of considerable excitation near young stellar objects. Some of the molecules seen in maser emission are OH, SiO, H_2O, and methanol (Elitzur 1999). The excitation of the maser emission can be achieved by radiation, collisions, or both.

The sundry spectroscopic observations discussed above show that the gaseous matter in dense interstellar clouds is overwhelmingly molecular in nature. Table 2 contains a list of the more than 120 molecules detected in interstellar and circumstellar sources with high resolution spectroscopic techniques. Many isotopomers have been detected, especially containing deuterium, carbon-13, and oxygen-18. Ranging in size up to 13 atoms, the molecules in dense clouds are all trace constituents compared with the dominant species - H_2. For example, the molecule with the second highest abundance, CO, has a concentration $\leq 10^{-4}$ that of hydrogen. Although chemistry plays a major role in the abundances of individual molecules, the dominant elemental abundance of hydrogen is critical (see Table 1). The fractional abundances of molecules in dense interstellar clouds are not homogeneous but depend on density and temperature; near star-forming regions, for example, the abundances change in a variety of ways depending on the mass of the star that is forming, the stage of evolution of the process, and the amount of energy that is being emitted. Even diffuse clouds show structure, and there are so-called translucent clouds, which appear to contain dense central cores surrounded by more diffuse material.

The molecules in Table 2 are mainly organic (carbon-containing) and tend to be unsaturated in nature except in unusual regions near newly formed high mass stars. Approximately half are well-known species by terrestrial standards, but about half are quite unusual and consist of positive molecular ions (e.g. HCO^+, H_3^+, H_3O^+), radicals (e.g. C_nH through n=8), three-membered rings (e.g. c-C_3H, c-C_3H_2), and isomers (e.g. HNC). Despite the detection of molecular ions, it would appear that the fractional ionization in dense clouds is considerably lower than in diffuse clouds, a not totally surprising observation since the penetration of ionizing radiation is less in the denser sources.

Table 2 Gas Phase Interstellar/Circumstellar Molecules Detected By High Resolution Spectroscopy

H_2	CH	CH^+	NH	OH
HF	C_2	CN	CO	CSi
CP	CS	NO	NS	SO
HCl	NaCl	KCl	AlCl	AlF
PN	SiN	SiO	SiS	CO^+
SO^+	H_3^+	CH_2	NH_2	H_2O
H_2S	C_2H	HCN	HNC	HCO
HCO^+	HOC^+	HN_2^+	HNO	HCS^+
C_3	C_2O	C_2S	SiC_2	SO_2
CO_2	OCS	MgNC	MgCN	N_2O
NaCN	CH_3	NH_3	H_2CO	H_2CS
HCCH	$HCNH^+$	H_2CN	C_3H (lin)	$c\text{-}C_3H$
HCCN	HNCO	$HOCO^+$	HNCS	C_2CN
C_3O	H_3O^+	C_3S	CH_4	SiH_4
CH_2NH	H_2C_3(lin)	$c\text{-}C_3H_2$	CH_2CN	NH_2CN
CH_2CO	HCOOH	C_4H	HC_2CN	HCCNC
HNCCC	C_4Si	H_2COH^+	C_5	CH_3OH
CH_3SH	C_2H_4	CH_3CN	CH_3NC	HC_2CHO
NH_2CHO	HC_3NH^+	H_2C_4 (lin)	C_5H	C_5N
CH_3NH_2	CH_3CCH	CH_3CHO	C_2H_3CN	$c\text{-}CH_2OCH_2$
C_6H	HC_4CN	C_7H	$HCOOCH_3$	CH_3COOH
CH_2OHCHO	CH_3C_2CN	H_2C_6 (lin)	C_8H	CH_3OCH_3
C_2H_5OH	C_2H_5CN	CH_3C_4H	HC_6CN	CH_3COCH_3
$CH_3C_4CN?$	$NH_2CH_2COOH?$	HC_8CN	$HC_{10}CN$	isotopomers

Spectral lines contain a wealth of information besides the identity and abundance of the species responsible for them. In the gas, the frequencies of the lines and the line shapes are due to the Doppler effect, and so tell us about the velocities of the clouds and the large scale motions inside them. In the main, clouds appear to be turbulent rather than purely thermal in nature. Since most interstellar molecules in dense clouds show many rotational emission lines, the relative intensities of the lines yield information on the rotational temperatures and, indirectly, on the kinetic temperature of the gas (Hollenbach & Thronson 1987; Wall et al. 1999). The existence of collisionally-excited rotational lines emanating from the dense cloud gas can be used to determine the gas density given knowledge of cross sections for rotationally inelastic collisions with the dominant species - H_2. Weakly polar species such as CO are excited at very low densities (10^2 - 10^3 cm^{-3}), especially for low-lying rotational transitions, when the collisional excitation rate equals the slow radiative relaxation rate, whereas polar species such as CS, which emit more rapidly, are excited only at higher densities, found in especially dense areas of the clouds known as "cores", which are often precursors to star formation. These cores are also detectable by changes in the fractional abundances of molecules compared with less dense regions. Once cores are discovered, they can be studied with high spatial resolution by using numbers of radio telescopes in interferometric mode.

Infrared spectroscopic features arising from the dust particles are broad and so more ambiguous to interpret than their gas-phase brethren. Nevertheless, absorption spectra due to silicate cores and icy mantles have been assigned by careful laboratory studies of analogous material. The major components of the ice are water, carbon dioxide, carbon monoxide, and methanol. In cold and dense environments, the amount of material tied up as ices can be large, with grain mantles of up to 100 monolayers. Laboratory reproduction of the spectra yields not only the carriers but their likely mantle environments (e.g. polar or non-polar) (Ehrenfreund & Schutte 2000). Not all dust particle cores are composed of silicates; a bump in the extinction curve at 2175 Å is customarily interpreted in terms of carbonaceous particles (Witt 2000).

The size distribution, shape, and porosity of interstellar dust grains are still only partially determined from spectroscopic and scattering data. Regarding size, the general viewpoint is that the particles exhibit a distribution that peaks around a radius of 0.1 μ but ranges to significantly larger and smaller sizes than this peak value. It is becoming clearer, however, that there is a significant component of the dust consisting of particles small enough to be more profitably regarded as large molecules such as PAHs. Although the surface chemistry to be discussed in this article will be based on 0.1 μ grains, the effective surface area of the small grain population is sufficiently large that future treatments must consider their chemistry as well. The 0.1μ grains considered here have roughly 10^9 atoms, a surface of 10^6 sites, and an abundance ratio to the overall gas density of 10^{-12}.

1.   WHERE ARE INTERSTELLAR MOLECULES FORMED?

The material in interstellar clouds derives from earlier generations of stars, especially in their later stages (Abell 1982). Whether explosively or gently, both gas and dust in stellar atmospheres are blown into the interstellar medium and eventually condense to form the current generation of clouds. The majority of stellar atmospheres contain more elemental oxygen than carbon (Table 1); the dust emanating from these stars is silicaceous in nature. Stars in which there is more carbon than oxygen are the producers of carbonaceous dust. Although gas-phase molecules are formed in stellar atmospheres, especially carbon-rich objects (Millar et al. 2000; Frenklach & Feigelson 1989), the radiation field due to the stellar background in unshielded interstellar space is intense enough to dissociate small molecules rather quickly. For example, the lifetime of CO in the unshielded interstellar medium is about 100 years (Van Dishoeck 1988). Lifetimes of sufficiently large species such as PAHs, however, may be long enough to withstand the destructive nature of the pervasive radiation so that these species can survive until they are incorporated into clouds. It is certainly true that dust particles will survive the journey. Thus, as clouds form, the matter in them consists of atomic gas, the cores of dust particles, and molecules large enough to withstand the interstellar radiation field. Once inside dense clouds, all matter is shielded from such radiation to a great extent. Inside diffuse clouds, on the other hand, photodissociation is often the dominant destructive mechanism for molecules (Van Dishoeck & Black 1988).

Small interstellar molecules must then be formed under the conditions of interstellar clouds, by synthesis from atoms, by sputtering off of dust particles, or by some degradation processes from large molecules present in the initial gas. Since any large molecules present are there because of their photostability, their fragmentation must occur via energetic particles such as cosmic rays (to be discussed below). Both sputtering and degradation have been discussed in the context of the diffuse interstellar bands seen in diffuse clouds (Ruffle et al. 1999). But, based on the success of models to be discussed in this article, it would appear that certainly the majority of species seen in dense clouds as well as the small molecules detected in diffuse clouds are formed by synthetic processes.

2.   GAS-PHASE REACTIONS

Synthetic processes in interstellar clouds can take place in the gas and on the surfaces of dust particles. In the gas, densities are sufficiently low even in the most dense regions that only two-body processes can occur. Even two-body collisions are sufficiently rare that the only important reactions are those occurring with high efficiency; at low temperatures such gas-phase reactions must be exothermic (energy-releasing) processes without activation energy. Consider a simple chemical reaction of the type

$$A + B \longrightarrow C + D \qquad (1)$$

where A and B are atoms or molecules. The rate of this reaction is given by the expression

$$\frac{d[C]}{dt} = \frac{d[D]}{dt} = k[A][B], \tag{2}$$

where the symbols $[\ldots]$ stand for concentration, and k is the rate coefficient. Note that $[\ldots]$ is used in the chemical literature; a more common astronomical notation for concentration is $n(\ldots)$. We utilize the chemical convention here to avoid confusion with surface abundances. Rate coefficients can normally be expressed by the Arrhenius equation:

$$k(T) = A(T)\exp(-E_a/k_B T) \tag{3}$$

where $A(T)$ is governed by both the long-range nature of the potential energy surface for the reaction and certain aspects of the short-range potential, and E_a, the activation energy, is the minimum energy needed for reaction. There is a possible tunneling correction to the Arrhenius equation, which is not included here. For endothermic reactions, the activation energy must be at the least the energy needed to form the products, but even for exothermic reactions, there is often a short-range barrier in the potential surface that is on the order of eV for most reactions involving neutral species. Consequently, at low temperature $E_a \gg k_B T$, and the exponential factor in the Arrhenius expression reduces the rate coefficient to near zero.

A simplistic explanation for the existence of activation energy is that old chemical bonds must be broken before new ones can be formed. That this explanation is simplistic is apparent when one considers that there are classes of exothermic reactions that do not possess activation energy barriers. These processes will dominate gas-phase interstellar chemistry except perhaps in regions in the proximity of young stellar objects. If the activation energy is zero, the Arrhenius expression shows that the rate coefficient is simply the so-called pre-exponential factor $A(T)$, which possesses at most a weak temperature dependence.

Of processes without activation energy, the most important are ion-molecule reactions; viz.,

$$A^+ + B \longrightarrow C^+ + D, \tag{4}$$

because these reactions have large rate coefficients due to the strongly attractive nature of their long-range potentials. Of course, ion-molecule reactions would still not be of importance in interstellar clouds unless there are sources of ions despite the low temperatures. In diffuse clouds, ions are formed by UV photons and by bombardment by high energy particles known as cosmic rays. These particles, which are high-energy protons and heavier nuclei travelling at relativistic speeds (Spitzer 1978) have sufficiently great penetrating power that, unlike photons, they are able to penetrate dense clouds as well. From the measurement of such particles above our own atmosphere as well as from the concentrations of interstellar ions, it is possible to estimate the rate of cosmic ray ionization to be $\zeta \sim 10^{-17}$ s^{-1} (Geballe & Oka 1996). This number is to be interpreted in the following manner. Consider the ionization of a species A:

$$A + \text{cosmic ray} \longrightarrow A^+ + \text{cosmic ray} + e. \tag{5}$$

The rate of this ionization is then given by the expression

$$\frac{d[A^+]}{dt} = \zeta[A]. \tag{6}$$

Actually the ionization is caused both directly by cosmic rays and indirectly by secondary electrons.

The rate coefficient k_{i-m} for ion-molecule reactions involving a non-polar neutral reactant is often found to agree with a simple expression first derived by Langevin (Rowe et al. 2000):

$$k_{i-m} = 2\pi e\sqrt{\alpha/\mu} \sim 10^{-9} \text{ cm}^3 \text{ s}^{-1} \tag{7}$$

where e (esu) is the electronic charge, α (cm^3) is the polarizability of the neutral reactant, and μ (gm) is the reduced mass of the reactants. This expression is independent of temperature, and the pre-exponential factor is 10-100 greater than the largest values for reactions involving neutral species. Ion-molecule reactions with a polar reactant occur even more rapidly, especially at low temperatures.

Not all neutral-neutral reactions possess activation energy; there are especially reactive species, known as radicals, that have an unpaired electron. Reactions involving two radicals, or a radical and an atom, occur rapidly even at low temperature. Recent experimental work has shown that, at least for some systems, reactions between radicals or atoms and normal neutral species can also be rapid at very low temperatures (Sims & Smith 1995). These reactions are now known to play crucial roles in the synthesis of certain classes of interstellar molecules such as the cyanopolyynes ($HC_{2n}CN$).

There are some unusual classes of reactions that also occur in interstellar clouds. Radiative association is a process in which two species literally stick together to form one product, which is stabilized against redissociation by the emission of energy in the form of a photon:

$$A^+ + B \longrightarrow C^+ + h\nu. \tag{8}$$

These reactions are often inefficient because it is more likely that the colliding species come apart, or react in a normal sense to form more than one product, than that they stick. Theoretical treatments show that radiative association can become rapid when the reactants are large so that the temporary collision "complex" AB^+ has many degrees of freedom and finds it difficult to dissociate along the path that produces it (Bates & Herbst 1988). Even for smaller systems, radiative association can be competitive in interstellar chemistry if one of the reactants is very abundant (e.g. H_2) so that the overall rate is large despite a relatively small rate coefficient.

Finally, there are reactions between molecular positive ions and electrons known as dissociative recombination processes since they are much more likely to produce neutral fragments than to produce parent neutral species. These processes typically have room temperature rate coefficients on the order of 10^{-6} cm^3 s^{-1} with a slight $(T^{-1/2})$ inverse dependence on temperature. Until recently, the products of these reactions were not known, but new measurements with both flowing afterglows and large storage rings have begun to remedy this deficiency (Vejby-Christensen et al. 1997; Larsson 2000). An example of a simple dissociative recombination reaction is that of the ion H_3^+:

$$H_3^+ + e \longrightarrow H + H + H; H_2 + H \tag{9}$$

where the fragmentation into three hydrogen atoms dominates.

This reaction is exceptional in that its overall rate coefficient has been the matter of a long dispute, with flowing afterglow results showing an unusually small rate coefficient and storage ring results showing a more typical, large rate coefficient (Larsson 2000). The discrepancy between high and low values appears to show no signs of being resolved. Theoretical treatments have up to now not helped the situation.

In addition to gas-phase chemical reactions, gas-phase species can undergo photodissociation and photoionization, even in dense clouds. These processes are initiated both by external photons (especially in diffuse clouds and in border regions of dense clouds) and by photons generated indirectly by cosmic rays. In the latter, cosmic rays first produce secondary electrons, which then excite the dominant interstellar molecule H_2 (Prasad & Tarafdar 1983). The subsequent relaxation of this species produces ultra-violet radiation even in the most dense clouds, where few external photons penetrate. The flux of such radiation is relatively small, however, compared with the flux in unshielded space. A typical neutral species in a dense cloud will be destroyed by reactions with ions more rapidly than it is photodissociated (Herbst & Leung 1986).

3.   REACTIONS ON GRAIN SURFACES

In addition to gas-phase processes, interstellar molecules can be produced on the surfaces of dust particles. The most important such process is the formation of molecular hydrogen from two hydrogen atoms (Hollenbach & Salpeter 1971), but there are many other significant processes as well, most of which appear to involve atomic hydrogen as a reactant.

Before such reactions can occur, gas-phase species must stick, or adsorb, to the grain surface. The rate of adsorption for a species A onto a grain is given by the equation

$$\frac{dN_A}{dt} = k_{ads}[A], \tag{10}$$

where N_A is the number of species A on a grain surface, and k_{ads} is the adsorption rate coefficient, which in turn is given by

$$k_{ads} = s\sigma v. \tag{11}$$

In this equation, s is the sticking probability, σ is the grain cross section, and v is the thermal velocity of A. Sticking probabilities are known to be near unity at low temperatures (Williams 1993), while the cross section for a grain of radius 0.1μ is 3×10^{-10} cm^2. The average velocity for H atoms at 10 K is 4.6×10^4 cm/s so that with $s = 1$, the adsorption rate coefficient for H at 10 K is 1.4×10^{-5} s^{-1}. Thus, if the initial gas composition in a dense cloud contains a concentration of 10^4 cm^{-3} of H atoms, the adsorption rate onto a given grain is 0.14 s^{-1}, so that an H atom sticks to a grain every 7 seconds.

Before one can consider the possibility of surface reaction, one most consider the rate of evaporation, which is clearly a competitive process. The rate coefficient for evaporation k_{evap}, which can be thought of as an inverse time scale, is given by the equation

$$k_{evap} = \nu_0 \exp(-E_D/T), \tag{12}$$

where ν_0 is a characteristic surface vibrational frequency, typically in the range 10^{12-13} s^{-1} for weakly adsorbed species, and E_D is the binding energy of the adsorbate to the surface, expressed in units of temperature (Tielens & Allamandola 1987). Although binding energies depend strongly on the surface material (carbon, silicates, ice), typical values for most adsorbates are sufficiently high that evaporation does not occur at 10 K. The exceptions are very light species, such as H, He, and H$_2$. The generally accepted minimal binding energy of H on surfaces for which it does not form a strong bond (e.g. silicates) is ~ 350 K; with this value and a typical vibrational frequency of 3×10^{12} s^{-1} (Hasegawa et al. 1992), the evaporation rate coefficient for H at 10 K is 1.9×10^{-3} s^{-1}. Thus, for H atoms to react on such a surface, they must do it before they evaporate in ~ 500 s. This is a longer time scale than the interval between adsorbing events for a dense cloud gas consisting mainly of atomic hydrogen (see above). For surfaces on which H is more strongly bound, such as amorphous carbon ($E_D = 511$ K; Katz et al. 1999), evaporation does not occur efficiently until higher temperatures. The actual rate of evaporation for any adsorbed species is given by the equation

$$\frac{dN_A}{dt} = -k_{evap}N_A. \qquad (13)$$

If reactions involving H atoms do not occur and one balances the evaporation (with $E_D = 350$ K) and adsorption rates of H atoms in a 10 K dense cloud with initial concentration 10^4 cm^{-3}, the average number of H atoms on a grain will be 75. But reaction does occur under these conditions. At later cloud stages, when the gas is mainly H$_2$ rather than H, the average number of H atoms on a grain is considerably less than unity, leading to problems, as we shall see.

The mechanism most often considered for surface reactions is diffusive in nature (Tielens & Allamandola 1987; Pickles & Williams 1977; Hasegawa et al. 1992; Hasegawa & Herbst 1993a), and is termed the Langmuir-Hinshelwood process. In this mechanism, two particles adsorbed weakly on the grain surface undergo a two-dimensional random walk among binding sites, or sites of minimum potential, before finding one another. If we label the energy needed to surmount the barrier between binding sites E_b, the rate coefficient, or inverse time scale, k_{hop} for diffusion of a species from one binding site to an adjacent one is given by an equation similar to that for the evaporation coefficient, but with a different energy:

$$k_{hop} = \nu_0 \exp(-E_b/T). \qquad (14)$$

If there are N binding sites on the surface of a grain, the rate coefficient k_s for sweeping over the entire grain is given by

$$k_s = \frac{k_{hop}}{N} = \frac{1}{N} \times \nu_0 \exp(-E_b/T), \qquad (15)$$

due to the two-dimensional nature of the surface (Biham et al. 2001). In addition to hopping from site to site, it is possible that quantum mechanical tunneling can be competitive. The rate of tunneling under a barrier is exponentially sensitive to the height and width of the barrier. If a rectangular barrier of width a and height E_b is assumed, the rate coefficient, or inverse time scale, for tunneling k_{tunn} is

$$k_{tunn} = \nu_0 \exp[-(2a/\hbar)(2mE_b)^{1/2}], \qquad (16)$$

where m is the mass of the species. In an imperfect lattice, the quantum mechanical tunneling rate over the entire grain is incoherent and is given by dividing k_{tunn} by N. Depending on the values of the parameters a and E_b, tunneling for light species such as H can be faster than classical diffusion. It is often written that for regular surfaces (Tielens & Allamandola 1987; Hasegawa et al. 1992)

$$E_b \approx 0.3 E_D \qquad (17)$$

but the generality of this relation is not apparent. If we assume it to be correct for H, $E_b \approx 100$ K if $E_D = 350$ K. Then, if $a = 1$ Å, the tunneling rate k_s for H atoms over an entire grain is 5×10^4 s^{-1} whereas the normal diffusion rate at 10 K is only 142 s^{-1}. Recent experiments show, however, that tunneling does not appear to occur for H atoms on olivine (polycrystalline silicate) and amorphous carbon surfaces at low temperatures (Katz et al. 1999). Moreover, Katz et al. (1999) found that for these two irregular surfaces, the relation between E_b and E_D does not hold. For olivine, while E_D was measured to be 373 K, near the previously assumed value, E_b was found to be 287 K, or 77 % of the binding energy. The higher value for the barrier against diffusion reduces the diffusion rate for H atoms at 10 K over an entire grain to $\ll 1$ s^{-1}, with important consequences for reactions involving this species.

Now let us consider a reaction to form products P between two species A and B that are diffusing on the surface of a grain. For reaction to occur, the species must be at the same binding site at the same time. From the point of view of species A, the rate of reaction is the rate of hopping or tunneling to an adjacent site multiplied by the probability that species B occupies that site. This probability is given by the expression N_B/N, where N_B represents the number of binding sites occupied by species B and N is the total number of sites. Recognizing that the hopping or tunneling rate of A to an adjacent site divided by N is the rate over the entire grain, and noting that species B also diffuses, we arrive at the rate law for the reaction:

$$\frac{dN_P}{dt} = \kappa(k_{s,A} + k_{s,B}) N_A N_B, \qquad (18)$$

where $k_{s,I}$ is the faster of the hopping and tunneling rates over the entire grain for species I, and κ is an efficiency factor that expresses the probability of reaction once the reactants are at the same binding site. The rate equation above does not include a so-called "rejection" term for saturated surfaces ($N_A, N_B \sim N$) because it is assumed that reaction can occur both on the original core and atop any mantle that is produced over time. The applicability of eq. (18) when N_A or $N_B > N$

is questionable since it implies that diffusive reactions can occur under the surface. A simple fix is to limit N_A and N_B to at most N when using this rate law. A more detailed rate law in which layers in the mantle are distinguished has been formulated by Hasegawa & Herbst (1993b). If it is desired to replace the numbers of species per grain with the volume concentration of surface species (cm^{-3}) in eqs. (10), (13), and (18), the N_I values must be multiplied by the grain concentration, as must the right-hand side of eq. (10). Also, in eq. (18), the diffusive rate coefficients given here must be divided by the grain concentration (Hasegawa et al. 1992). The disadvantage of this latter approach is that it requires changing the rate coefficients if the overall density of the source being modelled changes.

For reactive species such as atoms and radicals, $\kappa = 1$, while for unreactive species for which an activation energy barrier exists, κ can be much less than unity. Under interstellar conditions, the only possibility for reaction in this latter case is for quantum mechanical tunneling under the activation energy barrier; unlike the gas-phase case, this possibility cannot be ruled out for small but non-zero barriers since the reactants may have many opportunities to tunnel while they are still at the same binding site. Tunneling probabilities can be calculated with a similar equation to that used for quantum diffusion except that the activation energy barrier must be used. It should be noticed that we have only considered neutral species in our analysis. The surface lifetimes of positive ions that strike grains will probably be short because of the existence of electrons on the surface (Spitzer 1978; Aikawa et al. 1999).

What are the products of surface diffusion reactions? It is generally thought that, unlike the situation in the gas phase, association reactions dominate if there is a significant binding energy to form the product. Unlike the gas phase, the initial collision product of the two reactant species can give off its energy to the grain, which will rise slightly in temperature. It is even possible that some of the energy of reaction will be deposited in the adsorbate-surface mode, which will enable immediate desorption to take place. Such a situation was discovered by Katz et al. (1999) in their study of the formation of hydrogen molecules on low temperature surfaces.

Rate laws such as the above equation (Pickles & Williams 1977) , when coupled with adsorption and evaporation terms and integrated, yield average numbers of species per grain, which are not integers. If there are large numbers of surface species, then the distinction between integers and non-integers is unimportant, but if there are very small numbers of reactive surface species, especially fractional numbers well below unity, the concept of an average number of species is a dubious one. For example, if one computes that the average number of H atoms on a grain is 0.1, should one accept this number or should one replace it by zero or by unity? Under these circumstances, which can pertain in the interstellar medium, the rate equation approach to diffusive surface chemistry may only be an approximation, and other methods have been promoted.

3.1.   ALTERNATIVE APPROACHES TO DIFFUSIVE SURFACE CHEMISTRY

Once the gas phase of an interstellar cloud is mainly molecular hydrogen, the accretion rate of reactive atoms and radicals onto a grain is so tiny that on average there is a small number of these species on the surface. In this limit, known as the accretion limit, it makes sense to formulate an approach to surface kinetics in which the rate of accretion is more important than the rate of diffusion. In the early approach of Allen & Robinson (1977), based on the existence of grains small enough that the exothermicity of chemical reactions could supply the energy needed to desorb the reaction products, the rates of formation of surface species were equated, with approximations, to the rates of accretion of their constituent atoms or radicals, under the assumption that diffusion would enable the species to find one another. If one wishes to extend this approach to larger grains in which neither reaction nor desorption is guaranteed, a Monte Carlo procedure can be applied, where one drops randomly chosen species onto a grain at the accretion rate, and then follows what happens to two species dropped successively, choosing among the possibilities of reaction, evaporation (for light species), and neither of these. This approach, formulated initially by Tielens & Hagen (1982), is not a true stochastic method (see, e.g., Charnley 1998) because, among other reasons, the particles are dropped onto a grain at fixed rather than random intervals, and one follows only the accreting particles, not those remaining on the grain. The latter problem means that the procedure works only in the accretion limit, where one does not have to worry about the build-up of reactive particles. Of course, in the limit that large numbers of reactive particles exist on a grain surface, the normal rate equations for diffusive reaction should be adequate.

Despite the attractive nature of the Monte Carlo approach, it has not yet been possible to formulate a detailed model of the gas-phase and surface chemistry of interstellar clouds with it. In principle, this could be accomplished by stopping the gas-phase chemistry at specific intervals and then allowing the Monte Carlo procedure to determine what happens to the grain surface given fixed gas-phase abundances if the time scale of the surface processes is much smaller than that of the gas reactions. Unfortunately, this dichotomy holds only for rapid surface reactions, not reactions involving activation energy such as those that may explain the formation of methanol from CO (Charnley et al. 1997).

The difficulties of including the Monte Carlo procedure into a detailed model led Caselli et al. (1998) to formulate a set of semi-empirical corrections to the rate equations to enable the results of these equations to approach the Monte Carlo results in the accretion limit. The advantage of the approach of Caselli et al. (1998) is that it can easily be used in large gas-grain chemical models of dense clouds. The approach was compared with the Monte Carlo method by studying a series of simple, rather artificial systems at 10 K in which a small number of gas-phase species are considered and fixed in abundance. The simplest of these simple systems, which is similar to one discussed in a lecture by Tielens, consists of hydrogen and oxygen atoms in the gas phase, and the production of the three molecules H_2, OH, and O_2 on the surfaces of dust particles. The gas-phase O atom abundance is maintained at 1 cm^{-3} while the H atom abundance is varied from 0.1 cm^{-3} to 10^4 cm^{-3}. These numbers are based on the results of gas-phase models that show the O atom abundance in dense clouds to be rather constant but the H atom abundance to decrease with time. Unfortunately, Caselli et al. (1998) used an incorrect Monte Carlo procedure, an error that has only recently been corrected (Stantcheva, Caselli, & Herbst, in preparation). With a corrected Monte Carlo procedure, Stantcheva et al. have been able to reduce the number of modifications to one for the simple O, H system. The remaining modification is to slow down the diffusion rates of reactive species to the larger of their evaporation or accretion rates, if necessary. In this way, at most one molecule can be produced per accretion event. The agreement between the new modified rate approach and the Monte Carlo approach is satisfactory over a wide temperature range if the accretion limit pertains.

Although utilized in current gas-grain models of dense clouds, the modified rate method suffers from being semi-empirical in nature. Within the last year, several groups have reported a master equation method to study simple systems such as the O, H system (Green et al. 2001) or just the single reaction in which molecular hydrogen is formed (Green et al. 2001; Biham et al. 2001). For the O,H system, this new approach is in good agreement with the Monte Carlo and modified rate methods at 10 K. Whether or not this approach can be extended to large models with many surface reactions remains to be seen. The basic idea is to use differential equations similar to the diffusive rate equations to compute the probabilities that an integral number of atoms or molecules of a species is on a grain at a given time. For example, instead of computing the average number of hydrogen molecules N_{H_2}, one computes the probabilities $P_{H_2,i}$ that $i = 0, 1, 2, \ldots$ molecules are present. The average number of species can then be obtained from the sum over i; viz.,

$$N_{H_2} = \sum_{i=0}^{\infty} i \times P_{H_2,i}. \qquad (19)$$

In the accretion limit, the sum over i can be restricted to small integers only, so that a limited number of differential equations need to be incorporated into models. If, for example, one only has to include $i = 0, 1, 2$, then the model will contain only three times the number of differential equations as in the normal rate approach. Although this sounds like a simple extension of current models, the unanswered problem so far is whether or not one can compute independent probabilities for all surface species or whether the probabilities are correlated. The latter is obviously a more complex alternative.

3.2.   THE ELEY-RIDEAL AND HOT ATOM MECHANISMS

In addition to the complications discussed regarding the diffusive mechanism for surface chemistry, there is the problem that this is not the only mechanism. Suppose the binding energy between adsorbate and surface is high enough that at a given temperature no diffusion occurs at all within a reasonable time. Then, the surface will eventually become saturated with reactive but immobile species and the possibility that a gas-phase species striking a grain will come into contact with a reactive surface particle becomes high. If the gas-phase species strikes the adsorbed species directly, one refers to the "Eley-Rideal" mechanism, while if the non-thermalized adsorbing species must move on the surface first, one refers to the "hot atom" mechanism. Either mechanism can occur under interstellar conditions depending on the nature of the grain surface and the temperature. Qualitatively speaking, for any given surface and gas density, there will be a temperature range over which the diffusive mechanism can occur. Below the lower temperature limit, diffusion will no longer be competitive, and the Eley-Rideal and hot atom mechanisms will dominate. Above the upper temperature limit, evaporation will occur before diffusive reaction. Of course, if the binding energy between adsorbate and surface is large, the temperature range for diffusive reaction may occur at such large temperatures as to be of no importance. For example, Farebrother et al. (2000) have calculated that on graphite, the formation of molecular hydrogen occurs via an Eley-Rideal mechanism at all reasonable temperatures.

Although to the best of our knowledge, no model of gas-grain interstellar chemistry includes the Eley-Rideal mechanism, we are currently working to include this mechanism via both rate equations and a Monte Carlo approach.

3.3.   PHOTOCHEMISTRY ON SURFACES

Since normal stable species will not be reactive on cold grains via any of the mechanisms so far discussed, various investigators have suggested that photons can initiate a chemistry involving these species (Watson & Salpeter 1972; Gerakines et al. 1996). Indeed, laboratory groups have actually shown that photon bombardment of assorted mixes of stable species on cold surfaces leads to a complex photochemistry (Gerakines et al. 1996) in which many different species can be synthesized, even leading to the production of complex refractory material (Greenberg et al. 1995). The mechanism is the initial break-up of stable molecules into radicals, and the subsequent reactions involving these radicals. In these experiments, a heavy photon flux is applied for a short period of time, and the results compared with the interstellar situation in which a weak photon flux strikes grains over a very long period of time. It is not obvious that one can equate the laboratory and interstellar processes, both because of the different time scales and the interaction between interstellar gas and the dust. Consider, for example, the existence of H atoms in the interstellar gas. Even in dense clouds, the residual hydrogen atom abundance in the gas phase of dense clouds is sufficient to land on grains and hydrogenate radicals before they can react to form more complex species. For example, if water on the surface is broken into H and the radical OH, additional H atoms landing on the grain surface can simply reverse the process. On the other hand, if a molecule such as methanol is broken up into methyl and OH radicals, then H atoms will not reverse the process but will associate with the radicals to form the stable species methane and water. In either case, the existence of radicals may be short-lived.

We have attempted to include surface photochemistry into gas-grain chemical models of interstellar clouds in a rather simple manner (Ruffle & Herbst 2001a). Our basic approach is to equate the photodissociation rates of surface species with the better-measured rates appropriate to gas-phase species, based on the assumption that weakly-bound surface molecules can be thought of as individual entities. The photons in dense clouds come from the excitation of molecular hydrogen, the dominant species, by secondary electrons from cosmic ray ionization. Our results, which are admittedly approximate, indicate that surface photochemistry is only appreciable over very long time scales in dense interstellar clouds.

3.4.   NON-THERMAL DESORPTION

Except for light species such as H, H_2, and He, evaporation from dust particles does not occur at cold interstellar temperatures. To see this, consider a species that binds to a surface with a binding energy E_D of 1000 K, a reasonable value for heavy adsorbates. (A list of representative binding energies to silicate and ice surfaces is given in Hasegawa et al. (1992)). With a surface vibrational frequency of 10^{12} s^{-1}, we can calculate (see eq. 12) the rate of evaporation to be 4×10^{-32} s^{-1}, which is essentially zero. The view that evaporation is negligible is in agreement with the observation that ices build up on grains in cold dense clouds. There is a problem, however, in that it is difficult to see how a gas phase with any heavy species at all can remain much beyond the accretion time. The rate of accretion for heavy species with atomic mass ≈ 10 amu can be estimated from eq. (11) and the grain density ($\sim 10^{-8}$ cm^{-3} in a cloud of gas density 10^4 cm^{-3}) to be $\sim 10^{-6}$ yr^{-1}. This estimate contains the assumption that all grains are 0.1μ in radius. The accretion time, which is the inverse of the rate, is sufficiently small that many clouds should be older than this, which necessitates some alternative mechanism for desorption.

A number of possibilities have been suggested, common to which is the deposition of energy onto the surface in excess of the thermal value. What remains unclear, however, is whether or not the energy can actually be used efficiently to drive species off of the surface. Some specific mechanisms for non-thermal desorption are: photodesorption, cosmic ray-induced desorption, grain-grain collisions, both direct and indirect utilization of the energy of chemical reactions, and chemical explosions of grains induced by the reactions of free radicals when grains are heated above their ambient temperatures. Measurements exist for the direct ejection of H_2 following chemical reaction (Katz et al. 1999), chemical explosions Greenberg 1976), and photodesorption. Measurements of the efficiency of photodesorption are inconsistent. Although most mechanisms have been incorporated into selected model calculations at one time or another (see, e.g. Gwenlan et al. 1997), a general set of rate coefficients is available only for the cosmic ray-induced process (Hasegawa & Herbst 1993a) based on the assumption that cosmic rays heat up grains in a transitory manner and that some evaporation occurs before thermal relaxation. Bringa, in this volume, reports some detailed molecular dynamic calculations on the details of cosmic ray excitation of grains.

4.   THE SYNTHESIS OF INTERSTELLAR MOLECULES

Starting from a gas of atoms, how are interstellar molecules synthesized? Although gas-phase reactions can produce many of the observed gas-phase molecules, they cannot produce large amounts of molecular hydrogen, because the only process that might be able to convert H into H_2 efficiently under interstellar conditions is radiative association, and this process is inefficient for two H atoms (Gould & Salpeter 1963). Molecular hydrogen can be produced in the gas efficiently under different conditions from the interstellar medium via H^+ or H^- intermediates:

$$H^+ + H \longrightarrow H_2^+ + h\nu \tag{20}$$

$$H_2^+ + H \longrightarrow H_2 + H^+ \tag{21}$$

or

$$H + e \longrightarrow H^- + h\nu \tag{22}$$

$$H^- + H \longrightarrow H_2 + e \tag{23}$$

but these processes convert only a small portion of the interstellar gas into molecular hydrogen. The process that does produce H_2 from H atoms in clouds occurs on the surfaces of dust particles, either by a diffusive or Eley-Rideal chemistry. The diffusion-controlled formation of H_2 from H atoms on olivine and amorphous carbon grains has now been measured by Katz et al. (1999 and references therein) and the temperature range over which the diffusive mechanism operates efficiently on interstellar grains to produce H_2 estimated using classical rate equations. The range, under typical dense cloud conditions, appears to be very narrow - only a few degrees - for either grain surface, with olivine efficient around 8-10 K and carbon efficient at higher temperatures. Since molecular hydrogen is so widespread in dense clouds, it is unlikely that it can only be formed in narrow temperature ranges. Of course, the surfaces of interstellar dust particles are motley rather than pure, which will tend to broaden the range of temperatures for diffusion. Moreover, under the temperature range of efficient diffusive formation, the Eley-Rideal mechanism can be operative. Much more work remains, obviously, before a complete understanding of the formation of H_2 on interstellar grains is achieved.

4.1. SOME GAS-PHASE PATHWAYS

Once H_2 is formed on grains and ejected into the gas, a complex gas-phase chemistry occurs based mainly on ion-molecule reactions (Herbst & Klemperer 1973; Leung et al. 1984; Smith 1992). The initial process is the ionization of hydrogen molecules by cosmic rays to form the molecular hydrogen ion, as in reaction (5). The H_2^+ ion then reacts with neutral molecular hydrogen to produce the polyatomic ion H_3^+ via the well-studied ion-molecule reaction

$$H_2^+ + H_2 \longrightarrow H_3^+ + H. \tag{24}$$

The H_3^+ ion does not react with ubiquitous H_2 but is depleted by a variety of gas-phase reactions. In diffuse clouds, the major destruction mechanism is probably dissociative recombination with electrons (see reaction (9)), while in dense clouds, H_3^+ is destroyed more rapidly by reactions with heavy species present initially such as the atoms C and O and molecules formed via gas-phase chemistry such as the molecule CO.

The abundance of the ion H_3^+ in dense clouds is directly related to the cosmic ray ionization rate coefficient ζ and its measured abundance can be used to estimate this coefficient. The rate law for H_3^+ in dense clouds is given by the expression

$$\frac{d[H_3^+]}{dt} = \zeta[H_2] - k_{i-m}[X][H_3^+] \tag{25}$$

where X refers to heavy species that react with H_3^+, and it has been assumed that the ionization of molecular hydrogen leads eventually to H_3^+. Making the assumption that the abundance of H_3^+ reaches steady state so that the time derivative of its concentration can be set to zero, we obtain for the fractional abundance f of this species with respect to molecular hydrogen the expression

$$f(H_3^+) = \zeta/(k_{i-m}f(X)[H_2]). \tag{26}$$

With the Langevin value for the ion-molecule rate coefficient and a fractional abundance of 2×10^{-4} for the heavy species (CO possesses a typical fractional abundance of 10^{-4} by itself), we obtain that

$$f(H_3^+) \approx 5 \times 10^{-9} \tag{27}$$

for a cloud with hydrogen concentration 10^4 cm^{-3} if the standard value for ζ is assumed. The computed fractional abundance for H_3^+ is in excellent agreement with what is detected (Geballe & Oka 1996).

The reactions between heavy species X and H_3^+ are of the type

$$H_3^+ + X \longrightarrow XH^+ + H_2, \tag{28}$$

resulting in protonation of X. If the ion XH^+ can react with H_2 exothermically, it will do so quickly via a process in which a hydrogen atom is exchanged:

$$XH^+ + H_2 \longrightarrow H_2X^+ + H. \tag{29}$$

Subsequent H-atom transfer reactions with ubiquitous molecular hydrogen can also occur, until a molecular ion that does not react with H_2 is produced. In this manner, the ions H_3O^+, CH_3^+, and HCO^+ are synthesized. The hydronium ion – H_3O^+ – which has been detected in dense clouds, is depleted by a dissociative recombination reaction with electrons:

$$H_3O^+ + e \longrightarrow H_2O + H; OH + 2H; OH + H_2; O + H_2 + H. \tag{30}$$

There is still uncertainty regarding the product branching fractions for this reaction, with a flowing afterflow result indicating little or no water product (Williams et al. 1996), while the two latest storage ring results indicate 19% and 24% water, respectively (Larsson, private communication; Jensen 2000). The products of the reaction are themselves depleted by gas-phase reactions: water by ion-molecule reactions and the hydroxyl radical (OH) by both ion-molecule and neutral-neutral reactions. The methyl ion – CH_3^+ – is depleted by both dissociative recombination and by radiative association with H_2:

$$CH_3^+ + H_2 \longrightarrow CH_5^+ + h\nu. \tag{31}$$

This latter reaction occurs because the normal ion-molecule reaction between the methyl ion and molecular hydrogen is endothermic and cannot proceed rapidly at low temperatures. The radiative association has been studied in the laboratory (Gerlich & Horning 1992) and proceeds on one out of every 10^4 collisions at low temperatures. The dissociative recombination reaction between CH_5^+ and electrons leads mainly to the methyl radical CH_3 (Semaniak et al. 1998); methane is formed more efficiently by a reaction involving CO:

$$CH_5^+ + CO \longrightarrow CH_4 + HCO^+. \tag{32}$$

Another important radiative association reaction is that between atomic carbon ions and molecular hydrogen. Carbon ions are present in high abundance in diffuse clouds and to a lesser extent in dense clouds. The reaction

$$C^+ + H_2 \longrightarrow CH^+ + H \tag{33}$$

is endothermic by 0.4 eV and does not occur. The radiative association

$$C^+ + H_2 \longrightarrow CH_2^+ + h\nu, \tag{34}$$

is known to occur on roughly one out of every 10^{6-7} collisions according to experiment and theory (Gerlich & Horning 1992; Herbst 1982). Despite this low efficiency, the fact that molecular hydrogen is a reactant means that the overall rate of reaction is competitive in both dense and diffuse clouds.

Once OH is produced, it can lead to the synthesis of molecular oxygen via the neutral-neutral reaction

$$O + OH \longrightarrow O_2 + H \tag{35}$$

and to the synthesis of NO and molecular nitrogen by analogous neutral-neutral reactions:

$$N + OH \longrightarrow NO + H \tag{36}$$

$$N + NO \longrightarrow N_2 + O. \tag{37}$$

Molecular nitrogen can be broken up by reaction with atomic helium ions, formed via cosmic ray bombardment, to yield atomic nitrogen ions:

$$N_2 + He^+ \longrightarrow N^+ + N + He, \tag{38}$$

which can react with molecular hydrogen despite the fact that the reaction

$$N^+ + H_2 \longrightarrow NH^+ + H. \tag{39}$$

is weakly endothermic because of non-thermal excitation of the reactants. A series of hydrogen atom transfer reactions then ensues until the product NH_4^+ is produced. Dissociative recombination of this ion with electrons produces ammonia (NH_3) and smaller radicals (Vikor et al. 1999). Interestingly, the protonation of neutral nitrogen atoms by the ion H_3^+ does not lead to the production of NH^+ since the reaction is endothermic.

The abundant species CO has a variety of pathways leading to its synthesis, both ion-molecule and neutral-neutral. For example, once CH_2^+ is formed and converted by reaction with H_2 to the methyl ion, this latter species can react with oxygen atoms to produce carbon monoxide via the HCO^+ ion:

$$O + CH_3^+ \longrightarrow HCO^+ + H_2 \tag{40}$$

$$HCO^+ + e \longrightarrow CO + H. \tag{41}$$

Neutral-neutral reactions leading to CO include the reactions of O atoms with the radical CH and C atoms with molecular oxygen:

$$O + CH \longrightarrow CO + H \tag{42}$$

$$C + O_2 \longrightarrow CO + O. \tag{43}$$

Larger species are produced via chains of ion-molecule, dissociative recombination and neutral-neutral reactions. Ion-molecule reactions instrumental in the production of complex hydrocarbons include carbon insertion reactions; viz.,

$$C^+ + CH_4 \longrightarrow C_2H_3^+ + H; C_2H_2^+ + H \tag{44}$$

$$C^+ + C_2H_2 \longrightarrow C_3H^+ + H, \tag{45}$$

and condensation reactions between hydrocarbon ions and neutrals; viz.,

$$C_2H_2^+ + C_2H_2 \longrightarrow C_4H_3^+ + H. \tag{46}$$

The major neutral-neutral pathway in the production of hydrocarbons consists of reactions between neutral atomic carbon and hydrocarbons that are similar to the carbon insertion ion-molecule processes. An important example is

$$C + C_2H_2 \longrightarrow C_3H + H, \tag{47}$$

which not only produces the radical C_3H efficiently at low temperatures, but apparently produces both the linear and cyclic isomers, both of which are detected in dense interstellar clouds. Interestingly, however, there is no obvious neutral-neutral pathway to the far more abundant species c-C_3H_2, which is derived from the cyclic ion precursor c-$C_3H_3^+$. Ion-molecule and neutral-neutral reactions such as these have been shown to produce interstellar hydrocarbons as large as fullerenes, although the laboratory evidence for such processes is not well characterized for species of this great size (Bettens & Herbst 1996).

If one replaces the atomic carbon in reaction (47) with the well-known interstellar radical CN, the reaction

$$CN + C_2H_2 \longrightarrow HC_3N + H, \tag{48}$$

produces cyanoacetylene, an abundant interstellar molecule. Indeed, it now appears that this process dominates over ion-molecule chains such as

$$N + C_3H_3^+ \longrightarrow HC_3NH^+ + H, \tag{49}$$

$$HC_3NH^+ + e \longrightarrow HC_3N + H. \tag{50}$$

A weakness in this and analogous ion-molecule chains of reactions is that the products of the dissociative recombination reactions have not generally been studied and may, in this instance, include a variety of isomers of cyanoacetylene (e.g. HNCCC).

Radiative association reactions may also be of importance in the synthesis of complex species. Although this process has been studied in the laboratory by several groups, many specific reactions of interstellar interest have not been looked at. One example is part of the only known gas-phase synthesis of methanol:

$$CH_3^+ + H_2O \longrightarrow CH_3OH_2^+ + h\nu. \tag{51}$$

The protonated methanol ion can then possibly form methanol via dissociative recombination.

4.2.   SURFACE PATHWAYS TO POLYATOMIC MOLECULES

The extent of surface chemistry depends crucially on the mechanism for reaction and its efficiency. Whether the chemistry occurs by the diffusive or Eley-Rideal mechanism, and whether or not the accretion limit is reached, however, certain reactions are likely to be of importance. We have already discussed the formation of molecular hydrogen on surfaces; the existence of this reaction suggests strongly that rapidly migrating atoms of hydrogen can react with other atoms and radicals. In this way, atoms such as C, O, N, and S can be hydrogenated by successive reactions with H until additional atoms can no longer be added, leading to the production of surface water:

$$O + H \longrightarrow OH \tag{52}$$

$$OH + H \longrightarrow H_2O \tag{53}$$

and also to the formation of methane (CH_4), ammonia (NH_3), and hydrogen sulfide (H_2S). The efficiency of chains of this type depends on the products not being totally ejected from the surface through the exoergicity of reaction. That some ejection occurs is clear from the detection of the radical NH in diffuse clouds despite the lack of a gas-phase pathway to produce this species. There is an alternative hypothesis for the formation of saturated icy mantles on interstellar grains - a high-temperature gas-phase chemistry occurring during periodic shock waves, when reactions involving H_2 are possible despite their activation energy (Bergin et al. 1999), followed by deposition on the dust particles.

A more controversial sequence of hydrogenation reactions is thought to convert carbon monoxide landing on a grain surface into methanol:

$$H + CO \longrightarrow HCO \tag{54}$$

$$H + HCO \longrightarrow H_2CO \tag{55}$$

$$H + H_2CO \longrightarrow H_3CO \tag{56}$$

$$H + H_3CO \longrightarrow CH_3OH. \tag{57}$$

The problem here is that the first and third reactions are known to possess small amounts of activation energy, although the exact values are unknown. Current estimates are in the range 1000-2000 K (0.1-0.2 eV; Charnley et al. 1997; Ruffle & Herbst 2000). Laboratory evidence for the formation of methanol via this sequence is currently ambiguous, and quantum chemical treatments to determine the exact activation energies would be of great use.

The possibility of surface reactions involving two heavy atoms or radicals depends critically on the rates of diffusion. With the barriers against diffusion used by assorted astrochemists before the experiments of Katz et al. (1999) on H_2 formation, a variety of such processes are of importance. A large list can be found in Hasegawa et al. (1992), many of which derive from earlier papers by Tielens & Allamandola (1987), Tielens & Hagen (1982), and Allen & Robinson (1977). With the diffusive rates used by Hasegawa et al. (1992), surface processes can produce large molecules via sequences of reactions such as

$$C + C_n \longrightarrow C_{n+1}. \tag{58}$$

But the results of Katz et al. (1999) indicate that, at least for atomic hydrogen, the barrier against diffusion is more than twice what was envisaged by astrochemists. What does this mean for heavy species? If one assumes that, as for H, the barrier against diffusion is 0.77 of the adsorption energy instead of the previously estimated 0.30, and scales the barriers up accordingly, the diffusion of heavy species becomes so slow at typical interstellar temperatures that reactions between heavy species are no longer competitive (Ruffle & Herbst 2000, 2001a,b). Then, the only important surface reactions involve atomic H. To appreciate the problem, let us consider atomic carbon. The values of E_D and E_b given in Hasegawa et al. (1992) are 800 K and 240 K, respectively, and the rate of diffusion over an entire grain surface at 10 K is 4.9×10^{-5} s^{-1}. With the same adsorption energy and a higher barrier against diffusion of 616 K, the diffusion rate becomes 3×10^{-23} s^{-1}! To recover the old rate of diffusion at 10 K requires a much higher temperature of 28 K, which is probably not representative of cool clouds.

The chemistry that does occur on interstellar grain surfaces may not be too strongly coupled with the gas-phase chemistry in cold clouds unless non-thermal desorption occurs. The mechanisms considered here are not sufficient to maintain a gas phase indefinitely, nor does it appear, except in certain instances, that they can be competitive with gas-phase production of gas-phase species. Yet, non-thermal desorption can certainly prolong the lifetime of the gas phase against condensation, and the interplay between gas and surface does result in phenomena that do not occur in gas-phase models alone, such as secondary abundance peaks at late times (Ruffle & Herbst 2001a; Ruffle et al. 1997). Moreover, the surface chemistry is intimately related with the evolution of large mantles of icy species which are observed in cold sources.

The existence of a surface chemistry can also be inferred from the study of hot molecular cores, which are warm areas in the vicinity of young, high-mass stellar objects. In hot cores, the gas-phase molecules tend to be much more saturated (hydrogen-rich) than in ambient regions. Molecules such as methanol, ammonia, and hydrogen sulfide are far more abundant than elsewhere, and molecules such as ethanol (C_2H_5OH), dimethyl ether (CH_3OCH_3), and methyl formate ($HCOOCH_3$) are only seen in hot cores. The dominant school of thought on the reason for this change in the chemical make-up is that warm temperatures associated with star formation evaporate the mantles of dust particles, such that the hydrogen-rich molecules formed on previously cold surfaces are dispersed into the gas. These molecules can then undergo gas-phase chemical processes (Millar & Hatchell 1998; Charnley 1997; Viti and Williams 1999). Which molecules are formed on grain surfaces and which are produced via gas-phase chains of reactions from precursor evaporated species are still not well determined. Although most models consider only the gas-phase chemistry during the warm phase with assumed initial abundances, a gas-grain model of hot cores during both the previous cold and current warm eras has been published by Caselli et al. (1993). There is also the possibility that the saturated molecules on the cold grain surfaces during the earlier era were themselves produced at least partially in the gas during shock waves (Bergin et al. 1999).

5. GAS-PHASE CHEMICAL MODELS OF QUIESCENT CORES AND OTHER OBJECTS

Given the uncertainties associated with surface chemistry, most models of the past thirty years have only contained gas-phase reactions in the formation and destruction of molecules. Molecular hydrogen, a necessary precursor to gas-phase chemistry, is either persent as an initial condition or is produced via a simple approximation in which the frequency with which H atoms strike grains is multiplied by an efficiency factor. Less frequently, adsorption onto the grains and, possibly, desorption from the grains is considered but no surface reactions are included.

Most rarely, there are gas-grain models, where both surface chemistry and gas-phase chemistry are included.

In gas-phase models, the concentrations of species are determined by individual rate laws, or kinetic equations, such as eq. (25) for H_3^+. Solutions of these coupled equations as a function of time and possibly varying physical conditions are undertaken by use of modern computer algorithms such as that of Gear. Initial abundances for dense cloud cores consist either of atoms or of atoms and molecules representative of a diffuse cloud. In the latter case, a significant amount of H_2 is assumed. Elemental abundances are typically chosen to reflect the depletion of heavy elements from average stellar, or "cosmic" values, onto the grains, although there is no unanimity on the exact values for these abundances. It is typical, however, to have more oxygen than carbon, and to deplete metallic abundances to an extent either equal to that found in diffuse clouds ("high" metal abundances) or even more strongly than found in such clouds ("low" metal abundances; see Table 1). For some sets of elemental abundances not shown in Table 1, two solutions of the coupled differential equations can be obtained (Lee et al. 1998). It is still unclear whether both of these sets of solutions are real or artifacts of incomplete reaction networks.

Under constant physical conditions, the solutions of the coupled kinetic equations eventually reach a steady state, in which there is no time dependence. Although some species reach this condition rather quickly, others take up to 10^7 yr to do so in dense clouds. Before the advent of modern computers, steady-state models were all that was available. Even now, steady-state models are still often used for diffuse clouds, where the added complexities of radiative transfer of penetrating external photons and density inhomogeneities are often given precedence to the treatment of time dependence. The use of the steady-state approximation in such models is reasonable if one starts with a fixed fractional abundance of molecular hydrogen. But, if one actually considers the formation of H_2, albeit with a simplified approach, then the approach to steady state takes too long a time to be ignored, especially in regions with some extinction.

Current model networks of the gas-phase chemistry of interstellar clouds contain ≈ 4000 reactions involving ≈ 400 molecules and a wide assortment of different elements (Terzieva & Herbst 1998; Millar et al. 1997). Most of the reactions utilized have not been studied theoretically or in the laboratory, although many of the most significant ones have been studied. In addition, ion-molecule reactions have similar rate coefficients, and can be grouped into families so that the products are often estimable. The biggest uncertainties are the products of dissociative recombination reactions and the extent of neutral-neutral chemistry, particularly those reactions between atoms or radicals and stable species. Although a small subset of these systems has been measured at low temperatures and found to be rapid (Sims & Smith 1995), others possess activation energy barriers, so that it is presently unclear how to generalize these results. Our best comparisons with observation often come from our so-called "new standard model", to which neutral-neutral reactions involving one stable species are only added if they are studied in the laboratory or are very close cousins of studied systems. This network is similar to but not identical with the UMIST network (Millar et al. 1997).

When applied to dense cloud cores, gas-phase models reproduce the qualitative aspects of the chemistry correctly; that is, they provide a rationalization for the unusual species that are detected. Molecular ions are present because of cosmic rays and ion-molecule reactions; radicals and isomers are present because, among other reasons, dissociative recombination reactions produce these species as well as normal ones; the chemistry is mainly unsaturated because few large ions actually react with molecular hydrogen unless they contain very few hydrogen atoms; and isotopic fractionation occurs. The topic of isotopic fractionation, particularly regarding deuterium, is interesting enough to be discussed individually.

In dense clouds, HD is the major carrier of deuterium and possesses a fractional abundance of $\approx 10^{-5}$. For trace molecular species, however, the abundance ratio of the singly deuterated isotopomer to the normal species can be much larger than this, an effect known as fractionation. As an example, the abundance ratio of DCO^+ to HCO^+ can approach 0.10 in cold sources (Millar et al. 1989). Even doubly deuterated species such as D_2CO and NHD_2 have been detected (Tiné et al. 2000). How does the fractionation take place? Consider the well-studied reaction system

$$H_3^+ + HD \rightleftharpoons H_2D^+ + H_2. \tag{59}$$

The left-to-right reaction is exothermic by about 230 K due to the difference in zero-point vibrational energies between products and reactants and the Pauli Exclusion Principle, which forbids the ground rotational state of H_3^+. Although the system cannot reach equilibrium in most circumstances because of side reactions, the fact that the back reaction is endothermic and therefore slow for $T \ll 230$ K allows the abundance of H_2D^+ to build up dramatically relative to that of H_3^+ (Millar et al. 1989). Several other important "exchange" systems analogous to reactions (59) occur and also cause fractionation. The enhanced deuterated ions can then react with other species to spread the deuteration around the network of reactions. For example, DCO^+ can be produced via

$$H_2D^+ + CO \longrightarrow DCO^+ + H_2. \tag{60}$$

To include deuterated species, model networks must be increased significantly in size, and assorted approximations made regarding reaction rates and products (Millar et al. 1989). With such models, the deuterium fractionation can be used to estimate the ionization rate ζ in a large sample of dense cloud cores. A smaller degree of fractionation exists and has been studied for the isotope ^{13}C, while fractionation of ^{15}N appears to be minimal.

5.1.   CASE STUDIES: TMC-1 AND L134N

Two of the best studied dense cloud cores are known as TMC-1 and L134N. Over 40 molecules have been unamibiguously detected in the former and almost 30 molecules have been seen in the latter. Neither of these sources is homogeneous; in TMC-1 the abundances of different groups of molecules peak at different places. According to Pratap et al. (1997), the abundance variations stem from small changes in gas density and parameters such as ζ and elemental abundances, while according to Markwich et al. (2000), these variations are caused by a complex sequence of events related to star formation. Assuming the former interpretation to be at least partially applicable, it is still interesting to determine what percentage of the molecular abundances can be reproduced by simple homogeneous gas-phase models with fixed physical conditions. Such models have been formulated for many years, starting with the original steady-state models of Herbst & Klemperer (1973). In our recent homogeneous models (Terzieva & Herbst 1998; Terzieva 2000), we have utilized a standard temperature of 10 K and an overall gas density $n = n(H) + 2nH_2$ of 2×10^4 cm^{-3}. The "low metal" elemental abundances (Table 1) were used initially, but the C and O abundances were then varied, both by changing the C/O ratio and by common depletions of both elements. Our standard chemical network was employed, as were additional networks richer in unstudied neutral-neutral reactions.

If the criterion for agreement is to reproduce the observed abundances to within an order of magnitude, we can reproduce $\approx 85\%$ of the abundances in TMC-1 for a wide range of times and C and O elemental abundances (Terzieva & Herbst 1998; Terzieva 2000) with our standard network of reactions. With the normal oxygen-rich conditions (C/O $\approx$ 0.4), best agreement is achieved at the so-called early time of 10^{5-6} yr and worsens considerably thereafter, while increasing the C/O elemental abundance ratio or depleting the C and O equally increases the temporal interval over which equal agreement can be achieved. TMC-1 is known for its large abundances of complex hydrocarbons and cyanopolyynes; with the standard "low metal" abundances, these molecules have peak abundances at early time, then decrease in concentration as steady state is reached. Increasing the C/O ratio, or simply decreasing both C and O equally, tends to perpetuate higher abundances of these species, in agreement with observation. Some molecules are badly underproduced at all times, no matter what elemental abundances are used; these include oxygen-containing organic molecules such as methanol and acetaldehyde (CH_3CHO), which are prime candidates for formation on grain surfaces followed by non-thermal desorption.

The source L134N is not as rich in complex organic molecules as is TMC-1. Our best agreement, in which once again ≈ 85% of the molecular abundances are reproduced to within an order of magnitude, is obtained at an early time with a network rich in neutral-neutral reactions and with common depletions of 5 from the low-metal abundances for C and O. Our standard network yields an agreement of 80% with these parameters; the agreement is somewhat worse because the calculated abundances of complex molecules are occasionally too large. Removing the additional depletions for C and O slightly worsens the level of agreement.

From these two sources, we can conclude that homogeneous gas-phase models of interstellar chemistry can reproduce many but by no means all observed molecular abundances to within the observational uncertainties, but that problems remain. Nor is it obvious why different networks and depletions are needed to achieve best agreement in the two sources. For TMC-1, following the reasoning and analysis of Pratap et al. (1997), it is likely that a more detailed consideration of the heterogeneity and time-dependence of physical conditions would improve agreement somewhat, although the oxygen-containing organic molecules appear to be produced on grains or from granular precursors, as, for example. considered by Markwich et al. (2000).

5.2.   OTHER TYPES OF OBJECTS STUDIED

Gas-phase reaction networks have been used to model a variety of regions distinct from quiescent dense cloud cores, most associated in one way or another with star formation. A fairly ubiquitous type of object, known as a photon-dominated or photodissociation region (Hollenbach & Tielens 1997), occurs when a strong source of radiation, such as that from newly-born high-mass stars, impacts and heats the surrounding dense molecular gas and dust. As the radiation penetrates deeper into the material and weakens, the abundances and temperature tend towards the standard interstellar values. PDR models must treat the radiative transfer carefully and delineate the region into sub-areas characterized by differing abundances. For example, in the areas nearest to the source of photons, the dominant form of carbon is C^+, while somewhat further removed from the radiation source neutral atomic carbon achieves a high abundance. At greater depths, the dominant form of carbon is CO.

Star formation is also often associated with shock waves which, as they traverse cool neutral material, heat it up to temperatures as high as 4000 K. The physical details of the shock wave depend critically on the direction and size of the small interstellar magnetic fields in the area (Shull & Draine 1987). The shocked gas gradually cools, but before the pre-shocked conditions return, a significant change in the chemistry obviously happens, as exothermic reactions with barriers and even endothermic reactions occur, and material is sputtered off of grain surfaces. Shock models, with and without magnetic fields, have been used to explain abundances in a variety of objects with varying degrees of success (Bergin et al. 1999). In addition to photo-excitation (PDRs) and shock excitation, modellers have also begun to consider the type of excitation caused by intermittent turbulence in attempting to explain assorted regions of high excitation (Pety & Falgarone 2000).

Various stages in the cool collapsing cores of protstellar regions can be studied by analyzing molecular abundances in these objects. An object that is one of the earliest indicators of star formation is the starless, or prestellar, core. Here there is as yet no indication of a central star or even a warm central object, but Doppler analysis shows that molecules are falling in to the center. Some species have peak abundances at the center whereas others have holes there. These facts can be reproduced by chemical models with collapse dynamics (Aikawa et al. 2001). The radial distributions of molecules in these models are determined by the interplay of gas-phase chemical, adsorption, and dynamical time scales.

A much later manifestation of star formation is the protoplanetary disk, in which a disk of high density gas and dust rotates around a young "T Tauri" star for perhaps a million years. Protoplanetary disks are of great interest since their inner portions resemble the solar nebula, which existed before the solar system was formed. Observations of molecules in such disks indicate that gas-phase abundances are typically lower than in dense cloud cores, partially because the the time scale for condensation onto the grains is rather short. Detailed two-dimensional chemical models must be used for protoplanetary disks because the gas density and temperature are strong functions of distance from the central star and of height above and below the midplane of the disk. T Tauri stars are also strong emitters of X-rays, which can cause ionization, particularly in the inner portion of the disk. The gas-phase chemistry is, however, mainly powered by cosmic ray bombardment and bears a strong similarity to that in quiescent dense interstellar cloud cores. Models predict that most gas-phase molecules are to be found in regions intermediate in height between the midplane, where the gas phase is strongly depleted at typical T Tauri lifetimes, and the upper and lower edges, where photodissociation can occur, especially for disks that are no longer embedded inside the interstellar clouds where they were born (Aikawa & Herbst 1999; Willacy & Langer 2001).

6. GAS-GRAIN MODELS OF QUIESCENT CORES

A small group of intrepid modellers have attempted to solve the gas-phase and surface chemistry that occur in dense clouds simultaneously. Following early treatments by Watson & Salpeter (1972), Allen & Robinson (1977), Pickles & Williams (1977), and d'Hendecourt et al. (1985), Hasegawa et al. (1992) (see also Hasegawa & Herbst 1993a) undertook detailed model calculations and predicted time-dependent gas-phase and surface abundances. The abundances were obtained by solution of simultaneous rate equations for both gas-phase and surface species, coupled via adsorptive, evaporative, and non-thermal desorptive terms. Consider a neutral species A with a gas-phase concentration [A] and a surface population N_A. The rate equations used for the gas-phase concentration and surface population are as follows:

$$\frac{d[A]}{dt} = -\zeta'[A] + \sum\sum k_{B-C}[B][C] - \sum k_{A-B}[A][B] +$$
$$k_{\text{desorp,A}} N_A n_{\text{gr}} - k_{\text{ads,A}}[A] n_{\text{gr}} \tag{61}$$

$$\frac{dN_A}{dt} = -k_{\text{desorp,A}} N_A + k_{\text{ads,A}}[A] + \sum\sum (k_{s,B} + k_{s,C}) N_B N_C -$$
$$\sum (k_{s,A} + k_{s,B}) N_A N_B \tag{62}$$

where species B and C in the gas include ions and electrons, n_{gr} is the grain concentration, ζ' stands for the rate of direct and indirect cosmic ray ionization/dissociation, and the rate coefficients for desorption in our work include the non-thermal process of cosmic ray bombardment. The units used by Hasegawa et al. (1992) and Hasegawa & Herbst (1993a,b) for the surface populations are different from those shown here; these authors preferred to compute surface concentrations rather than number of species per grain. A more detailed surface rate equation than eq. (62), incorporating surface populations within individual monolayers of the surface, was formulated by Hasegawa & Herbst (1993b) but has not been extensively used.

The detailed results of these models show that the gas-phase abundances are mainly unaffected by the inclusion of grain chemistry until rather late times and that grain mantles grow in cold regions with some of the expected species dominating. It is pointless to discuss the results in great detail since the diffusive rates and rate equation method used for the surface chemistry are questionable at best, as discussed above.

After Caselli et al. (1998) had published their modified rate treatment to account semi-empirically for the defects of the rate equation approach in the accretion limit, Shalabiea et al. (1998) published a gas-grain model calculation of cold dense cloud cores incorporating the modifications. These authors found discrepancies between results obtained with modified rates and unmodified rates but also found that the discrepancies depend on the initial form of hydrogen. When hydrogen is initially molecular, the accretion limit always pertains, and significant discrepancies persist for up to $\sim 10^6$ yr, during which hydrogenated surface species (e.g. NH_3) tend to be lower in abundance and non-hydrogenated surface species (e.g. O_2) higher in abundance when the modified rates are used. The reason for this is that the modifications slow down the diffusion rate and therefore the effectiveness of atomic hydrogen dramatically. When the initial form of hydrogen is atomic, there is so much atomic hydrogen that changing the diffusion rate of this species has little affect.

The diffusive rates used by Shalabiea et al. (1998) before modification are the rapid ones of Hasegawa et al. (1992, and references therein). After the paper was published, the work of Katz et al. (1999) appeared, in which the experimental results for the surface formation of molecular hydrogen on cold olivine and amorphous carbon were analyzed in terms of much slower diffusion rates for H due to higher values of E_b. Ruffle & Herbst (2000, 2001a,b) have now developed new gas-grain models of cold, quiescent dense-cloud cores by generalizing the results of Katz et al. (1999) in several ways. In their first paper (Ruffle & Herbst 2000), they constructed two new models, labelled M1 and M2, respectively, with chemistry occurring on an olivine-like surface along with gas-phase chemistry in a 10 K core. In both M1 and M2, they changed the binding energies of H and H_2 from their previous values of 350 K and 450 K to the newly measured values of 373 K and 315 K. Regarding the more important barriers against diffusion, in M1 they altered the barrier for H from 100 K (Hasegawa et al. 1992) to 200 K and excluded tunneling, while keeping fixed the barriers for all other species. The barrier for H was not raised higher than 200 K because to have done so would have meant slowing the diffusive rate of H below that of heavier species, a non-physical transformation. The diffusive rates for heavier species were not changed in M1 because there is no direct experimental evidence concerning these rates. In M2, the barrier against diffusion was raised to the full value of 287 K determined by Katz et al. (1999) and the barriers of all other species were raised proportionately to $0.77 \times E_D$. In M2, the diffusive rates are so slow that no modifications of the type advocated by Caselli et al. (1998) are necessary. Put another way, the accretion limit does not pertain. In M1, the diffusive rate of H is rapid enough that the modifications are marginally needed. So, although the generalization of the results of Katz et al (1999) is sufficiently unclear as to lead to two very different models, an advantage in generalizing the results of Katz et al. (1999) is that the artificial reduction in diffusion rate of Caselli et al. (1998), or any other alteernative method such as the Monte Carlo approach or master equation technique, is no longer really needed, at least at 10 K.

The major results of models M1 and M2 for a dense cloud core at 10 K, gas density $n = n(H) + 2n(H_2) = 2 \times 10^4$ cm^{-3}, and low-metal abundances for the initial gas can be summarized as follows. The surface and gas-phase abundances are functions of whether H (Case B) or H_2 (Case A) is used as the initial form of hydrogen up to a time of $\sim 10^{6-7}$ yr. The gas-phase abundances begin to differ appreciably from their values in purely gas-phase models after 10^5 yr, with many small species maintaining surprisingly large gas-phase abundances through 10^7 yr and organic molecules such as the cyanoopolyynes tending to show a second maximum in their abundance at 10^7 yr. The agreement with gas-phase abundances in the well-studied source TMC-1 is still best at early time but does not worsen as quickly as in purely gas-phase models. The surface results show that mantles composed mainly of the ices of water, CO, and, to a lesser extent, methanol and CO_2 grow at a rate such that by 10^6 yr, a mantle of 100 monolayers has developed around each grain, and much of the elemental abundance of O is consumed as water ice. The large abundance of water ice cannot be accounted for by purely gas-phase models at 10 K followed by adsorption onto the grains. In M1, surface reactions between two heavy species are still competitive, but in M2 only hydrogenation reactions involving H occur. In this latter model, surface CO is not formed by recombination of C and O atoms but by deposition from the gas. Surface carbon dioxide can be formed in M1 by the surface reaction

$$CO + O \longrightarrow CO_2, \qquad (63)$$

although this reaction is thought to possess some activation energy. Other surface reactions that lead to carbon dioxide in M1 include

$$OH + CO \longrightarrow CO_2 + H \qquad (64)$$

and

$$O + HCO \longrightarrow CO_2 + H. \qquad (65)$$

In M2, surface CO_2 is mainly deposited from the gas.

The surface results are best compared with sources in which the infrared spectrum arises from the absorption of the radiation of background field stars rather than internal cloud sources, which also lead to higher temperatures than 10 K. The best-studied source with a field star lies in the direction towards Elias 16, a star behind the Taurus molecular cloud. The worst disagreement with observations of this source for all four models tested (M1/A, M1/B, M2/A, M2/B) occurs for CO_2, which is vastly underproduced at all times. All models produce lots of water, while models M1/A and M2/A reproduce the CO abudance and upper limit to methanol well for a reasonable interval of time. Unfortunately, only one gas-phase species – CO – has been studied along the line of sight so that the models cannot be constrained by gas-phase observations in this direction.

In their next paper, Ruffle & Herbst (2001a) investigated the changes that photochemistry on grain surfaces could make to the results of M1 and M2. The photochemical models (see Section 3.3) are labelled P1 and P2. In dense cloud cores, only cosmic ray-induced photons play a significant role. With this small flux of photons, salient differences between M1 and M2 and P1 and P2 occur only for cloud ages longer than 10^6 yr. Whether or not a surface species is strongly affected by photochemistry depends on its dissociation products (see also Section 3.3). Species such as water, which dissociate mainly into OH and H, are quickly reproduced on grain surfaces by reactions involving H atoms from the gas that land on the surfaces. Hence their abundances are not changed much. Species such as methanol, which dissociate into radicals (e.g. CH_3 and OH), can be reduced in abundance because these radicals are hydrogenated to form new stable species rather than the parent one. Species such as CO, which are photoproducts of reactions, increase in abundance. The abundance of surface carbon dioxide is little affected. In the gas-phase, calculated abundances tend to increase at late times when surface photochemistry is included, improving the agreement with the TMC-1 abundances at this time. In general, the effects of photochemistry do not appear to play a major role for the surface chemistry, at least according to the model of Ruffle & Herbst (2001a) although they do tend to increase the time interval over which theory and observation are in reasonable agreement. We (Nguyen et al., in preparation) are currently trying to extend the models to sources with more radiation. It would also be of some interest to use the model network, or extensions of it, to simulate some of the laboratory photochemical experiments that utilize bright radiation sources for small periods of time.

In their most recent paper, Ruffle & Herbst (2001b) considered the problem of surface carbon dioxide, which was underproduced compared with observation in Elias 16 and many other sources in all of their earlier gas-grain models of quiescent cores. They took several approaches to boosting the surface CO_2 abundance:

- increasing the density of the gas,

- increasing the temperature of the gas,

- reducing the activation energy of the surface $CO + O$ reaction,

- using a grain surface of amorphous carbon.

The first three modifications were performed on the P1 and P2 models, while the last required two new analogous models, labelled P3 and P4.

Ruffle & Herbst (2001b) found that the observed large surface abundance of CO_2 as well as the other major constituents of the mantles could be accounted for over reasonable time intervals with suitable conditions but only if the atomic H barrier against diffusion is at most raised to a value (200 K) somewhat below that measured by Katz et al. (1999) and the other barriers are left unchanged; i.e., for the P1 and P3 cases. Put alternatively, gas-grain models with diffusive chemistry can reproduce the abundances detected on cold icy mantles *only* if surface diffusion of heavy species such as atomic oxygen occurs efficiently to form CO_2. Specifically, with olivine as the reactive surface, model P1/A works well at a density of 5×10^5 cm^{-3} and a temperature of 10 K, or at the standard density and a temperature of 12 K. It also works well if the activation energy for reaction (63), thought to be ≈ 1000 K, is reduced. Model P3/A works well at a temperature of 20 K.

7. CURRENT UNCERTAINTIES AND FUTURE PROGRESS

Although gas-phase chemistry can explain the abundances of most of the smaller interstellar molecules detected in the gas, it is clear that surface chemistry is needed to produce molecular hydrogen, and, although there are alternatives based on shock chemistry, it is likely that surface chemistry is needed to produce the large abundance of water ice detected in most cold dense clouds. Surface chemistry appears to be necessary for the production of methanol as well as the large amounts of carbon dioxide ice, although there is little experimental evidence that the several critical reactions between heavy species that form surface CO_2 can take place at low temperatures. The saturated gas-phase molecules detected in hot molecular cores are best understood as emanating at least partially from a previous era of surface chemistry at low temperatures followed by desorption during the process of star formation.

Given the need for surface chemistry, some investigators have included diffusive surface reactions into their chemical models of interstellar regions. Besides the recent quiescent core models of Ruffle & Herbst (2000, 20001a,b) discussed here, the protoplanetary disk model of Willacy & Langer (2001) should be mentioned. Current investigators are working in an atmosphere of considerable uncertainty, since few experimental results relating to low temperature surface chemistry have been published. The main exception to this statement concerns photochemistry, but in this instance there is still a large gap between laboratory experiments and the one detailed model (Ruffle & Herbst 2001a) that even attempts to include such processes.

Among the many interrelated difficulties that afflict modellers of grain chemistry are the following, most of which have been discussed in this article:

- uncertain rates of diffusion on cold interstellar-like surfaces for all reactive species of interest other than atomic hydrogen,

- the uncertain size distribution and topological nature of interstellar grains (spherical vs. porous and fluffy) and how they affect surface chemistry,

- the uncertain and time-dependent chemical nature of grain surfaces (silicates, amorphous carbon, dirty ices),

- alternative mechanisms of condensed-phase chemical reactions to the commonly assumed surface diffusive process (Eley-Rideal, hot atom, internal, photochemical),

- mathematical problems associated with the rate equation approach to diffusive chemistry in the accretion limit of small particles,

- the role of and mechanisms for non-thermal desorption.

Despite the gravity of these difficulties, there are many experiments, theoretical treatments, and observational programs planned and in progress to seek answers to most of the above questions. It is to be hoped that grain chemistry will eventually be treatable with the same exactness that is now possible for gas-phase chemistry.

Once gas-grain models become more reliable, they can be used with more confidence for an increasing number of environments in the interstellar medium including prestellar cores, collapsing cores, protoplanetary disks, and hot molecular cores, as well as diffuse and translucent clouds. Nor is the interstellar medium the only environment for which gas-grain models are suitable. The extended envelopes of old stars, especially carbon-rich sources such as IRC+10216, are replete with organic molecules that are formed via both gas phase and surface chemistry.

Dust particles are ubiquitous throughout the universe. They are critical in extinguishing harsh UV radiation from stars so that interstellar gas-phase molecules can grow in dense clouds, and they are critical as sites for a not yet fully understood surface chemistry. As we comprehend more about the dust, we will understand more about the surface chemistry occurring on it. Although the many uncertainties concerning dust particles and surface chemistry may seem overwhelming at times, it is reassuring to remember how much has been learned in the past half century, and that much will be learned in the near future. Given a reasonable rate of progress in our knowledge, the next fifty years should lead to a much more detailed understanding of interstellar grains, what happens on them, and what the information tells us about the sources where they are located.

Acknowledgments

The support of the National Science Foundation (US) for my research program in astrochemistry is gratefully acknowledged.

REFERENCES

Abell, G.O. (1982), *Exploration of the Universe*, Saunders College Publishing, Philadelphia.

Aikawa, Y. and Herbst, E. (1999), *A&A*, **351**, 233.

Aikawa, Y., Herbst, E. and Dzegilenko, F. (1999), *ApJ*, **527**, 262.

Aikawa, Y., Ohashi, N., Inutsuka, S.-I., Herbst, E. and Takakuwa, S. (2001), *ApJ*, **552**, 639.

Allamandola, L.J., Tielens, A.G.G.M. and Barker, J. (1989), *ApJSS*, **71**, 733.

Allen, M. and Robinson, G.W. (1977), *ApJ*, **212**, 396.

Bates, D.R. and Herbst, E. (1988), in Millar, T.J. and Williams, D.A. (eds.), *Rate Coefficients in Astrochemistry*, Kluwer Academic Publishers, Dordrecht, p.17.

Bergin, E.A. (2000), in Minh, Y.C. and van Dishoeck, E.F. (eds.), *Astrochemistry: From Molecular Clouds to Planetary Systems*, Sheridan Books, Chelsea, Michigan, p.51.

Bergin, E.A., Neufeld, D.A. and Melnick, G.J. (1999), *ApJ*, **510**, L145.

Bettens, R.P.A. and Herbst, E. (1996), *ApJ*, **468**, 686.

Biham, O., Furman, I., Pirronello, V. and Vidali, G. (2001), *ApJ*, **553**, 595.

Caselli, P, Hasegawa, T.I. and Herbst, E. (1993), *ApJ*, **408**, 548.

Caselli, P, Hasegawa, T.I. and Herbst, E. (1998), *ApJ*, **495**, 309.

Charnley, S.B. (1997), *ApJ*, **481**, 396.

Charnley, S.B. (1998), *ApJ*, **509**, L121.

Charnley, S.B., Tielens, A.G.G.M. and Rodgers, S.D. (1997), *ApJ*, **482**, L203.

d'Hendecourt, L.B., Allamandola, L.J. and Greenberg, J.M. (1985), *A&A*, **152**, 130.

Ehrenfreund, P. and Schutte, W.A. (2000), in Minh, Y.C. and van Dishoeck, E.F. (eds.), *Astrochemistry: From Molecular Clouds to Planetary Systems*, Sheridan Books, Chelsea, Michigan, p.135.

Elitzur, M. (1999), in Wall, W.F. et al. (eds.), *Millimeter-Wave Astronomy: Molecular Chemistry & Physics in Space*, Kluwer Academic Publishers, Dordrecht, p.127.

Farebrother, A.J., Meijer, A.J.H.M., Clary, D.C. and Fisher, A.J. (2000), *Chemical Physics Letters*, **319**, 303.

Frenklach, M. and Feigelson, E.D. (1989), *ApJ*, **341**, 372.

Geballe, T.R. and Oka, T. (1996), *Nature*, **384**, 334.

Gerakines, P.A., Schuttte, W.A., and Ehrenfreund, P. (1996), *A&A*, **312**, 289.

Gerlich, D. and Horning, S. (1992), *Chemical Reviews*, **92**, 1509.

Gould, R.J. and Salpeter, E.E. (1963), *ApJ*, **138**, 393.

Green, N.J.B., Toniazzo, T., Pilling, M.J., Ruffle, D.P., Bell, N. and Hartquist, T.W. (2001), *A&A*, **375**, 1111.

Greenberg, J.M. (1976), *Astrophysics and Space Science*, **39**, 9.

Greenberg, J.M., Li, A., Mendoza-Gómez, C.X., Schutte, W.A., Gerakines, P.A. and de Groot, M. (1995), *ApJ*, **455**, L177.

Gwenlan, C., Ruffle, D.P., Viti, S., Hartquist, T.W. and Williams, D.A. (2000), *A&A*, **354**, 1127.

Hasegawa, T.I. and Herbst, E. (1993a), *MNRAS*, **261**, 83.

Hasegawa, T.I. and Herbst, E. (1993b), *MNRAS*, **263**, 589.

Hasegawa, T.I., Herbst, E., and Leung, C.M. (1992), *ApJSS*, **82**, 167.

Herbst, E. (1982), *ApJ*, **252**, 810.

Herbst, E. (1999), in Wall, W.F. et al. (eds.), *Millimeter-Wave Astronomy: Molecular Chemistry & Physics in Space*, Kluwer Academic Publishers, Dordrecht, p.329.

Herbst, E. and Klemperer, W. (1973), *ApJ*, **185**, 505.

Herbst, E. and Leung, C.M. (1986), *MNRAS*, **222**, 689.

Hollenbach, D. and Salpeter, E.E. (1971), *ApJ*, **163**, 155.

Hollenbach, D. and Thronson, H. A. Jr. (eds.) (1987), *Interstellar Processes*, Reidel, Dordrecht.

Hollenbach, D. and Tielens, A.G.G.M. (1997), *Annual Reviews of A&A*, **35**, 179.

Jensen, M. et al. (2000), *Euro Summer School on Dynamics of Molecular Collisions Relevant to the Evolution of Interstellar Matter*, Weizmann Institute of Science, Rehovot, Israel. Contact mariejj@ifa.au.dk

Katz, N., Furman, I., Biham, O., Pirronello, V. and Vidali, G. (1999), *ApJ*, **522**, 305.

Larsson, M. (2000), *Philosophical Transactions of the Royal Society: Mathematical, Physical & Engineering Sciences*, **358**, 2433.

Lee, H.-H., Roueff, E., Pineau des Forêts, G., Shalabiea, L. M., Terzieva, R. and Herbst, E. (1998), *A&A*, **334**, 1047.

Leung, C.M., Herbst, E., and Huebner, W.F. (1984), *ApJSS*, **56**, 231.

Lucas, R. and Liszt, H.S. (2000), *A&A*, **358**, 1069.

Maier, J.P., Lakin, N.M., Walker, G.A.H. and Bohlender, D.A. (2001), *ApJ*, **553**, 267.

Markwick, A.J., Millar, T.J., and Charnley, S.B. (2000), *ApJ*, **535**, 256.

Melnick, G.J. et al. (2000), *ApJ*, **539**, L77.

Millar, T.J., Bennett, A. and Herbst, E. (1989), *ApJ*, **340**, 906.

Millar, T.J., Farquhar, P.R.A. and Willacy, K. (1997), *A&A Supp.Ser.*, **121**, 139.

Millar, T.J. and Hatchell, J. (1998), *Faraday Discussions*, **109**, 15.

Millar, T.J., Herbst, E., and Bettens, R. P. A. (2000), *MNRAS*, **316**, 195.

Pety, J. and Falgarone, E. (2000), *A&A*, **356**, 279.

Pickles, J.D. and Williams, D.A. (1977), *Astrophysics and Space Science*, **52**, 443.

Prasad, S.S. and Tarafdar, S.P. (1983), *ApJ*, **267**, 603.

Pratap, P., Dickens, J.E., Snell, R.L., Miralles, M.P., Bergin, E.A., Irvine, W.M., and Schloerb, F.P. (1997), *ApJ*, **486**, 862.

Rowe, B.R., Rebrion-Rowe, C. and Canosa, A. (2000), in Minh, Y.C. and van Dishoeck, E.F. (eds.), *Astrochemistry: From Molecular Clouds to Planetary Systems*, Sheridan Books, Chelsea, Michigan, p.3237.

Ruffle, D.P., Bettens, R.P.A., Terzeiva, R. and Herbst, E. (1999), *ApJ*, **523**, 678.

Ruffle, D.P., Hartquist, T.W., Taylor, S.D., and Williams, D.A. (1997), *MNRAS*, **291**, 235.

Ruffle, D.P. and Herbst, E. (2000), *MNRAS*, **319**, 837.

Ruffle, D.P. and Herbst, E. (2001a), *MNRAS*, **322**, 770.

Ruffle, D.P. and Herbst, E. (2001b), *MNRAS*, **324**, 1054.

Semaniak, J. et al. (1998), *ApJ*, **498**, 886.

Shalabiea, O.M., Caselli, P., and Herbst, E. (1998), *ApJ*, **502**, 652.

Shull, J.M. and Draine, B.T. (1987), in Hollenbach, D.J. and Thronson, H.A. Jr. (eds.), *Interstellar Processes*, Reidel, Dordrecht, p.283.

Sims, I.R. and Smith, I.W.M. (1995), *Annual Review of Physical Chemistry*, **46**, 109

Smith, D. (1992), *Chemical Reviews*, **92**, 1473.

Snow, T.P. and Witt, A.N. (1996), *ApJ*, **468**, L65.

Spitzer, L. (1978), *Physical Processes in the Interstellar Medium*, John Wiley & Sons, New York.

Terzieva, R. (2000), PhD thesis, Chemical Physics Program, The Ohio State University.

Terzieva, R. and Herbst, E. (1998), *ApJ*, **501**, 207.

Tielens, A.G G M. and Allamandola, L.J. (1987), in Hollenbach, D.J. and Thronson, H.A. Jr. (eds.), *Interstellar Processes*, Reidel, Dordrecht, p.397.

Tielens, A.G.G.M. and Hagen, W.F. (1982), *A&A*, **114**, 245.

Tielens, A.G.G.M. and Snow. T.P. (eds.) (1995), *The Diffuse Interstellar Bands*, Kluwer Academic Publishers, Dordrecht.

Tiné, S., Roueff, E., Falgarone, E., Gerin, M. and Pineau des Forêts, G. (2000), *A&A*, **356**, 1039.

Van Dishoeck, E.F. (1988), in Millar, T.J. and Williams, D.A. (eds.), *Rate Coefficients in Astrochemistry*, Kluwer Academic Publishers, Dordrecht, p.49.

Van Dishoeck, E.F. and Black, J.H. (1988), in Millar, T.J. and Williams, D.A. (eds.), *Rate Coefficients in Astrochemistry*, Kluwer Academic Publishers, Dordrecht, p.209.

Vejby-Christensen, L., Andersen, L.H., Heber, O., Kella, D., Pedersen, H.B., Schmidt, H.T. and Zajfman, D. (1997), *ApJ*, **483**, 531.

Vikor, L. et al. (1999), *A&A*, **344**, 1027.

Viti, S. and Williams, D.A. (1999), *MNRAS*, **305**, 755.

Wall, W.F., Carraminana, A., Carrasco, L. and Goldsmith, P.F. (eds.) (1999), *Millimeter-Wave Astronomy: Molecular Chemistry & Physics in Space*, Kluwer Academic Publishers, Dordrecht.

Watson, W.D. and Salter, E.E. (1972), *ApJ*, **174**, 321.

Willacy, K. and Langer, W.D. (2000), *ApJ*, **544**, 903.

Williams, D.A. (1993), in Millar, T.J. and Williams, D.A. (eds.), *Dust and Chemistry in Astronomy*, Institute of Physics, Bristol, p.143.

Williams, T., Adams, N.G., Babcock, L.M., Herd, C.R. and Geoghegan, M. (1996), *MNRAS*, **282**, 413.

Winnewisser, G., Herbst, E. and Ungerechts, H. (1992), in Rao, K.N. and Weber, A. (eds.), *Spectroscopy of the Earth's Atmosphere and Interstellar Medium*, Academic Press, New York, p.423.

Witt, A.N. (2000), in Minh, Y C. and van Dishoeck, E.F. (eds.), *Astrochemistry: From Molecular Clouds to Planetary Systems*, Sheridan Books, Chelsea, Michigan, p.317.

HOW TO IDENTIFY DIFFUSE BAND CARRIERS?

Jacek Krełowski

Center for Astronomy, N. Copernicus University Gagarina 11, Pl-87-100 Toruń, Poland

Keywords: Interstellar medium – clouds – diffuse interstellar bands – identification

Abstract   The longest standing unsolved problem in all of spectroscopy – identification of the carriers of diffuse interstellar bands (DIBs), is considered from the observer's point of view. Limitations caused by the available apparatus are discussed together with those following our location in the Milky Way. Two possible methods of DIB identification are proposed. One of them requires very high S/N spectra in a very broad range of wavelengths which may allow to select subsets of DIBs matching wavelengths and relative intensities of spectra originating in certain species, known from laboratory experiments. The second one requires a very high resolution and S/N profiles of individual features. Already observed high resolution profiles of several DIBs contain substructures, resembling those of known (from laboratory experiments) molecular bands, but with unresolved rotational structure. The substructure pattern may be characteristic for a given species and thus a match of laboratory and observational profile may be a sufficient argument to identify a known molecule as a DIB carrier. It is emphasized that a criterion of identification must meet the requirements of both: experimentalists and observers.

Introduction

For most of their history DIBs have not been observed in individual ISM clouds, in spite of the fact that since the observations of Beals (1938), who discovered the Doppler splitting in interstellar atomic lines, it is commonly accepted that the interstellar medium is not homogeneous, but contains many separate clumps. The Doppler components are usually observed in high resolution profiles of strong interstellar absorption lines, such as NaI D_1 and D_2, CaII H and K or KI 7698 Å feature

V. Pirronello et al. (eds.), Solid State Astrochemistry, 147–174.

along sightlines towards distant, reddened OB stars. Since its discovery, this fact created the question whether all these individual clouds can be characterized by identical spectra and physical parameters.

The question remained open until Krełowski & Walker (1987) and Krełowski & Westerlund (1988) have demonstrated convincingly that spectra of interstellar clouds are not identical. The strength ratios of many different pairs of DIBs vary substantially from object to object (Fig. 1) revealing varying physical conditions inside the intervening clouds and different origin of the simultaneously observed features. These variations seem to be related to the behaviour of another spectral characteristics of ISM clouds: intensities of atomic and molecular features (Fig. 2) or shapes of interstellar extinction curves (Krełowski et al. 1992).

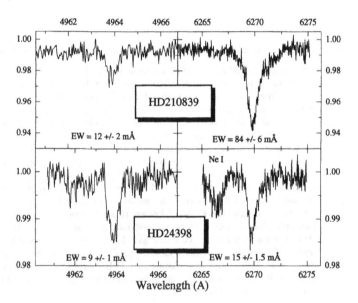

Figure 1 Varying strength ratios of diffuse interstellar bands in high resolution spectra from Terskol spectrometer installed at 2m Zeiss telescope. The plot shows clearly that the two features assumed to be originated in C_7^- by Tulej et al. (1998) are not of common origin.

Herbig & Soderblom (1982) demonstrated convincingly that the profiles of DIBs, as well as those of interstellar atomic lines, can be modified by the Doppler splitting (Fig. 3). The same was shown later by Westerlund & Krełowski (1988) where only the broadening of some DIB profiles, caused by the Doppler effect, was presented. The discussion was, how-

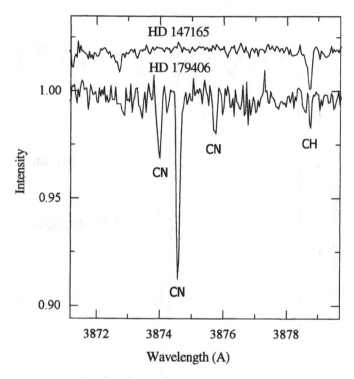

Figure 2 Complicated relations between simple molecules in the spectra characterized either by low ratio of 5780/5797 DIBs (HD179406) or by the same ratio higher
by an order of magnitude (HD147165). Spectra from the old coudé spectrometer fed
by CFHT.

ever, extended to the features broader than those considered by Herbig
& Soderblom which can hardly be observed as Doppler–splitted. These
observations raised the problem of intrinsic profiles of DIBs i.e. profiles which are not modified by the Doppler effect. Such profiles can be
observed only in individual clouds i.e. in the cases of low to moderate
reddening (single clouds of very high reddening are not common). It
is to be emphasized that only such profiles can be compared to those,
found (for certain species) in laboratory experiments. Heavily reddened
stars, i.e. the objects in which DIBs have been discovered, are usually obscured by several clouds and the Doppler splitting is very likely
to modify their profile shapes and – to a certain extent – their wavelengths. The strength ratios of individual Doppler components can vary
from cloud to cloud as shown by Krełowski & Westerlund (1988).

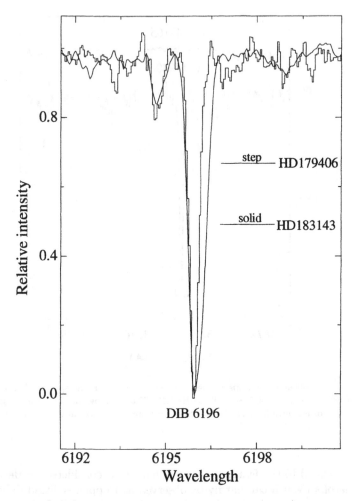

Figure 3 The narrow 6196 DIB broadened by the Doppler splitting in the spectrum of HD 183143. The second target, HD 23180, shows narrow and symmetrical interstellar atomic lines. The spectra from the Musaev echelle spectrometer fed with the 2m Terskol telescope.

To the present day no reliable identification of a carrier of any of the DIBs has been found. Almost all conceivable forms of matter, ranging from dust grains, to free molecules of very different sizes and structures, to even the hydrogen negative ion, have been proposed as the possible carriers; for a review see Snow (1995). It is commonly believed now that

DIBs originate in some large, carbon–bearing molecules. The latter can be responsible both for DIBs and some segments of the interstellar extinction curve. Their internal structure can be very complex and the possible presence of a mixture of many species in every cloud makes the identification of DIBs very difficult.

During the decade of 90–ties the list of unidentified spectral features, found in the interstellar medium, has grown up to nearly 300 entries (Galazutdinov et al. 2000). About 90% of them are weak and narrow features, barely seen even in high S/N spectra. A majority of these diffuse bands is concentrated in the red and near infrared spectral ranges (Fig. 4). The wealth of observed bands should make the task of identification of their carriers simpler but seemingly the observed clouds contain many different species and their relative abundances vary from cloud to cloud. Moreover, the probability of DIB blending grows with the number of discovered features.

During the last decade it was convincingly demonstrated that the profiles of diffuse bands are not gaussians but show some internal structure which may be intrinsic to certain molecules (Sarre et al., 1995; Kerr et al. 1998; Walker et al. 2000). However, despite many efforts, none of the possible carriers have been identified yet. Fig. 5 shows how the substructures of the strong 5797 DIB are getting revealed with the growing resolution of spectra.

The current review discusses the problems of the identification of diffuse band carriers from the observer's point of view. The possible ways of completing this task as well as the difficulties which make identifications uncertain are presented. The future observational projects are outlined and their possible limitations discussed.

1. POSSIBLE RELATIONS BETWEEN DIFFUSE BANDS

The standard way of spectroscopic identification is based on comparison of two sets of spectral features: one obtained at a laboratory (or calculated theoretically) and one observed. Both: wavelengths and relative intensities should match each other. The more features meet the above conditions the more certain is the identification. To complete this task one needs to acquire high quality spectra of reddened stars allowing precise wavelength and intensity measurements (i.e. free of Doppler splitting). The whole sets of features originated in any of complex organic molecules cover usually broad spectral ranges and the observed spectra should meet the same condition.

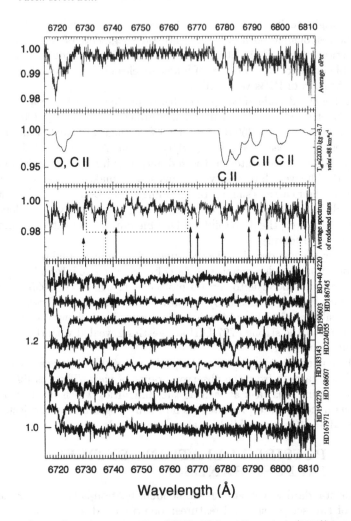

Figure 4 A page from the recent atlas of DIBs (Galazutdinov et al. (2000)) based on spectra from the Terskol Observatory. The presented wavelength range is extremely densely populated with weak, unidentified interstellar features marked with arrows and the rectangle.

The analysis made recently by Moutou et al. (1999) showed the general lack of strict correlation between almost any of possible pairs of strong diffuse bands i.e. those listed already by Herbig (1975). However, some of the pairs of DIBs correlate reasonably well, some of them show

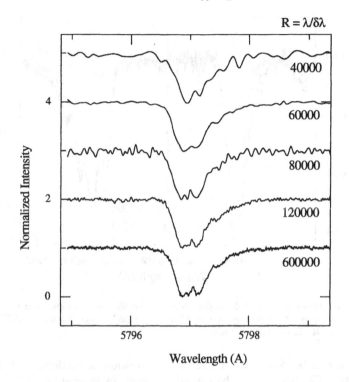

Figure 5 The effect of growing resolution. The 5797 DIB observed in the spectrum of HD 24398 using different resolutions. From top to bottom: CFHT (old coudé spectrometer), McDonald (Sandiford spectrometer), Terskol (echelle spectrometer of F. Musaev), CFHT (Gecko), UHRF (courtesy of P. Sarre).

practically no correlation. Only the relation between the strong 6614 DIB and the narrow 6196 one, may suggest a close relation, possibly a common origin. The relation between 6614 and 6196 looks attractive but their profiles differ seriously in the width and shape; also their strength ratio is not always exactly the same (Fig. 6) which makes unlikely their common origin. One may conclude that every strong DIB is caused by another carrier which the fact forces observers to search for correlations between strong and weak DIBs. Let's mention that spectra of single carbon–bearing molecules are composed of one strong and a couple of weak features each.

An analysis of mutual correlations between pairs of strong DIBs is quite easy as measurements of either their central depths or equivalent

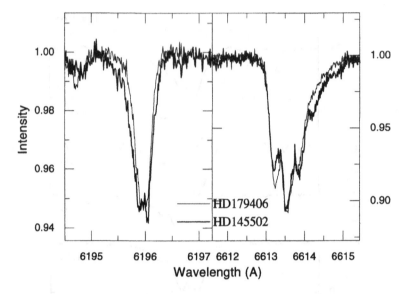

Figure 6   6196 and 6614 DIB profiles observed towards two stars obscured by single clouds. Note that their profiles vary in different fashion. Spectra from ESO CES high resolution (R=220,000 spectrometer.

widths can be done with a reasonable precision in modern, high S/N spectra. The scatter may be caused, however, by several factors:

- the lack of proper telluric line divisors; at the moment I can recommend only two for the Northern hemisphere: HD116658 (*Spica*) and HD120315 (*η*UMa). In other stars, used as such standards (e.g. *δ*Per or Vega), some weak interstellar or stellar features can be traced inside DIB profiles which makes the measurements unreliable, especially in cases of low reddenings

- continuum setting, which may be done in different ways by two different observers; this effect may be especially strong in cases of broad DIBs and affects equivalent widths more than central depths

- separation of blended features – it is usually up to the observer's experience

- instrumental effects. The 6196 DIB intensity, observed in the spectrum of HD179406 with the aid of the Gecko spectrometer fed with the CFH telescope (Fig. 7), is lower then in those recorded with different instruments. Most probably it follows improperly subtracted scattered light.

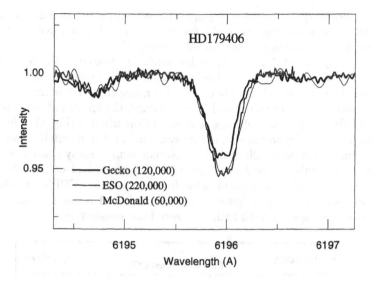

Figure 7 Profiles of the narrow 6196 DIB in three spectra of HD179406. The spectra from McDonald Sandiford echelle and ESO-CES apparently coincide while that from from Gecko CFHT shows the DIB weaker. The intensity of the neighbour, 6194.5 Å DIB is identical in all spectra.

The laboratory spectra of complex molecules are known to consist of one strong feature and several others, much weaker ones (for details – see the paper by J. Fulara in this volume). If the diffuse bands are of molecular origin, the observed spectrum should be a mixture of several strong bands and many much weaker features. Generally this expectation is confirmed by the observations (Jenniskens & Désert 1994, Krełowski, Sneden & Hiltgen 1995, Tuarisg et al. 2000, Weselak et al. 2000, Galazutdinov et al. 2000): the number of weak unidentified interstellar features is by an order of magnitude larger than that of the well–known, strong DIBs. In such a case we can expect that every strong diffuse band may be a "dominant" member of a "family" which includes several weak features and forms a spectrum of single species.

The task of retrieving spectra of single carriers (expected to consist a few features each) from the general spectrum of nearly 300 DIBs is a very complicated problem. The central depths of most of the observed features are of the order of 1% of the continuum and thus it is difficult to achieve a high S/N ratio inside their profiles. This makes precise measurements of either central depths or equivalent widths uncertain. Theoretically the problem can be solved while using spectra of very high S/N. The practice is, however, more complicated as will be shown later.

Any laboratory spectrum of a single species, suspected to carry some DIBs, covers a spectral range of a few thousand of Å which requires the astrophysical spectra to cover the same range(s).

The majority of diffuse interstellar bands is situated in the red and near infrared spectral ranges. The popularity of back illuminated CCD's makes their investigations difficult. In this range the effect of fringing is substantial. Even while divided by a standard, the spectra of reddened stars suffer the presence of some remnants of the fringing (Fig. 8). These remnants are comparable with some weak unidentified interstellar bands and thus can strongly affect their measurements. In many cases it may even be difficult to decide whether some feature is real or not. It is possible to avoid this problem using front illuminated CCD's but this reduces sensitivity by a factor of two or more and makes the task of recording the spectra with high S/N very time–consuming.

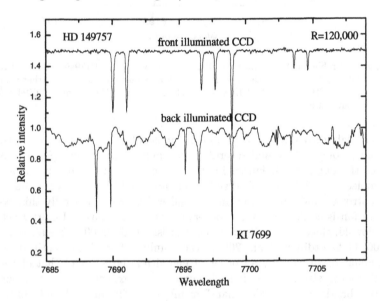

Figure 8 The effect of fringing in back illuminated CCDs. The upper spectrum was recorded with a less sensitive front illuminated CCD.

Krelowski, Schmidt & Snow (1997) proposed a special method of relating strong and weak DIBs. High resolution spectra can be re-scaled to match the depth of a certain feature. If the chosen, strong feature is related to some weak DIBs, their depths should be the same in all spectra which have been scaled down or up to make the strong feature identical. This method can be a more efficient tool for finding DIB families than the direct measurements of e.g. equivalent widths of both strong and weak DIBs; especially the latter may be substantially contaminated with noise. While the spectra are overlayed it is much easier to decide whether the intensities (and sometimes – profiles) are identical or not.

A collaboration with experimentalists should be more than welcome. If any of the strong DIBs originates in some species formerly investigated in a laboratory, the wavelengths and relative intensities of the other features originating in the same species can be predicted providing the laboratory conditions match reasonably well those of the interstellar space.

2. HIGH RESOLUTION PROFILES OF DIFFUSE BANDS

High resolution, high S/N profiles of strong DIBs may allow identification of carriers. The vibrational–rotational transitions, while taking place in complex molecules, are very unlikely to show the rotational features separated. Large molecules are typically characterized by small rotational constants. The rotational structures should thus blend and form band-heads which allow only to recognize the P, R, and Q branches inside a profile. However, even in such a case profile shapes may be specific to given molecules.

The first description of the DIB profiles was given by Herbig (1975). It was based on photographic spectra of heavily reddened stars. Despite the S/N enhancement made by means of averaging of 8 – 10 spectra of each star the profiles could hardly show any fine structure inside as the targets were apparently reddened by several clouds each and thus the Doppler splitting masked possible substructures.

The problem was then raised by Snell & Vanden Bout (1981) in the case of the broad 5780 DIB only. Their high resolution and high S/N spectra did not show any substructures inside the broad profile. They have not considered simultaneously profiles of any atomic and/or molecular features in spectra of the same targets and thus were unable to reveal any Doppler splitting along the chosen sightlines.

Westerlund & Krełowski (1988) made a more systematic survey of the intrinsic profiles of five DIBs. Their spectra contained the interstellar NaI lines which allowed to select for the profile analysis only the targets which are very likely to be obscured by single, individual clouds. The analysis showed that all DIB profiles are non–gaussian and that some substructures may exist inside the relatively narrow DIBs (5797 and 6379).

The recent survey of Krełowski & Schmidt (1997) proved that substructures are very likely to be present inside profiles of almost all of the strong diffuse bands. The profiles showing the internal structure are very likely to be molecular bands, originating in large molecules of low rotational constants which the factor makes impossible any further resolving of their rotational profiles.

The highest resolution DIB profiles, published until now are those of Sarre & al. (1995) and Kerr et al. (1998). The Ultra High Resolution Facility allowed to see some very fine details of the 5797 and 6614 profiles. It is characteristic that the 5797 profile contains three features which could be the P, R and Q branches of a molecular band. It is of importance that the possible Q branch becomes undetectable when the resolution is as low as 60,000 (Fig. 5). In laboratory spectra of complicated molecules this branch is easily seen with the resolution of 150,000 (T. Motylewski, private communication) which may be thus close to an optimum value as the growing resolution extends substantially the exposure time. Fig. 9, based on the UHRF spectra, demonstrates the striking similarity of the 5797 and 6614 profiles in the spectra of single clouds which may follow a similarity of their carriers.

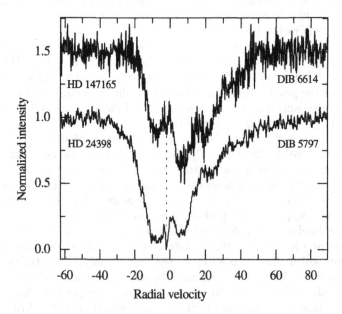

Figure 9 The surprising similarity of the ultra high resolution profiles of two DIBs (R=600,000). The DIB cores resemble laboratory spectra of complex molecules. Courtesy of P. Sarre.

The complicated DIB profiles can also follow the isotopic abundances of carbon (Walker et al. 2000). As the wavelength differences between the substructures, seen inside certain DIBs (5797, 6614) are nearly constant, it seems possible that the observed profiles are composed of the features (of nearly gaussian profiles) originating in molecules containing 0, 1, 2 etc. ^{13}C atoms. The presence of ^{13}C causes a wavelength shift which is directly proportional to the number of ^{13}C atoms. The proposed mechanism can work if the numbers of C atoms in DIB carriers are comparable with the $^{12}C/^{13}C$ ratio which is about 60. This may put some constraints on the sizes of molecules being DIB carriers. It must be, however, mentioned that the very narrow feature, marked in Fig. 9 cannot be interpreted this way.

High resolution profiles of some DIBs may be so characteristic to a given carrier that their comparison with high resolution laboratory spectra (acquired in gas phase) may allow identification. It is to be emphasized, however, that the spectra used to determine the profiles must be of very high resolution (R>100,000) and very high S/N (>1000). This requirement makes the observational programmes very time consuming. The procedure requires also a very careful analysis of possible stellar contaminations which could have very likely avoided any attention while spectra of lower S/N have been analyzed.

Another problem is the lack of proper targets. Only a few sightlines towards reasonably nearby stars in the whole sky intersect single dark clouds of reasonable optical depth which allow an analysis of intrinsic DIB profiles. An extension of the possible sample towards slightly reddened objects, which are very likely to be obscured by single clouds only, leads to the necessity of very high S/N ratios which are currently not available.

Westerlund & Krełowski (1988) claimed that the widths of the observed diffuse bands are constant in every single cloud, assuming Doppler splitting to be the only mechanism of the observed broadening. Porceddu et al. (1991) found some physically broadened profiles inside the Ori OB1 association. Krełowski & Greenberg (1999) proved also that these broadened DIBs are red-shifted in comparison with interstellar atomic lines. However, many DIBs in this association are extremely weak and are thus hardly observable. The broadening, apparently caused by physical conditions, can be also observed in very high resolution spectra (Fig. 6).

The profiles of diffuse bands may differ from object to object, not only in width or wavelength but also in the pattern of weak substructures seen in the bottom of some broad features (Fig. 6). This pattern was suggested as variable by Krełowski & Schmidt (1997) and by Seab et al. (1998) at least in the case of the 6614 DIB. In other DIBs the same phenomenon is suggested (5797) but far not as well evidenced as in the case of 6614.

3.   DIFFUSE BANDS VS. OTHER INTERSTELLAR ABSORPTIONS

The dark interstellar, HI clouds, produce several kinds of interstellar absorptions. Some of them are well–identified (atomic or molecular lines), some others are reasonably well–understood (dust grains) – only diffuse interstellar bands still remain unidentified. The whole spectrum of a dark, interstellar cloud may contain:

- continuous extinction – the selective attenuation of the light of stars shining through ISM clouds; the most interesting part of the extinction curve is situated in the vacuum–UV. For a review see Mathis (1990) or Krełowski & Papaj (1993)

- polarization – closely related to extinction and probably caused by the same dust grains but partially aligned, most probably by the galactic magnetic field

- atomic lines of FeI (3719.937, 3859.913 Å), CaII (H – 3968.468 & K – 3933.663 Å), CaI (4226.728 Å), NaI (D_1 – 5895.9236 & D_2 – 5889.9504 Å), LiI (6707.83 Å), KI (4044.136, 4047.206 & 7698.959 Å) – see Crawford (1992) and Morton (1975). In many cases these features show several Doppler components because of different radial velocities of clouds along the same sightline; most of the atomic lines (not listed above) is situated in far–UV

- molecular features of CN (3873.37, 3873.998, 3874.608, 3875.763 Å), CH (3878.768, 3886.41, 3890.213 & 4300.321 Å), CH^+ (4232.54 Å & 3957.70 Å), C_2 (8750 – 8775Å); in the vacuum–UV strong bands of CO are observed. For a review see Snow (1992).

- numerous unidentified, diffuse interstellar bands. The recent surveys (Jenniskens & Désert 1994, Krełowski, Sneden & Hiltgen 1995, Galazutdinov et al. 2000) have grown their number to about 300. Most of these interstellar features are barely visible even in high S/N spectra. For a review see Herbig (1995).

The absorption spectra of ISM HI clouds reveal the complexity of the physical processes inside them. The behaviour of the spectral features reflects variations of physical parameters of the clouds (Krełowski et al. 1992) which cause the observed variations of DIB intensities and profiles. Relating the DIB behaviour to that of the identified interstellar features one can try to understand how the DIB carriers react to variations of temperature and/or density. To do this it is necessary to search for targets obscured by only one cloud to avoid a confusion caused by ill–defined averages. A reliable description of the physical processes inside the ISM should be restricted to such individual clouds.

The latter requirement is, in fact, very severe. Usually heavily red-dened stars are observed through several clouds. Any physical parameter averaged along such a line of sight hardly makes any sense. On the other hand the observed correlation between colour excesses and DIB intensi-ties makes many DIBs hardly detectable in spectra of slightly reddened stars. It is very difficult to record spectra in which noise does not con-taminate seriously profiles of weak features. This situation forces us to observe just a bunch of moderately reddened stars ($E_{B-V} \sim 0.3$). Nobody knows how typical are these objects in our Milky Way.

3.1. RESONACE ATOMIC LINES AND DIFFUSE BANDS

The ground–based observations can cover only a few spectral lines, originating in neutral interstellar atoms. Among them the NaI lines are usually saturated and thus their intensities cannot be directly cor-related with those of diffuse interstellar bands as the latter are far from the saturation level. The line which is always observed but very rarely saturated is that of KI (7699 Å). Krełowski, Galazutdinov & Musaev (1998) proved that narrow diffuse bands, such as 5797 or 6379 are well–correlated with this line. The relation between DIBs and KI suffers some scatter but seems evident (Fig. 10). It is, however, easy to demonstrate also that while the 5797 DIB is of identical intensity in two spectra, the KI line may be different (HD 24534 vs. HD23180). The same result may be suggested in the case of CaI, but its line (4226 Å) is usually very weak and measurements cannot be done with a high precision. The observed relation may suggest that intensities of some DIBs depend on the ionization conditions inside certain clouds.

The above suggestion cannot be, however, blindly applied to every case. The observed correlation may give us only some very general guidelines concerning the conditions which facilitate the formation of certain DIB carriers. The broader features (5780 or 6284) show a rather poor correlation with KI.

It is difficult to say whether the intensities or profiles of certain DIBs depend on the interstellar depletions. The abundances of elements in interstellar clouds are uncertain (Snow & Witt 1996) and no systematic

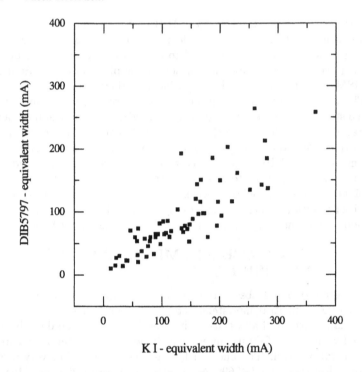

Figure 10 The apparently good correlation between the intensities of interstellar potassium line and the narrow 5797 DIB.

survey of column densities of many elements has been performed until now.

3.2.   TWO–ATOM MOLECULES AND DIFFUSE BANDS

Currently it is rather commonly accepted that diffuse bands originate in some interstellar molecules, most probably complex, carbon–bearing ones. It may be thus interesting to relate the DIB intensities to those of the features originating in some well–identified, simple molecules, such as CH, CN or CO.

The recent compilation by Krełowski et al. (1999) demonstrates that the narrow DIBs (related to KI as mentioned above) can be also related to the CH molecule (Fig. 11). The relation is not very tight and thus it can be only a suggestion of a general coincidence of physical conditions facilitating the formation (preservation) of some DIB carriers and the

CH molecule. The molecular ion CH$^+$ is not correlated to any of the DIBs. Apparently its formation requires some conditions which neither facilitate nor forbid the DIB carriers formation.

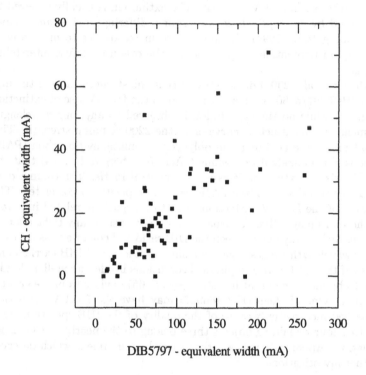

Figure 11 The relation between the intensity of interstellar CH and the 5797 DIB. The DIB carrier seems to be related more closely to KI than to CH.

Sikorski et al (1998) proved that two other simple interstellar molecules, CN and CO behave in a very similar fashion to CH. Most probably they are also related to narrow DIBs but the sample of the CN and CO observations is relatively scarce (based on vacuum ultraviolet spectra which are difficult to get at least with a high resolution and S/N) and this conclusion does not seem to be very well evidenced. CO is usually considered as the tracer of the most abundant molecule, H$_2$, and thus a more detailed examination of the behaviour of CO may shed some light on the possible relations between the hydrogen molecule and DIB carrier(s).

3.3.   INTERSTELLAR EXTINCTION AND DIFFUSE BANDS

Sneden et al. (1991) suggested that the behaviour of diffuse bands can be closely related to the shape of interstellar extinction curve, especially in its vacuum–UV segment. The extinction is usually believed to be caused by 2 – 3 populations of grains, differing in sizes and shapes (Greenberg & Li, 1994). It does not seem reasonable to make any of these grain populations responsible for the origin of diffuse interstellar bands.

Megier et al. (2001) demonstrated that the strength ratio of the major DIBs 5797/5780 is generally related to the far–UV rise of extinction (Fig. 12) while no similar relation is observed to any long wavelength segment of the extinction curve or to the 2200 Å bump strength. This result makes the DIB origin in polycyclic aromatic hydrocarbon (PAH) cations quit unlikely if we assume (after Greenberg & Li 1994) that the far–UV rise of extinction is caused by neutral PAHs. The column densities of neutral and ionized PAHs cannot be positively correlated. The carriers of the far–UV extinction seem to be spatially related in a way to those of many DIBs. As usually we do not observe any tight relation but the lack of any one between the DIB ratio and colour excess ratios in other wavelength ranges suggests a similar reaction of DIB carriers and those of the far–UV rise to physical conditions inside interstellar clouds. It is to be mentioned that five stars (out of 65) deviate from the general relation. One of them (HD 166937) may have the far–UV extinction measured incorrectly because of the quality of the IUE spectrum. Out of other four remaining objects three belong to the nearby Per OB2 association. Apparently it is a rather special environment which deserves further investigations.

4.   DISCUSSION

Generally in the course of investigations of the diffuse bands the list of invented DIB carriers which proved NOT to cause any of the observed features keeps growing. The lack of identification follows partially the lack of gas phase spectra of many complex molecules caused by substantial difficulties that experimentalists face. However, we can also indicate certain astrophysical reasons i.e. the lack of proper observational data which seem necessary to identify DIBs.

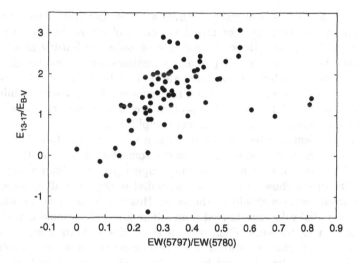

Figure 12 The correlation between far–UV extinction and the strength ratio of the major diffuse bands. The DIB ratio does not correlate at all with colours characterizing other segments of the extinction curve.

4.1. REST WAVELENGTHS

Diffuse interstellar bands remain unidentified and thus their rest wavelengths can be determined only by comparison with those of the well–identified interstellar spectral features, presumably narrow ones. A bunch of atomic and molecular lines can be observed from ground based observatories (see Sec. 4) and, especially in echelle spectra, they allow to shift the wavelength scale to that of the zero radial velocity of the intervening clouds.

One problem which we face while trying to determine rest wavelengths is that, already mentioned in the case of Ori OB1 association (Krełowski & Greenberg 1999). In this particular case we clearly observe a red–shift of diffuse bands in comparison to interstellar atomic lines. Perhaps it is due to the sticking of large, carbon–bearing molecules to surfaces of interstellar dust particles, although an alternative mechanism (e.g clustering) can also be proposed. Anyway the result creates doubt – diffuse bands do not necessarily share the radial velocities of atomic lines.

A similar doubt may be created by an analysis of the Doppler splitting observed in spectra of distant and heavily reddened stars (Galazutdinov et al. 2000a). Apparently the radial velocities of components seen in different atomic or molecular features are not identical. Once again it creates the warning: the radial velocities of diffuse bands may be slightly different than those of the atomic or molecular features chosen to measure Doppler shift(s). Apparently sometimes the rest wavelengths of DIBs, while determined as outlined above, may be incorrect. Moreover, we cannot be sure whether such a shift is common for all observed diffuse bands. Let's add that the DIB profiles are usually asymmetric which makes the wavelength measurements complicated.

The problem may be not so severe when we observe no Doppler splitting in interstellar atomic lines towards some target (e.g. in the case of HD 23180). In such a case we may expect just one cloud along the sightline of the chosen target and the radial velocities of all considered interstellar features should be the same. However, even in such a situation the radial velocities derived from different atomic or molecular lines are not exactly the same (Galazutdinov et al. 2000). At the moment it is difficult to say what is responsible for this phenomenon but the problem of DIB rest wavelengths must be considered with a proper caution.

It must also be added that among weak diffuse interstellar bands there are many features which can be traced only in spectra of heavily reddened stars. They are not observed in any single cloud cases regardless the S/N ratio achieved. The striking example is the case of two features in near infrared, close to 9577 and 9632 Å (Fig. 13) respectively, which are tentatively attributed to the C_{60}^+ molecule (Foing & Ehrenfreund 1994, 1997). The features are difficult to be observed because of the strong telluric and stellar contamination. Anyway, their interstellar origin was shown beyond a doubt (Galazutdinov et al. 2000b).

4.2.   DETAILS OF DIB PROFILES

As mentioned above two possible ways of identifying DIB carriers can be possible. One of them can be based on an analysis of the profile shapes which can match or not those obtained in laboratories for a given candidate. In principle the substructure patterns of certain DIBs may suggest their molecular origin. This was shown by means of simulations such as those of Schulz et al. (2000) – see Fig. 14. Apparently some species, of rotational constants characteristic for long chain molecules, can be responsible for some of the diffuse interstellar bands. Unfortunately, despite the striking similarity of the calculated and observed profiles the achieved result is still only a suggestion. The simulations use wavelengths as free parameters. Only after we find gas phase laboratory spectra of known species matching the diffuse band profiles one can consider them as identified. Let's mention that the calculated profile does

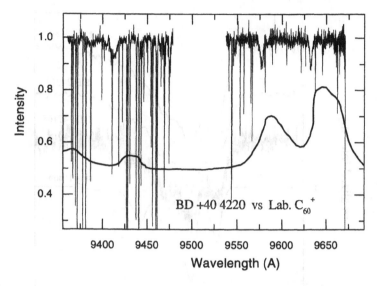

Figure 13 A comparison between the laboratory spectrum of the C_{60}^+ ion (matrix isolation) and the spectrum of the heavily reddened star BD +40 4220. Note the remnants of the telluric contamination in the stellar spectrum.

not contain the narrow peak, marked in high resolution spectra shown in Fig. 9.

From the point of view of an observer it is of basic importance to analyze a reasonable sample of profiles of every DIB. In the case of the 6614 feature, observed in very high resolution spectra, some variations of the substructure pattern from object to object have been reported (Krełowski & Schmidt 1997, Seab et al. 1998). Whether this substructure pattern is related to a behaviour of other interstellar absorptions such as: intensities of molecular or atomic features, strength ratios of other DIBs or shape of extinction curve – remains an open question. It cannot be solved before a reasonably large sample of profiles is collected. The observed objects must be known as free of any Doppler splitting in atomic interstellar lines – this may require an analysis of high resolution spectra from ground based observatories as well as high resolution of far–UV HST spectra from the range in which a majority of interstellar lines can be observed.

Only a big enough sample of high resolution spectra can answer the question of whether variations of the substructure pattern, first observed in 6614, can be found in other DIB profiles as well. The existing vari-

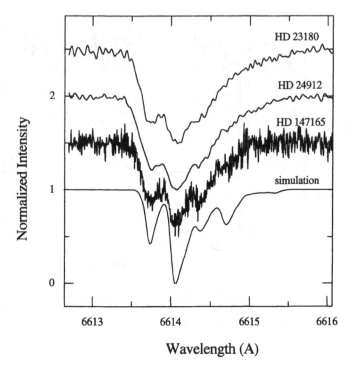

Figure 14   The observed CFHT and UHRF profiles of the 6614 DIB compared to the result of simulation for $C_{12}H_2$ by Schulz et al. (2000). The narrow feature, marked in Fig. 9 is absent in the calculated profile.

ations must be interpreted in terms of a proposed mechanism of the formation of any DIB before we may consider any feature identified.

High resolution spectra are also necessary while determining precisely strength ratios of some DIBs. Any of the diffuse features can be, in principle, blended with some weak stellar lines. The nicely–looking high resolution DIB profiles (Fig. 9) may contain "foreign" weak features. A separation of stellar components requires a sample of a high resolution and high S/N spectra of both reddened and unreddened stars as well as of synthetic spectra. However, the theoretically calculated spectra are still short of some observable features and require a deeper analysis.

Let's mention that blending is quite popular among strong DIBs: 5780 and 5778, 5797 and 5795, 6203 and 6205; blending among much more numerous weak DIBs may be more probable. Thus some of the profiles may contain in fact more than one feature and the only possibility to realize this fact may follow their variable intensity ratio discovered in a large enough sample of targets.

4.3.   OUR LOCATION IN THE GALAXY

Our location in the Milky Way also creates a problem. As already mentioned, to analyze precisely DIB profiles we need targets free of any Doppler splitting along the chosen sightlines. Otherwise possible substructures of DIB profiles are very likely to be smoothed by the Doppler effect. Another requirement is a high enough reddening ($E_{B-V} > 0.2$); otherwise the expected features are very weak which would require the application of extremely high S/N spectra very difficult. In most of cases the stars fulfilling the above condition are those, belonging to the so–called Gould Belt – the local structure tilted to the plane of the Galaxy. This tilt allows to observe these nearby objetcs on the "black" background: no more distant aggregates are observed in the same directions. Other OB associations are crowded around the galactic equator and the interstellar features observed towards them are usually "coctails" of many clouds, differing in both: radial velocities and physical parameters.

The Gould belt associations: Sco OB2, Per OB2, Ori OB1 and Lac OB1 are thus very attractive targets. We can believe that the observed interstellar features are originated in some remnants of their parent clouds. However I have to mention that the interstellar spectra observed towards these aggregates show practically all extremities (or peculiarities) found until now. In Sco OB2 we observe very strong 5780 DIB and a very high 5780/5797 ratio. Also a rather low far–UV extinction is observed. In Ori OB1 the physically broadened and red–shifted DIBs are observed (Krełowski & Greenberg, 1999; Jenniskens et al., 1993). The extinction curves, derived from the IUE spectra of Ori OB1 stars show very low 2200 "bump" and low far–UV extinction (Fitzpatrick & Massa 1990). Diffuse bands are extremely weak in relation to E_{B-V} in this aggregate. Per OB2 association shows typically the "average" extinction curve but also extremely strong 5797 feature as seen in Fig. 15.

The above outlined situation creates the problem of how far the environments of the Gould belt associations are typical for the Milky Way. If they are not we face a tremendous problem of dealing with the DIB profiles modified by the Doppler splitting while observed towards other associations.

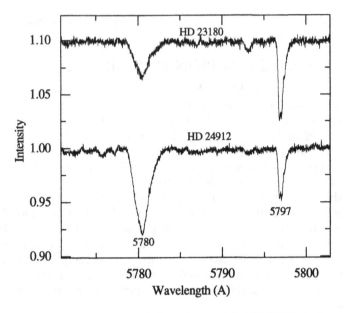

Figure 15   The varying DIB ratio in high resolution, high S/N CFHT (Gecko) spectra of the two stars from the Per OB2 association. The 5797/5780 strength ratio, observed towards HD 23180, is the extremely high one.

4.4.   VERY HIGH SIGNAL–TO–NOISE RATIO

The necessity of collecting high S/N spectra was mentioned already. The problem gets especially difficult while composed with that of acquiring high resolution as well. In such cases even very large telescopes need a lot of (hard to get) time to collect enough of light to meet the observers' requirements. OB stars are usually situated at rather long distances from our Sun. This makes their apparent magnitudes relatively faint.

High S/N is absolutely necessary while trying to determine strength ratios of strong and weak DIBs. The latter are typically very shallow which makes the S/N inside their profiles up to 100 times lower than in the continuum. As the spectral features of the same large molecule can be separated by as much as 1000 – 1500 Å, only echelle spectra can be used to search for such DIB "families" – believed to be spectra of single species.

High S/N spectra cannot be recorded in cases of many stars. With the stellar apparent magnitudes getting fainter the exposure time gets very long which reduces the possibility of acquiring large samples of high quality spectra. This fact also creates the question of how far the relatively nearby targets, which can be observed in high resolution and S/N, are "typical" for the Milky Way.

While trying to acquire a very high S/N one has to take care of possible remnants of fringes which are very likely not to be completely divided out even if the divisor is selected very carefully. Moreover, in the spectral ranges strongly contaminated with telluric lines it is practically impossible to get rid of fringes and telluric lines simultaneously. Front illuminated CCDs should be recommended despite their lower sensitivity. Very high S/N spectra can also reveal some very weak stellar lines which have never been observed (and identified) before. A sample of spectra broadened in different ways by stellar rotation can be very helpful while discriminating stellar features.

4.5. PHYSICAL CONDITIONS INSIDE INTERSTELLAR CLOUDS

Interstellar spectra, observed towards many OB stars, usually differ more or less from object to object. The only two targets in which the diffuse bands 5780 and 5797 seem to be identical are HD 23180 and HD 24534. But even in this case atomic and molecular features are of different intensities as well as E_{B-V}. This fact supports the idea of DIBs originating in many different carriers, not necessarily related either to simple molecules or to dust grains. However, if a set of DIBs can be selected as being of the same strength ratios in all considered spectra it may be considered as originated in the same carrier.

It must be emphasized that an attempt to identify a molecule as the carrier of some DIB requires an analysis of many spectra. DIB strengths, as measured in a single spectrum, give insufficient information to complete such a task. This is why the recent DIB atlas (Galazutdinov et al 2000) does not include DIB equivalent widths. Proposed spectral features (i.e. found as the spectrum of a molecular species) must match precisely wavelengths and intensity ratios in several observed spectra. It is of importance to select spectra in which the strength ratios of major DIBs (5780 and 5797) are strongly variable. A constant ratio of some other DIBs makes the case strong. An example is shown in Fig. 16 but even here only the ratio of central depths of 6196 and 6614 is the same in both targets, despite the evident difference of the 5797/5780 ratio. The ratio of equivalent widths is different as well.

The additional problem is the growing number of stellar features observed in colder stars. The frequently observed, heavily reddened star, HD 183143, is of the spectral type B7I and thus it's spectrum is full of

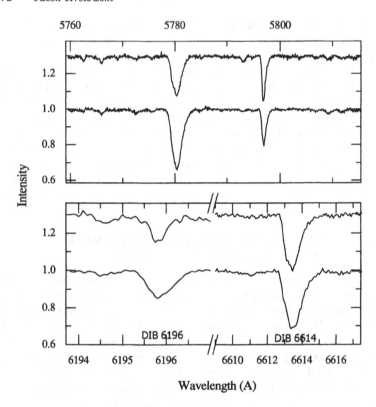

Figure 16 The spectra of two heavily reddened stars: HD 186745 (upper plots) and HD 183143 (lower plots); note the ratio of central depths of 5780 and 5797 strongly variable in contrast to that of 6196 and 6614. Spectra from the Musaev echelle spectrometer fed with the 2m Terskol telescope.

weak, unidentified stellar lines. Some of them form reasonably broad blends which are very likely to be misidentified as DIBs e.g 5747 – see Galazutdinov et al. (2000a). Much more reliable are narrow features observed in spectra of rapidly rotating O type stars where spectral lines are scarce because of the effective temperature. Such stars are, however, very uncommon and usually distant i.e. quite faint.

It seems of basic importance to create in any possible case a kind of criterion tailored both by observers and experimentalists which could give us a definite answer whether any proposed set of features, found in laboratory as the spectrum of a given molecule, can be identified as belonging to one species, present in the interstellar space. The conclusions must be satisfactory both from the point of view of these who investigate molecular structure and molecular spectra in laboratory as well as from the point of view of observers who compare the laboratory expectations to many different, observed spectra.

Acknowledgments

The author wants to express his gratitude to the Polish National Committee for Scientific Research for the financial support under the grant 2 P03D 019 23 and to Dr. Gazinur Galazutdinov for the very helpful discussions and suggestions.

REFERENCES

Beals, C.S. (1938), *ApJ*, **87**, 568.
Crawford, I.A. (1992), *MNRAS*, **254**, 264.
Fitzpatrick, E.L., and Massa, D. (1990), *ApJSS*, **72**, 163.
Foing B.H. and Ehrenfreund P. (1994), *Nature*, **369**, 296.
Foing B.H. and Ehrenfreund P. (1997), *A&A*, **317**, L59.
Galazutdinov, G.A., Krełowski, J., Musaev, F.A. (2000a), *MNRAS*, in press.
Galazutdinov, G.A., Krełowski, J., Musaev, F.A., Ehrenfreund, P. and Foing, B.H. (2000b), *MNRAS*, in press.
Galazutdinov, G.A., Musaev F.A., Krełowski J., Walker, G.A.H. (2000), *Publs Astr. Soc. Pacific*, in press.
Greenberg, J.M. and Li, A. (1994), in Greenberg, J.M. (ed.), *The Cosmic Dust Connection*, Kluwer, p.43.
Herbig, G.H. (1975), *ApJ*, **196**, 129.
Herbig, G.H. (1995), *Ann. Rev. A&A*, **33**, 19.
Herbig, G.H. and Soderblom, D. R. (1982), *ApJ*, **252**, 610.
Jenniskens, P. and Dèsert, X. (1994), *A&A Supp. Ser.*, **160**, 39.
Jenniskens, P., Ehrenfreund, P. and Foing, B. (1993), *A&A*, **281**, 517.
Kerr T.H., Hibbins R.E., Miles J.R., Fossey S.J., Sommerville W.B. and Sarre P.J. (1996), *MNRAS*, **283**, 105.
Krełowski, J., Ehrenfreund, P., Foing, B. H., Snow, T. P., Weselak, T., Tuairisg, S. ., Galazutdinov, G. A. and Musaev, F. A. (1999), *A&A*, **347**, 235.
Krełowski, J., Galazutdinov, G.A. and Musaev, F.A. (1998), *ApJ*, **493**, 217.
Krełowski, J. and Greenberg, J.M. (1999), *A&A*, **346**, 199.
Krełowski, J., Papaj, P. (1993), *Publs. Astr. Soc. Pacific*, **105**, 1209.
Krełowski, J., Schmidt, M. (1997), *ApJ*, **477**, 209.
Krełowski, J., Schmidt, M. and Snow, T.P. (1997), *Publs. Astr. Soc. Pacific*, **109**, 1135.
Krełowski, J., Sneden, C. and Hiltgen D. (1995), *Planetary Space Sci.*, **43**, 1195.
Krełowski, J., Snow, T.P., Seab, C.G. and Papaj, J. (1992), *MNRAS*, **258**, 693.
Krełowski, J., and Walker, G.A.H. (1987), *ApJ*, **312**, 860.
Krełowski, J. and Westerlund, B.E. (1988), *A&A*, **190**, 339.
Mathis, J.S. (1990), *Ann. Rev. A&A*, **28**, 37.
Megier, A., Aiello, S., Barsella, B., Casu, S. and Krełowski, J. (2001), *A&A*, **326**, 1095.
Morton, D.C. (1975), *ApJ*, **197**, 85.
Moutou, C., Krełowski, J., D'Hendecourt, L., Jamroszczak, J. (1999), *A&A*, **351**, 680.
Porceddu, I., Benvenuti, P., Krełowski, J. (1992), *A&A*, **260**, 391.

Sarre P.J., Miles J.R., Kerr T.H., Hibbins R.E., Fossey S.J., Sommerville W.B. (1995), *MNRAS*, **277**, 41.

Schulz, S.A., King, J.E. and Glinski, R.J. (2000), *MNRAS*, **312**, 769.

Seab, C. G., Sneden, C., Snow, T. P., Riedlinger, L., Bommer, J. and Lewis, D. (1998), *Amer. Astr. Soc. Meeting*, **193**, 65.22.

Sikorski, J., Krełowski, J., Gnaciński, P., Kaczmarczyk, G., Snow, T.P. and Krivova, N.A. (1998), in *Ultraviolet Astrophysics, Beyond the IUE Final Archive*, ESA SP-413, p.505.

Sneden, C., Woszczyk, A. and Krełowski, J. (1991), *Publs. Astr. Soc. Pacific*, **103**, 1005.

Snell, R.L. and vanden Bout, P.A. (1981), *ApJ*, **244**, 844.

Snow, T.P. (1992), *Australian J. Phys.*, **45**, 543.

Snow, T.P. (1995), in Tielens, A.G.G.M. and Snow, T.P. (eds.), *The diffuse interstellar bands*, Kluwer Academic Publ., p.325.

Snow, T.P. and Witt, A. N. (1996), *ApJ*, **468**, L65.

Tuairisg, S.., Cami, J., Foing, B. H., Sonnentrucker, P. and Ehrenfreund, P. (2000), *A&A Suppl. Ser.*, **142**, 225.

Tulej, M., Kirkwood, D.A., Pachkov, M., Maier, J.P. (1998), *ApJ*, **506**, L69.

Walker, G. A. H., Bohlender, D. A. and Krelowski, J. (2000), *ApJ*, **530**, 362.

Weselak, T., Schmidt, M., and Krelowski, J. (2000), *A&A Suppl. Ser.*, **142**, 239.

Westerlund, B., E. and Krełowski, J. (1988), *A&A*, **203**, 134.

MATRIX AND GAS - PHASE SPECTROSCOPIC STUDIES OF POSSIBLE DIB CARRIERS

Jan Fulara
Institute of Physics, Polish Academy of Sciences, Al. Lotników 32/46 PL- 02-668 Warsaw, Poland

Keywords: ISM: lines and bands – Molecules – Methods: laboratory

Abstract: The origin of diffuse interstellar bands (DIBs) is discussed in this review. Numerous astrophysical observations indicate that narrow DIBs are rotationally - unresolved electronic bands of free gas-phase molecules. Open - shell molecular systems based on the carbon skeleton are the best candidates for DIB carriers. Such chemical species are very difficult objects for spectroscopic examinations due to their very high reactivity. The best suited method for such purposes is matrix isolation of mass selected ions. The performance of this method is demonstrated for several linear carbon chains and general features of the electronic absorption bands of such molecules isolated in a matrix are discussed in this paper. Very sensitive detection methods used for spectroscopic examination of radical species in the gas phase are also discussed and comparison of the gas phase electronic spectra of specific molecules with the spectrum of DIBs is made.

INTRODUCTION

Diffuse Interstellar Bands (DIBs) are the spectroscopic absorption structures that are observed in the spectra of reddened stars. The first observation of two intense interstellar structures ($\lambda = 5780$, 5797 Å) comes from the early twenties (Heger 1922). Further strong DIBs have been discovered in the following years (Merrill 1936). Till early seventies, the spectra of reddened stars have been recorded by using photographic plates. The spectra acquired at such conditions had a low signal to noise ratio and

175

they revealed merely several tens of the strongest interstellar absorption features (Herbig 1975). The subsequent replacement of the photographic plates with electronic detectors led to a drastic improvement in the signal to noise ratio of recorded spectra. At the beginning of the nineties more than one hundred DIBs were known (Jenniskens & Désert 1994), and at present this number exceeds 270 (Galazutdinov et al. 2000). The reader can find a comprehensive review on DIBs in the paper by Krełowski in the same issue.

1.   MATRIX SPECTROSCOPIC STUDIES

1.1   MOLECULAR ORIGIN OF DIBS

The spectrum of known DIBs extends from the violet to the near infrared spectral region and it consists of numerous absorption structures that differ in their half widths as well as in their intensities. In the spectrum of DIBs we can distinguish very broad structures (FWHM>10 Å) e.g. the 4430 or 6177 Å DIBs and very narrow ones (FWHM ≤ 1 Å) e.g. the 6196, 6379 Å DIBs. The spectral position of any DIB remains almost constant irrespective of the line of sight. In contrast to the spectral position of DIBs the relative intensities of any pair of DIBs vary in an irregular manner from one cloud to another. A very important feature of DIBs is the profile that remains nearly unchanged irrespective of the direction of observation. The profiles of many DIBs measured at high - resolution conditions exhibit fine multiplet structures, as have been shown by Sarre et al. (1995) and Kerr et al. (1998) for the 5797 and 6614 Å DIBs.

The above - mentioned features of DIBs give us some suggestions about the origin of these bands. The spectral region, where DIBs occur (from the visible to the near infrared), is characteristic for the <u>electronic transitions</u> of chemical species. The constant positions and unchanged profiles of DIBs irrespective of the line of sight point to free gas -phase molecules as the carriers of these bands. If DIBs were to originate from chemical species frozen into dust grains the narrow interstellar absorption features would be absent in the spectrum of DIBs. The positions of the absorption peaks of molecules embedded into dust grains strongly depend on the chemical composition of the microenvironment of these molecules. The local chemical composition can vary from one grain to another so that in the interstellar clouds a broad absorption spectral feature is expected as a result of the superposition of absorption peaks of individual molecules subjected to a range of chemical shifts. The gas - phase hypothesis of the origin of DIBs is substantiated by the observation of the fine internal structures in the profiles of many DIBs. The multiplet structures seen in the profiles of many DIBs are the envelopes of unresolved individual rovibronic lines of gas - phase molecules.

An additional argument supporting the gas - phase hypothesis comes from the observations of several DIBs in emission in the Red Rectangle nebula (Warren-Smith, Scarrott & Murdin 1981). The emission bands lie at wavelengths, which are close to the positions of DIBs. However the half widths of the emission bands are larger than the half widths of the respective DIBs. As distance from the central star that illuminates the nebula increases, the emission bands become narrower and their positions approach the positions of DIBs (Sarre, Miles & Scarrott 1995). Such behaviour can be explained in terms of a variation of temperature across the nebula and the assumption that the emission originates from the free gas - phase molecules.

The variability of the relative intensities of DIBs from one cloud to another suggests that the bands originate from a great number of different gas - phase molecules and in the extreme case each DIB might originate from a different molecule.

1.2   WHAT FREE GAS - PHASE MOLECULES CAN BE CARRIERS OF DIBS?

Some suggestions concerning the chemical composition of the carriers of DIBs arise from the abundance of elements in the universe. The carriers of DIBs ought to be built of the most abundant elements: H, O, C, N, a small contribution of other elements also can not be excluded. Among this group of elements carbon is in exceptional position, since it can form a great number of stable compounds with a linear, planar and spherical structure. Addition to carbon of other elements from this group drastically increases a number of stable compounds. *Let us consider now whether ordinary molecules built of carbon and hydrogen atoms, which are stable in normal conditions, are suitable carriers of DIBs?*

The saturated and unsaturated hydrocarbon chains represent one group of such compounds. Aliphatic hydrocarbons have strong electronic transitions of the σ^*-σ type in the short - wavelength UV range. The unsaturated hydrocarbons absorb also in the UV spectral range (the electronic transitions of the π^* - π type), however the positions of their absorption peaks are shifted towards the red with respect to the saturated counterparts. The position of the origin band of unsaturated hydrocarbons depends on a number of π bonds in a molecule and it shifts towards the red as the number of π electrons increases. Sufficiently large π - electronic systems should absorb in the visible spectral range.

Another important group are polycyclic aromatic hydrocarbons (PAHs) that form planar π - electronic systems. PAHs have strong electronic absorption bands in the UV. Among PAHs regular behaviour is observed according of which the long-wavelength edge of the absorption band shifts towards the red as the size of a molecule increases. The position of the origin band of PAHs is not a simple function of a number of carbon atoms (the number of π electrons), but it depends also on the chemical structure of molecules. If we consider PAHs of a very similar chemical structure, the wavelength of the 0_0^0 band depends linearly on a number of carbon atoms that form the molecules. This is well illustrated in fig. 1 for PAHs that belong to the same series as benzene, naphthalene, and anthracene.

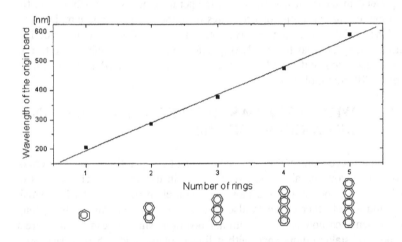

Figure 1. The wavelength of the origin band of PAHs (Birks 1970) that belong to the same series as benzene, naphthalene as a function of the number of benzene rings in a molecule.

Such linear dependence is predicted in the frame of the simplest quantum chemical model - the particle in a box, where particles are electrons occupying π orbitals, and the box has the dimensions of a molecule. PAHs larger than anthracene that belong to the same series have electronic transitions in the visible spectral range. However the intensities of absorption bands in the visible range are considerably smaller than the intensities of the UV bands of these molecules.

If we look for DIB carrier molecules that would have absorption bands in the visible or in the near infrared spectral range, we should choose a sufficiently large π -electronic molecular system. Apart from PAHs also fullerenes fall in this category. Fullerenes have absorption bands in the visible spectral range. In the case of C_{60} very weak absorption peaks in the visible range correspond to the electronic transitions that are forbidden for symmetry reasons (Gasyna et al. 1991). C_{70} has a lower symmetry than C_{60} and the absorption bands in the visible range are stronger than the bands of C_{60} (Fulara, Jakobi & Maier 1993). Nevertheless the intensities of the bands of C_{70} in the visible range are still two orders of magnitude weaker than the intensities of the bands in the UV range. The strongest absorption bands of fullerenes fall in the UV spectral range.

A question arises whether a large π - electronic molecular system would have strong electronic transitions in the visible or in the near infrared spectral region in the case when the size of such molecule increases up to "microscopic" dimensions? A negative answer to this question has been given by the results of the spectroscopic studies of nano- and micro- carbon particles isolated in low temperature matrices (Schnaiter et al. 1998). The carbon particles have a very strong absorption band in the UV spectral range and a long weak tail that extends to the visible spectral range.

In summary even though molecular systems such as unsaturated hydrocarbons, PAHs, fullerenes and nano- or micro- carbon particles have absorption bands in the visible or in the near - infrared spectral region they are not good candidates for the DIB carriers, since their visible or near - infrared bands are much weaker than the UV bands.

Among other stable molecules, dyes have very strong absorption bands in the visible or the near - infrared spectral regions. However due to a very low probability of formation of these molecules they should also be excluded from further consideration as DIB carriers.

Since ordinary molecules that are available under normal laboratory conditions are not good candidates for DIB carriers, we have to look for other molecular systems based on the carbon skeleton that would have strong electronic transitions in the visible or in the near infrared regions. A very broad group of such chemical species are open - shell molecular systems that possess one or two unpaired electrons. Good examples of such molecules are naphthalene and diacetylene cations that form from the neutral precursors by the removal of the valence electron. These cations, in contrast to the neutral precursors, have strong electronic transitions in the visible spectral range. Open - shell molecular systems can also form by a partial or total dehydrogenation of hydrocarbons and they can be generated by electrical discharge running through organic vapours. Carbon radicals can be generated in the laboratory by the evaporation of graphite at high temperature (above 2500 °C) or by the laser ablation of a graphite surface. A great number of neutral and charged radical species can be produced in such experiments.

The carbon radicals and partially hydrogenated carbon radicals are stable, but very reactive molecular systems. Their high reactivity is the reason why only a very low transient concentration of these species can be reached in laboratory gas - phase experiments. A low concentration, on the other hand, precludes direct spectroscopic measurements of these species in the gas phase. Inert gas matrix isolation is the most suitable method for the spectroscopic studies of such species. This method, developed more than fifty years ago, was successfully applied to spectroscopic examinations of many radical systems. When the chemical species generated in one of the above - mentioned ways are codeposited with the noble gas onto a cold substrate, a matrix is formed that contains a mixture of different molecules. The electronic absorption spectrum of the matrix recorded in such conditions is very complicated due to overlapping of the absorption bands of different molecules and very often it is not possible to assign the observed bands to a specific molecule.

1.3   METHOD OF MATRIX ISOLATION OF MASS SELECTED IONS

In order to obviate such difficulties, we have developed several years ago in Basel a method that permits the spectroscopic studies of molecules of any type. This method combines matrix isolation with mass spectrometry. The scheme of our apparatus, shown in fig. 2, illustrates how this method operates (Freivogel et al. 1994a). The molecules which are the object of the spectroscopic examination are generated in a charged form in the ion source. A simple unsaturated hydrocarbon - diacetylene is used as the precursor of ions. In the ion source many charged and neutral molecules are produced from diacetylene.

Because we are interested only in one specific molecule, we have to separate it from the others that would contaminate a matrix. In the first place all kinds of ions have to be separated from neutral molecules. This is achieved in the quadrupole deflector. The ion trajectories are curved in the quadrupole deflector, while neutral molecules go directly to the pump. Next the ions enter to the quadrupole mass filter. In an oscillating electric field only ions of a specific mass to charge ratio have stable trajectories and they are transmitted through the mass filter. Other ions hit the walls of the vacuum chamber and they are removed from the system. The ions of the specific m/e ratio are deposited with an excess of neon at 4.5 K on the cold finger of the cryostat.

The ion current that is required for the matrix isolation experiments exceeds by several orders of magnitude a typical ion current that is used in mass spectrometry. A large ion current can be supplied from the ion sources that are usually used in the ion implanting technique. Our discharge - type ion source can supply a large ion current and it can work with organic precursors. Many ion - molecule reactions take place in the source and very large polyacetylene cations form (see fig. 3) from diacetylene. The mass spectrum shown in fig. 3 was recorded at relatively high mass resolution. At such conditions the current of the transmitted ions was strongly discriminated. An ion current of the order of nA is needed for the matrix isolation experiments. In order to achieve such current the resolution of the mass filter has to be lowered. The cations that differ in the number of the hydrogen atoms have not been separated in the mass filter and several different ions have been deposited in a matrix.

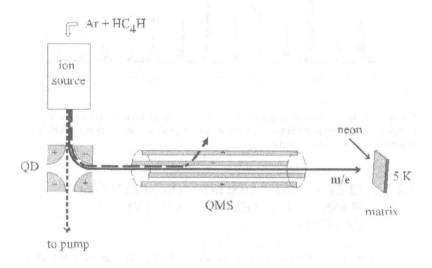

Figure 2. Schematic outline of the apparatus for matrix isolation of mass selected ions. QMS - quadrupole mass filter, QD- quadrupole deflector.

Molecules can be stored in a matrix as long as low temperature is maintained and all spectroscopic measurements can be carried out during this period. The electronic absorption spectra of mass-selected ions were recorded by using a single beam spectrometer that consists of a light source, a monochromator, lenses and a detector. The light from the monochromator is focused onto the entrance slit of the matrix substrate and the light passes through the matrix in the 'waveguide mode'. A matrix some one hundred μm thick was deposited onto a 2 x 2 cm superpolished copper substrate coated with rhodium. The light beam is guided parallel to the surface of the substrate. The light is reflected from one side from the rhodium surface and by total internal reflection from the other side. The light path in the matrix is longer than 2 cm, which considerably increases the detection limit of the apparatus.

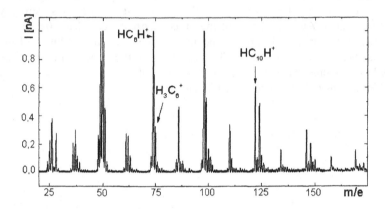

Figure 3. Mass spectrum of polyacetylene cations produced from diacetylene in the ion source. Note that the spectrum was recorded at much higher resolution than that at which the deposition of cations was carried out.

1.4   ELECTRONIC ABSORPTION SPECTRA OF MASS - SELECTED POLYACETYLENE CATIONS

The electronic absorption spectrum shown in fig. 4a was recorded after deposition of the polyacetylene cations of m/e ≈ 74 in a matrix. This spectrum extends from 460 to 610 nm and consists of several broad absorption bands that exhibit multiplet structure. We can distinguish two sets of bands that differ in the shape of bands. The origin band of the first band system lies at 604.6 nm and bands at 582.8, 563.2, 543.3, 538 nm also belong to this system. The second band set originates at 530.4 nm and is composed of several absorption bands that lie at shorter wavelengths than the origin.

When the matrix is irradiated with UV photons (λ>360 nm) from a medium pressure mercury lamp, the first band system vanishes entirely, while the peaks of the other system grow slightly in intensity. The absorption bands that vanish upon UV irradiation belong to the polyacetylene cation that contains six carbon atoms. The UV light - induced decay of the absorption bands of matrix isolated cations is a very well known phenomenon called photobleaching. In addition to the cations, negative ions are also present in a matrix that balance the positive charge of the cations. Anions are present in a matrix as the result of the capture of free electrons by impurity molecules from the vacuum system.

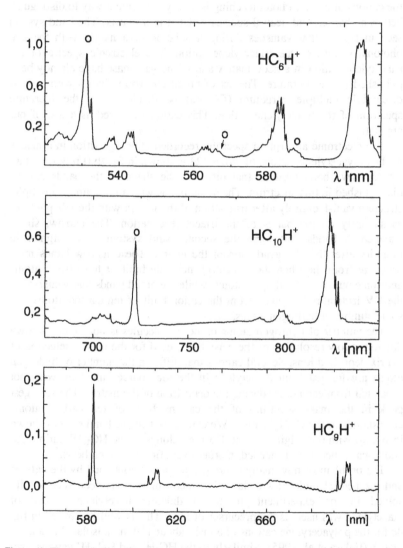

Figure 4. Electronic absorption spectra of mass - selected polyacetylene cations in neon matrices. The cations are indicated on the figures. Open circles mark the electronic absorption bands of neutral molecules C_6H, $C_{10}H$ and HC_9H from the top to the bottom trace respectively.

The anions and cations are spatially separated in a matrix. The UV photons destroy weakly bound anions and free electrons are liberated. Electrons can travel in a matrix over quite a large distance, and once they meet cations they neutralize them so that in effect, the absorption peaks of the cations diminish. Photobleaching is a very convenient way to distinguish between charged and neutral species isolated in a matrix. The band system seen in fig. 4a that vanishes during irradiation of a matrix with the UV photons belongs to the triacetylene cation. The electronic spectra of the triacetylene cation in a neon matrix and in the gas phase have already been published in the literature. The set of broad absorption bands with a well-discernible multiplet structure (fig. 4a) is identical with the literature spectrum of the triacetylene cation. This cation was used as a test of our apparatus.

The electronic absorption spectrum recorded after deposition in a matrix of the polyacetylene cations of m/e ≈ 122 is shown in fig.4b (Freivogel et al. 1994a). Two band systems that differ in the shape of the bands can be distinguished in this spectrum. The broad bands with a pronounced multiplet structure vanish entirely after irradiation of the matrix with the UV photons, as similarly do the bands of the triacetylene cation. The narrow, short-wavelength bands that form the second band system grow slightly in intensity after the UV irradiation of the matrix. These narrow bands must originate from the photostable neutral molecule built of ten carbon atoms and some number of hydrogen atoms, while the broad bands that vanish after the UV irradiation originate from the cation built of ten carbon atoms and some number of hydrogen atoms.

The number of hydrogen atoms in these molecules is not exactly known due to the low resolution of the mass filter, used for the mass - selection of matrix deposited ions. Several cations that differ in the number of hydrogen atoms are formed from diacetylene in the ion source, and they were not separated from each other during the deposition of the matrix. The strongest peak in the mass spectrum of the cations built of ten carbon atoms corresponds to the $C_{10}H_2^+$ cation. We can guess that the band system shown in fig. 4b with the origin band at 823 nm belongs to the $HC_{10}H^+$ cation. An additional experiment is needed in order to confirm this hypothesis.

The problem of how many hydrogen atoms are contained by the cations and the neutral molecules isolated in a neon matrix has been successfully solved in the experiment, in which dideuterodiacetylene instead of diacetylene was used as the precursor of ions. This is well illustrated in fig. 4c for the polyacetylene cations of a mass about 110 m. u. isolated in a neon matrix (Fulara et al. 1995). Similarly to the HC_6H^+ and $HC_{10}H^+$ cations, two band systems that differ in the shape of the bands are well-discerned in fig. 4c. The band system with a well resolved band structure (the origin band at

694.9 nm) vanishes entirely after irradiation of the matrix with the UV photons, while the second band system that is composed of the narrow bands, increases slightly in these conditions. The first band system that vanishes after UV irradiation originates from the polyacetylene cation built of nine carbon atoms and some number of hydrogen atoms.

The expanded view of the origin band of this cation is shown in fig. 5a. When dideuterodiacetylene was used as the precursor of cations all the bands of this cation were shifted to the violet, while the fine structure of the bands was remained unchanged. The expanded view of the origin band of the deuterated cation is shown in fig. 5c. The spectrum depicted in the middle trace in fig. 5 was recorded in the case when the 1:1 mixture of normal and dideuterodiacetylene was used for the generation of cations. The spectrum shown in fig. 5b comprises the overlapping origin bands of three isotopomers of this cation. Such spectral pattern of the origin and the vibronic bands can be expected only for the molecules that have two equivalent hydrogen atoms.

The substitution of normal hydrogen atoms with deuterium atoms in connection with mass selection enables unambiguous identification of the molecules trapped in a matrix. In this manner the neutral molecules C_6H, $C_{10}H$ and HC_9H have also been identified (Freivogel et al. 1995). Additional experiments that were used to identify the trapped in a matrix molecules are described in paragraphs 1.6, 1.7. The absorption bands of carbon hydride molecules are seen in the spectra shown in fig. 4a, b, c respectively (the narrow absorption features marked with the open circles). The neutral molecules are present in a matrix as a result of charge neutralization of deposited cations. As it has been found on the basis of radioastronomical observations the monohydrogenated carbon chain molecules appear to be very important constituents of matter in dense interstellar clouds (Suzuki et al. 1986; Guélin et al. 1987; Cernicharo & Guélin 1996; McCarthy et al. 1996).

What regularities we can see in the spectra of the polyacetylene cations and neutral molecules shown in fig. 4? All the absorption bands of the polyacetylene cations isolated in neon matrices are much broader than the absorption bands of neutral molecules of size similar to that of the cations. The absorption bands of cations reveal a complicated multiplet structure. In contrast to the gas phase, the shape of the bands in a matrix does not relate to the size and the structure of isolated molecules. Due to the solid environment the molecules can not freely rotate in a matrix. The width of the bands of molecules trapped in a matrix depends on the strength of interaction of the guest molecules with the host environment. The ions interact with noble gas atoms stronger than neutral molecules of similar size and this is the reason why the bands of cations are broader than the bands of neutral molecules.

Larger cations (e.g. $HC_{10}H^+$) due to deleocalization of their charge over a larger distance interact with the matrix less strongly than smaller cations (e.g. HC_6H^+) and, in effect, the absorption bands of $HC_{10}H^+$ are narrower than the bands of the HC_6H^+ cation. The multiplet structure of the absorption bands of the cations is a result of a slightly different energy of interaction of these cations with their closest microenvironment. The cations occupy several different sites in a matrix. The coupling of phonons with the electronic states of molecules isolated in a matrix can also lead to the multiplet structure of the electronic absorption bands.

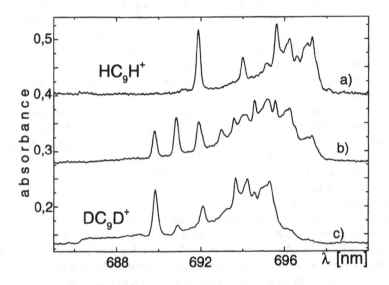

Figure 5. The expanded view of the origin band of the HC_9H^+ cation and its isotopomers (Fulara et al. 1995). Trace b) was recorded for 1:1 mixture of dideuterodiacetylene with diacetylene used for the generation of the cations.

The electronic absorption spectrum of a molecule isolated in a matrix has a simpler pattern than the gas - phase spectrum. Due to a very low temperature (around 5 K) only the lowest vibrational level of a molecule is populated in a matrix. The transitions from this level to several vibrational levels of the excited electronic state are observed in the electronic spectrum. The diagram in fig. 6 shows the potential energy curves for the ground and the excited electronic states of a molecule and explains the differences in appearance of the electronic absorption spectra in a matrix and in the gas phase. The lowest energy band in the spectrum of a molecule isolated in a matrix is the origin band. The spectrum consists also of the vibronic bands that lie at shorter wavelengths with respect to the origin band. In contrast to the matrix, the transitions between the excited vibrational levels of the ground and the excited electronic states are observed in the gas - phase spectrum.

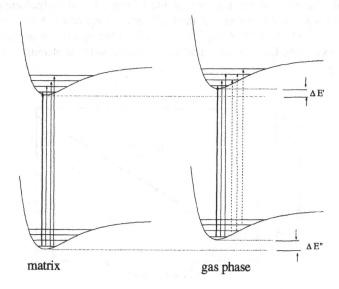

Figure 6. Potential energy curves and vibronic transitions of a molecule in the gas phase and a molecule isolated in a low temperature matrix. Thick arrow denotes the 0_0^0 transition. Dashed arrows denote transitions from the excited vibrational levels of the ground electronic state.

The energy levels of a molecule in a matrix are shifted with respect to their gas - phase positions by $\Delta E''$ and $\Delta E'$ cm^{-1} respectively in the ground and the excited electronic states. The shift of the wavelengths of the bands in a matrix with respect to their gas-phase positions depends on the interaction of a molecule with the solid microenvironment and on whether it is stronger in the ground or in the excited electronic state. If $\Delta E'' < \Delta E'$ the origin band in a matrix is redshifted with respect to the gas - phase position.

The polyacetylene $HC_{2n}H^+$ n = 2 - 8 cations have been mass - selected and deposited in neon matrices (Freivogel et al. 1994a). The electronic absorption spectra of these cations exhibit regular behaviour, according to which the relative intensities of the vibronic bands with respect to the origin band decrease when the size of the cations increases. The absorption bands of the larger cations, are shifted to the red with respect to the bands of the smaller cations. In a series of polyacetylene cations the wavelength of the origin band increases linearly with the number of carbon atoms in these cations (see fig. 7). By extrapolation of this linear dependence, the position of the origin band of even larger and yet unmeasured polyacetylene cations can be predicted. Similar regularities to these in the polyacetylene cations have also been observed in the spectra of the $C_{2n}H$ radicals (Freivogel et al. 1995). The strongest band in the spectra of these molecules is the origin band. The ratio of the intensity of the vibronic bands to the origin band becomes smaller as the size of the $C_{2n}H$ radicals increases. A vibronic band of energy of about 2000 cm^{-1} is apparent in the spectra of polyacetylene cations and the $C_{2n}H$ radicals, and it corresponds to the stretching of the $C{\equiv}C$ bonds.

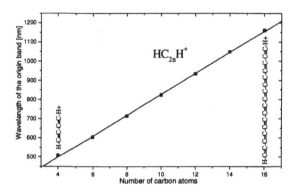

Figure 7. The wavelength of the origin band of polyacetylene cations as a function of the number of carbon atoms in a molecule.

1.5   ELECTRONIC ABSORPTION SPECTRA OF CYANOPOLYACETYLENE CATIONS

The linear cyanopolyacetylene chains are the largest molecules identified so far in the interstellar medium (ISM) (Bell et al. 1998). The cyano- and dicyanopolyacetylene cations should also be taken into account when we consider molecules that can be responsible for the occurrence of DIBs. The cyanopolyacetylene cations can be generated efficiently in a situation when cyanoacetylene is used instead of diacetylene as the precursor of ions. A low-resolution mass spectrum of cations produced from cyanoacetylene is shown in fig. 8. Due to the small difference in the mass of the CH group and that of the nitrogen atom the cyanopolyacetylene and dicyanopolyacetylene ions have not been separated from each other in the mass filter. In consequence, for example when the ions of mass of about 124 are selected the mixture of the $C_8N_2^+$ and HC_9N^+ ions is present in the matrix.

The electronic absorption spectrum recorded in such conditions is shown in fig. 9. Two band systems are apparent in this spectrum. The broad absorption peaks with multiplet structure vanish after the UV irradiation of a matrix. This result suggests that the bands originate from charged species. However we can not conclude which band system belongs to which cation.

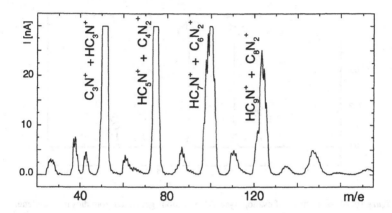

Figure 8. Mass spectrum of the cyanopolyacetylene cations generated from cyanoacetylene. The spectrum was recorded at the resolution typical for the deposition of mass - selected ions.

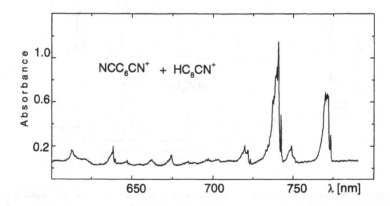

Figure 9. Electronic absorption spectrum of a mixture of cations of mass of about 124 isolated in a neon matrix. The cations were produced from cyanoacetylene as the precursor of ions.

The mass spectrum becomes much simpler when cations are produced from dicyanoacetylene (see fig 10). In this case the cations do not contain the hydrogen atoms. The electronic absorption spectrum recorded after deposition of $C_8N_2^+$ is shown in fig. 11b. All peaks with multiplet structure that constitute this spectrum belong to the $C_8N_2^+$ cation. When the electronic absorption spectrum of cations generated from dicyanoacetylene (fig. 11b) is compared with the spectrum of a mixture of ions shown in fig. 11a, the absorption bands of the HC_9N^+ cation can be identified.

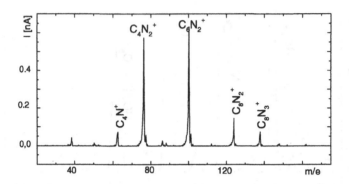

Figure 10. Mass spectrum of cyanopolyacetylene cations generated from dicyanoacetylene.

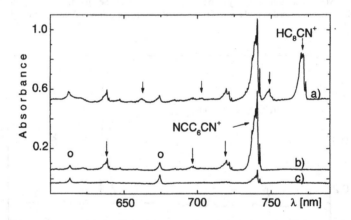

Figure 11. Electronic absorption spectra of mass selected cyanopolyacetylene cations isolated in a neon matrix (Forney et al. 1995). Trace a) spectrum from fig. 9, trace b) the spectrum of the $C_8N_2^+$ cation generated from dicyanoacetylene., trace c) the spectrum of the photostable neutral molecule, which forms in the matrix by neutralization of $C_8N_2^+$ cations.

Systematic spectroscopic studies of the cyanopolyacetylene cations have been performed (Forney et al. 1995a). A list of the cyanopolyacetylene cations studied so far in neon matrices is given in table 1. The cyanopolyacetylene cations form a homologue series of molecules that show similar regularities to the polyacetylene cations. In a homologue series of cyanopolyacetylene cations their electronic absorption spectra have a similar appearance. The ratio of the intensities of the vibronic bands to the intensity of the origin band decreases with increasing size of these molecules. As the size of cations increases by a single C_2 fragment the wavelength of the origin band of the cations shifts to the red by several tenths of cm^{-1}.

1.6 ELECTRONIC ABSORPTION SPECTRA OF CARBON ANIONS AND OF NEUTRAL CARBON MOLECULES

The method of matrix isolation of mass - selected ions also allows spectroscopic studies of molecular anions to be performed. In order to carry out such experiments we have to replace in our apparatus the cation source by the source that can produce negative ions and we have to reverse the polarity in the ion optics. The anions are produced from diacetylene in a discharge ignited by electrons emitted from a hot cathode. Many ion - molecule reactions take place in the source so that, a rich mixture of negative ions forms as a result of capture of slow electrons by the neutral molecules. The mass spectrum of anions produced from diacetylene is shown in fig. 12.

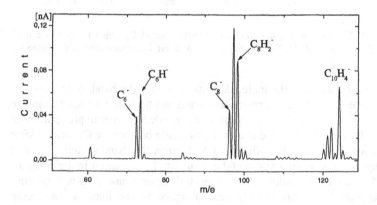

Figure 12. Mass spectrum of molecular anions generated from diacetylene in the anion source.

Several closely lying mass peaks of anions that differ in the number of hydrogen atoms are apparent in the spectrum shown in fig. 12. The depositions of the anions have been carried out at much lower mass resolution than the mass spectrum (fig. 12) was recorded. When for example the anions of mass of about 72 were deposited into a neon matrix the C_6^-

C_6H^- and $C_6H_2^-$ anions, have not been separated. The electronic absorption spectrum recorded after deposition of a mixture of these anions is shown in fig. 13. In order to establish whether the molecule responsible for the band system seen in fig. 13 contains hydrogen atoms or not the deposition of bare carbon anions of a mass 72 has been carried out (Forney et al. 1995b). The bare carbon anions were produced from graphite by bombarding the graphite surface with Cs^+ ions. The electronic absorption spectrum recorded in such conditions was identical with the spectrum shown in fig. 13.

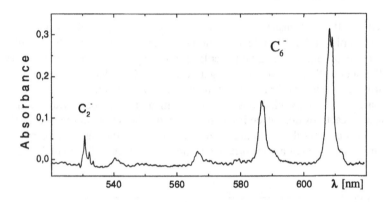

Figure 13. Electronic absorption spectrum of mass selected C_6^- anion isolated in a neon matrix (Forney et al. 1995). C_2^- is present in a matrix due to fragmentation of deposited C_6^- anions.

We can conclude that the molecule responsible for the band system seen in fig. 13 does not contain hydrogen atoms and that the spectrum must originate from C_6^- or C_6. An additional experiment is needed in order to prove whether the bands originate from the neutral C_6 molecule or from the C_6^- anion. After irradiation of the matrix with near UV photons this band system vanished entirely and a new band system (shown in fig. 14) appeared in the range of 450 -510 nm. In the middle infrared spectral region a new band at 1964 cm^{-1} also appeared that has already been assigned to the linear C_6 molecule (Kranze & Graham 1993). The carbon anions, during irradiation with the UV photons, lose electrons and form neutral carbon molecules. Thus, the band system shown in fig. 13 originates from C_6^-, while the C_6 neutral molecule is the carrier of the bands in the 450 -510 nm spectral range.

Systematic spectroscopic studies of the C_{2n}^- carbon anions (n = 2 - 10) isolated in neon matrices have been carried out (Freivogel et al. 1995). The electronic absorption spectra of the C_{2n}^- carbon anions have been measured. The wavelengths of the origin band of these anions are collected in table 1.

Many features common to all studied carbon anions are apparent in these spectra. The origin band is the strongest band in the spectrum of each of the C_{2n}^- anions. The relative intensities of the vibronic bands with respect to the intensity of the origin band decreases with increasing size of the carbon anions. The wavelength of the origin band shifts monotonically to the red with increasing number of carbon atoms in an anion. As the size of anions increases the width of their absorption peaks decreases. This is due to the fact that in the case of large anions an unpaired electron is delocalized over a larger distance and heavier anions interact more weakly with a neon matrix than lighter ones.

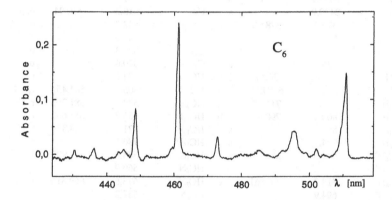

Figure 14. Electronic absorption spectrum of the neutral C_6 molecule isolated in a neon matrix (Forney et al. 1995). C_6 was produced from C_6^- by irradiation of the matrix with UV photons.

In addition to the carbon anions, the electronic transitions of the neutral carbon molecules (C_{2n} n = 2 - 5) have also been determined in this type of experiment (Freivogel et al.1995). The wavelengths of the origin band of the carbon molecules are collected in table 1. The electronic absorption spectra of neutral carbon molecules reveal many features common to the other carbon chains. The vibronic band of frequency of about 2000 cm^{-1} is also apparent in the spectra of neutral carbon molecules. The origin band of carbon molecules shifts monotonically to the red as the size of the molecule increases (see table 1).

Table 1. The wavelengths of the origin bands of carbon chain molecules isolated in neon matrices and in the gas phase.

molecule	0_0^0/Ne [nm]	0_0^0/gas [nm]	ref.	molecule	0_0^0/Ne [nm]	0_0^0/gas [nm]	ref.
C_4^-	456.7	457.09	a, b	C_3^-	403.1	404.04	m
C_6^-	607.6	606.87	a, c	C_5^-	495.1		n
C_8^-	773.2	768.42	a, c	C_7^-	626.8	627.02	n, c
C_{10}^-	967.3		a		492.3	492.8	n, c
C_{12}^-	1249.1		a	C_9^-	764.4	765.21	n, c
C_{14}^-	1460		a		607.3	607.45	n, c
C_{16}^-	1729		a	C_{11}^-	905.6		n
C_{18}^-	2069.1		a		719.1		n
C_{20}^-	2439.9		a	C_5	510.2	510.94	o, p
C_4	379.9	378.90	d, e	C_7	542.3		o
C_6	511.3		a	C_9	295.0		o
C_8	639.8		a	C_{11}	336.3		o
C_{10}	735.5		a	C_{13}	379.6		o
C_6H	530.4	526.576	a, f	HC_5H	434.2		r,
C_8H	631.0	625.866	a, g	HC_7H	505.8	504.453	r, s
$C_{10}H$	721.9	714.09	a, g	HC_9H	582.8	581.755	r, s
$C_{12}H$	800.5	790.45	a, h	$HC_{11}H$	655.7	653.632	r, s
$C_{14}H$	865.5		a	$HC_{13}H$	721.9	718.379	r, s
$C_{16}H$	924.1		a	$HC_{15}H$	780.5		r
HC_6H^+	604.6	600.214	i, j	HC_5N^+	583.8	581.927	l, h
HC_7H^+	599.5		i	HC_6N^+	569.9		l,
HC_8H^+	713.2	706.861	i, k	HC_7N^+	673.6	669.815	l, u
HC_9H^+	694.9		i	HC_8N^+	657.2		l
$HC_{10}H^+$	823.3		i	HC_9N^+	770.5		l
$HC_{11}H^+$	788.8		i	$HC_{10}N^+$	746.6		l
$HC_{12}H^+$	934.1		i	$HC_{11}N^+$	872.0		l
$HC_{13}H^+$	873.1		i	$HC_{12}N^+$	831.5		l
$HC_{14}H^+$	1047.1		i	$HC_{13}N^+$	973.1		l
$HC_{15}H^+$	959.2		i	NC_4N^+	598.1	595.774	l, h
$HC_{16}H^+$	1159.8		i	NC_6N^+	659.1	655.752	l, h
NC_7N^+	629.1		l	NC_8N^+	741.3		l
NC_9N^+	713.1		l	$NC_{10}N^+$	831.3		l
$NC_{11}N^+$	794.5		l	$NC_{12}N^+$	923.3		l

a) Freivogel et al.. 1995; b) Zhao, de Beer, Neumark 1996; c) Tulej et al. 1998; d) Freivogel et al. 1996; e) Linnartz et al. 2000; f) Kotterrer, Maier 1996; g) Linnartz, Motylewski, Maier 1998; h) Motylewski et al. 2000; i) Freivogel et al. 1994a, j) Sinclair et al. 1999; k) Pfluger et al. 2000; l) Forney et al. 1995a; m) Tulej et al. 2000; n) Forney et al. 1997; o) Forney et al. 1996; p) Motylewski et al. 1999; r) Fulara et al. 1995, s) Ball et al. 2000, u) Sinclair et al. 2000.

1.7   RELEVANCE OF CARBON CHAIN RADICALS TO THE ORIGIN OF DIBS

The hypothesis postulating that linear carbon molecules C_n (5<n<15) may be carriers of DIBs was formulated by Douglas in 1977. This hypothesis has been extended by Fulara et al. (1993) and Freivogel, Fulara & Maier (1994) who suggest that highly unsaturated hydrocarbons may be good candidates for the DIB carriers. The carbon chain molecules and their hydride analogues have strong absorption bands in the red or the near infrared spectral regions. The wavelengths of these bands, measured in neon matrices, are close to positions of DIBs. These systems are built of carbon or carbon and hydrogen atoms, and fulfil the criterion of the abundance of elements in the universe. The simplest carbon molecule, C_2, occurs quite commonly in diffuse interstellar clouds. The larger C_3 and C_5 molecules have also been discovered, through their high-resolution infrared absorption spectra, in the nebulae surrounding carbon-rich stars (Hinkle et al. 1988, Bernath et al. 1989). The monohydride carbon chains C_nH (n=2-8) have been identified in dark interstellar clouds by using their emission spectra in the microwave region (Travers et al. 1996, Cernicharo & Guélin 1996, Guélin et al. 1997).

The C_{2n}^{-} carbon anions that have strong electronic transitions in the visible or the near infrared region should also be taken into account when considering possible DIB carriers. Carbon anions with an even number of atoms have the highest electron affinities of all the affinity values known at present. The electron affinity of, e.g., C_8 is 4.38 eV (Zhao et al. 1996) and is close to the ionization potential of potassium (4.339 eV). The carbon anions, in contrast to alkali metals, should be more resistant to the strong UV radiation that penetrates interstellar clouds. This is due to the possibility of redistribution of the excitation energy of the anion between many available degrees of freedom.

Apart from anions and the neutral C_{2n} and $C_{2n}H$ chains also positive ions with skeleton built of carbon atoms should be considered as DIB carriers. In the ISM, cations can form by photoionization of neutral molecules. Special attention should be paid to cyanopolyacetylene cations, HC_nCN^+, whose precursors are the HC_nCN (n=0-10) molecules, the largest molecules identified in the ISM thus far (Bell et al. 1997). Symmetric dicyanopolyacetylenes, probably also abundant in the ISM, cannot be observed through their microwave spectra since they do not possess a dipole moment. The cyano- and dicyanopolyacetylene cations have strong absorption bands situated in the red or the near infrared spectral regions, where DIBs are particularly numerous. The absorption of the $HC_{2n}H^+$,

$HC_{2n+1}H^+$ polyacetylene cations isolated in neon matrices also falls in this spectral region.

Which features of DIBs can we predict from the knowledge of the electronic absorption spectra of linear carbon chain species? A typical electronic spectrum of a linear carbon chain molecule is composed of the strong origin band and of several weak vibronic bands that are located at shorter wavelengths with respect to the 0_0^0 band. In the electronic spectra of such species a characteristic vibronic band with frequency in the 1800 - 2200 cm^{-1} energy region is well discerned and it corresponds to the stretching of the $C{\equiv}C$ and $C{\equiv}N$ bonds. If a carbon chain molecule were responsible for the existence of a DIB, the strong diffuse band would be in correlation with a weak interstellar structure that would be shifted by ca. 1800 - 2200 cm^{-1} to the violet with respect to the origin band. Other weak interstellar structures that would lie at shorter wavelengths than the origin band could also accompany a strong DIB. The intensities of these structures could be as much as an order of magnitude lower with respect to the intensity of the 0_0^0 band. This imposes very sever requirements on the quality of astrophysical spectra acquired in order to detect such weak interstellar features because the central depths of most DIBs do not exceed 10% of the continuum level.

Another characteristic feature of carbon chain species is that small molecules absorb in the region shifted more towards short wavelengths than is the case with large molecules. The position of the 0_0^0 band depends almost linearly on the length of the carbon chain, as has been shown in Fig. 15 for polyacetylene cations, the carbon anions C_{2n}^- and the radicals $HC_{2n+1}H$, $C_{2n}H$. The absorption spectra of two successive molecules that belong to the same homologous series are displaced with respect to one another by 60 to 150 nm. If such chain species were responsible for DIBs, we should expect that 3 to 8 DIBs related to this homologous series of species would fall within the 600-1000 nm region. These DIBs would be accompanied by weaker interstellar structures situated on the short wavelength side of the stronger bands. This is very well illustrated in fig. 16 and 17 for the electronic absorption spectra of C_{2n}^- anions and the spectra of $HC_{2n+1}H$ molecules that have been compared with the synthetic spectrum of DIBs. In the case of carbon anions only four anions have their electronic transitions in the 400 - 1000 nm spectral range, whereas in order to cover the DIB spectrum with the spectra of the unsaturated hydrocarbons ($HC_{2n+1}H$) eight molecules are needed.

The carbon chain species possess an additional very important feature that the oscillator strength of their electronic transitions increases considerably with the size of the molecule. This feature may allow detection of the longer chains in the ISM.

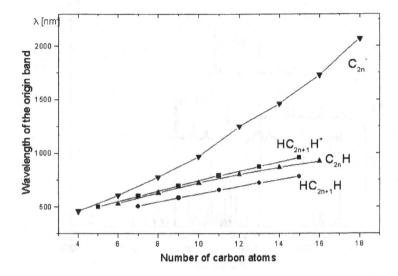

Figure 15. Dependence of the wavelength of the origin bands of carbon chain species on the number of carbon atoms that constitute the skeleton of a molecule.

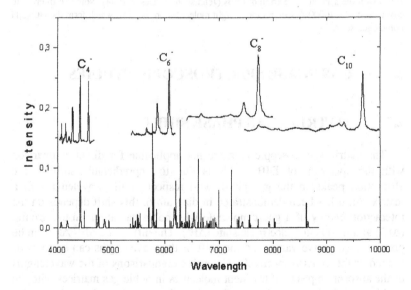

Figure 16. Comparison of the electronic spectra of the C_{2n}^- carbon anions isolated in a neon matrix (Freivogel et al. 1995) with the synthetic spectrum of DIBs (Jenniskens & Désert 1994). Note that in order to cover the 4000 - 10000 Å spectral range only four C_{2n}^- carbon anions are needed.

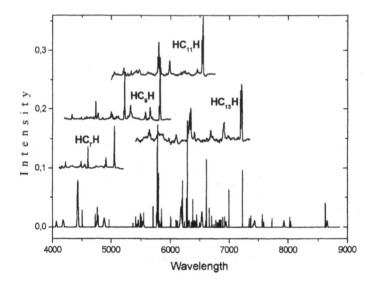

Figure 17. Comparison of the electronic spectra of the $HC_{2n+1}H$ polyacetylenes (Fulara et al. 1995) with the synthetic spectrum of DIBs (Jenniskens & Désert 1994). Note that in order to cover the 4000 -10000 Å spectral range eight molecules are needed, which form the $HC_{2n+1}H$ homologue series.

2.   GAS PHASE SPECTROSCOPIC STUDIES

2.1   MATRIX - GAS PHASE SHIFT

The matrix spectroscopic data are not applicable for direct comparison with the spectrum of DIBs. This is due to unpredictable shift of the absorption peaks in the gas phase with respect to the wavelengths in a matrix. As it has been demonstrated in diagram 6, this shift depends on the interaction energy of a guest molecule with noble gas atoms in the ground (ΔE") and the excited electronic state (ΔE'). The difference ΔE'- ΔE" can be positive or negative value, and the shift of the wavelengths can be towards the red or the violet respectively. Extensive comparisons of the wavelengths of the absorption peaks of transient molecules in noble gas matrices with the positions of the bands in the gas phase have been made by Jacox (1994) for the middle infrared spectral region. The shift of the wavelengths in a matrix shows a scatter around the gas - phase positions, which resembles a Gaussian curve. The positions of most known transient molecules and molecular ions in a matrix lie within ±1% of their gas - phase wavelengths. For a typical vibrational energy of about 2000 cm^{-1} the expected shift is about ± 20 cm^{-1}.

If we scale this regularity to the electronic transition of energy near 20000 cm^{-1} we can expect that the positions of the electronic absorption bands of most molecules in the gas phase should lie ± 200 cm^{-1} around the matrix wavelengths. Nevertheless the matrix spectroscopic data are invaluable for searching of the electronic bands of a specific molecule in the gas phase. The gas - phase spectroscopic measurements of transient molecules are a very difficult task due to very low concentration of such species attainable in the gas phase experiments. The reason for this is the very high reactivity of transient molecules. In order to study such molecular systems in the gas phase very sensitive detection methods are needed.

2.2   EMISSION METHOD

One of the most sensitive methods is emission. The first large transient molecule, which has been detected in emission, was the diacetylene cation that was observed about 50 years ago in the electrical discharge ignited in organic vapours (Callomon 1956). In following years the emission method has successfully been applied to other molecular cations. The cations were produced in the electron impact experiments. Collisions of electrons with energy of about 200 eV with neutral molecules produce cations in excited electronic states. The excited cations decay radiatively and their emission can be measured. In the mid - eighties the electronic spectra of about one hundred organic cations in the gas phase were known (Maier 1982, Bondybey & Miller 1983), but all these cations are of very similar structure. These are substituted acetylene, diacetylene or benzene cations. Although the emission is one of the most sensitive methods, it has a very serious limitation, because it is applicable only to molecules, which decay radiatively. Most transient molecules do not emit a detectable amount of photons.

2.3   MULTIPHOTON IONIZATION METHOD

Another very sensitive method, which can be used for the spectroscopic studies of transient molecules in the gas phase, is the resonance enhanced multiphoton ionisation (REMPI) (Hollas 1998). This method was developed for the spectroscopic studies of ordinary molecules. The principle of operation of this method is illustrated in fig. 18. A molecule absorbs a photon of energy $h\nu_1$ and an excited state is formed. The molecule then absorbs a second photon of energy $h\nu_2$, which leads to its ionisation. By measuring the current of electrons or of the positive ions as a function of the wavelength of the first photon (hv_1) the electronic absorption spectrum of a molecule is obtained.

The multiphoton ionisation method is very well suited for the gas - phase spectroscopic studies of molecular anions. The diagram in fig. 18 explains how this method works in the case of carbon anions. In the pulsed ion source a mixture of different carbon anions is produced. The anions are extracted form the source and are injected into the tube where the electric potential is applied. The lighter anions move faster than the heavier ones. After appropriate time delay the anions interact with the laser beam. Absorption of the first photon causes excitation of the anion, and the second photon photodetaches an electron from the anion. By measuring the electron current (or the beam intensity of the neutral molecules that are produced from the anions) as a function of the wavelength of the first photon the electronic absorption spectrum of the anions is recorded.

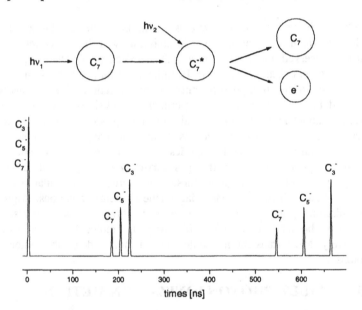

Figure 18. The principle of operation of the two photon photodetachement method in the case of carbon anions (upper diagram) and the principle of mass separation in the time of flight mass spectrometer (bottom trace).

Several scientific groups have recently undertaken the gas phase spectroscopic studies of carbon anions (Ohara et al. 1995, Zhao, de Beer & Neumark 1996, Tulej et al. 1998a). At present, the carbon anions are one of the best-known group of transient molecules in the gas phase. The wavelengths of the origin bands of the carbon anions in the gas phase are collected in table 1. In the case of C_6^- the origin band lies very close to the 6065 Å DIB. An amazingly good fit of the wavelengths and the intensities of four absorption bands of C_7^- with the spectrum of DIBs has recently been reported (Kirkwood et al. 1998, Tulej et al. 1998b). This is the first case, when all the absorption peaks of a molecule fit the wavelengths of DIBs exactly. Thus four DIBs (λ = 6270, 6065, 5748, 5610 Å) may already have been identified. However the situation does not look as good as presented above. If these four DIBs originated from C_7^-, the relative intensities of these DIBs would remain constant irrespective of the direction of observation. Galazutdinov and co-workers (1999) have shown that the relative intensities of these DIBs can vary from one line of sight to another. Moreover C_7^- has a second electronic transition in the short-wavelength range (400- 500 nm) of the spectrum (Forney et al. 1997). The intensity of the peaks that belong to this transition is similar to the intensity of the origin band of the first electronic transition. The positions of the peaks of the second electronic transition of C_7^- do not match the wavelengths of DIBs (Galazutdinov et al. 1999). This is a very strong argument against C_7^- as the carrier of four DIBs discussed - above.

The multiphoton ionization method is probably the best method for spectroscopic studies of negative ions in the gas phase. However, it is difficult to apply this method to neutral transient molecules and cations. In the case of cations the second ionization potential is much higher than the first one and it is not possible to ionize the cations in a one - photon process with the available UV lasers. However, instead of observing doubly charged ions, it is possible to monitor the fragment ions, but the fragmentation also needs high-energy photons. Moreover the fragment ions can differ very little in a mass relative to the parent ions. Such situation takes place e.g. in the case of hydrocarbon cations.

In order to overcome such difficulties a very clever method has recently been proposed by Bréchignac and Pino (1999). They demonstrated this method for the phenanthrene cation. The diagram in fig. 19 explains the idea of their method. In a supersonic expansion van der Waals aggregates of neutral phenanthrene with argon atoms are formed. Subsequently, the phenanthrene complex that contains one argon atom is ionised by an absorption of two photons of energy $h\nu_1$ and $h\nu_2$. The vibrationally cold complex of the phenanthrene cation with an argon atom is formed. This complex absorbs a third photon ($h\nu_3$) and dissociates into the phenanthrene cation and an argon atom. In order to distinguish between the phenanthrene cation and the complex cation a time of flight mass spectrometer has been used. By measuring the current of the phenanthrene cations (that are produced during laser fragmentation of the complex cations) as a function of the wavelength of third photon ($h\nu_3$), the absorption spectrum of the complex cation was recorded.

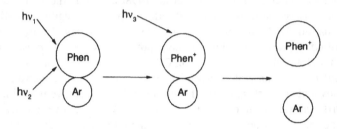

Figure 19. The principle of operation of the method developed by Pino and Bréchingnac (1999) for spectroscopic examination of PAH cations in the gas phase.

When a similar experiment was performed with the phenanthrene complex that contains two argon atoms the shift of the bands due to one argon atom was evaluated. By knowing this shift the positions of the bands of the complex cation can be extrapolated to the wavelengths of the free phenanthrene cation. The spectrum of the phenanthrene cation in the gas phase is shifted by ca. 80 cm^{-1} towards the violet with respect to the matrix one. This spectrum has been compared with the spectrum of DIBs. The positions of the phenanthrene cation bands in the gas phase do not match the wavelengths of any DIBs.

The method proposed by Bréchignac and Pino (1999) is quite general and can be applied to the spectroscopic studies of any cations in the gas phase.

2.4   CAVITY RING DOWN METHOD

Another very sensitive method, which can be used for the spectroscopic studies of any kind of gas - phase molecules is 'cavity ring down' (CRD). This method was proposed twelve years ago by O'Keefe and Deacon (1988) for the studies of very weak absorptions. The idea of this method is illustrated in the next scheme (fig 20a). The main part of the CRD apparatus is an optical resonator that is built of two highly reflecting mirrors. Light from a pulsed dye laser is injected into the resonator. The light oscillates back and forth in the resonator and a small portion of the light escapes the resonator through the mirrors. The intensity of the light in the resonator drops exponentially with time (see fig. 20b). The decay time depends on the length of the resonator and on the reflectivity of the mirrors. For mirrors of reflectivity of 99.99% and for cavity length of 1 m the decay time is of the order of several tens of microseconds. During this period the light travels a distance of several kilometres in the cavity.

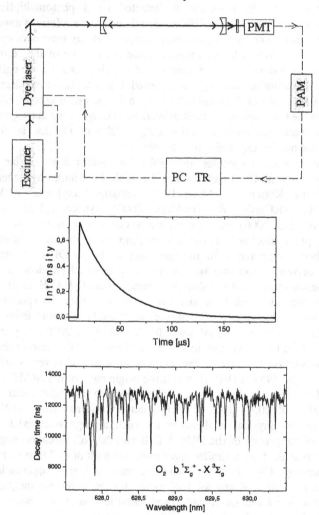

Figure 20. Schematic outline of the cavity ring down (CRD) apparatus (upper diagram). PMT - photomultiplier, PAM- preamplifier, TR - transient recorder. The middle figure shows the exponential decay of the light intensity within the cavity. The bottom figure shows the electronic spectrum of atmospheric oxygen around 630 nm recorded with the CRD apparatus built in Warsaw.

When a gas sample is placed into the resonator the decay time diminishes due to additional light losses caused by the absorption by the sample. The light that escapes the resonator is detected by a photomultiplier. A preamplifier is then used to amplify the signal and the waveform is averaged for several laser shots by a digital oscilloscope. A decay time is calculated for each laser wavelength. By measuring the decay time of the resonator filled with the gas sample as a function of the dye laser wavelength the absorption spectrum of the sample is recorded. Fig. 20c shows the electronic absorption spectrum of the molecular oxygen in the range around 628 nm recorded in the CRD apparatus built in Warsaw. The spectrum corresponds to a very weak electronic transition ($^1\Sigma_g^+$ - $^3\Sigma_g^-$) of O_2 that is doubly forbidden for multiplicity and symmetry reasons.

Several transient molecules and molecular cations have so far been studied by using the CRD method. The list (table 1) includes hydrogenated carbon chains (Kotterrer & Maier 1997; Linnartz, Motylewski & Maier 1998; Ball, McCarthy & Thaddeus 2000) cyanoacetylene cations (Motylewski et al. 2000) and the naphthalene cation (Romanini et al. 1999). The gas - phase electronic absorption spectrum of naphthalene cation has recently been compared with the spectrum of broad DIBs. A striking similarity between these two spectra has been reported (Krełowski et al. 2000). The wavelengths of the absorption bands of the C_6H C_8H, $C_{10}H$, $C_{12}H$ radicals in the gas phase have also been compared with the spectrum of DIBs. However the interstellar structures that would originate from these radicals have not been detected yet (Motylewski et al. 2000). A puzzling case is provided by the electronic spectrum of the NC_4N^+ cation in the gas phase. The rotational contour of the origin band of NC_4N^+ is very similar to the profile of the 5959 Å DIB. However the origin band of the NC_4N^+ cation is shifted to the red by 0.9 Å with respect to the interstellar band (Motylewski et al. 2000). The 5959 Å DIB can not originate from the NC_4N^+ cation since no physical explanation for such a large shift exists. The similarity of the profile of the 5959 Å DIB and the NC_4N^+ cation suggests that this DIB arises from a similar linear molecule built of 6 -7 heavy atoms.

Summarising, CRD is a very sensitive method that is applicable for spectroscopic studies of any kind of molecules. Nevertheless the lack of mass selections is a serious limitation and identification of absorption peaks measured with this method is based on the matrix data.

2.5   SEARCH FOR DIBS THAT WOULD PRESERVE THE CONSTANT INTENSITY RATIOS

A number of transient molecules for which the electronic spectra were measured in the gas phase continuously grows. In order to test a hypothesis that a specific molecule can be the carrier of certain DIBs good quality spectra of reddened stars are needed. If a molecule were to be responsible for the occurrence of several DIBs the wavelengths of all bands of this molecule would match the positions of DIBs. Apart from the positions, a very important feature of DIBs are their profiles, which should be measured in high - resolution conditions with as high as possible signal to noise ratio. The profiles of DIBs should be similar to the rotational contours of the bands of a molecule considered for the carrier of DIBs. The gas - phase electronic spectrum of a molecule recorded in the laboratory can differ slightly from the spectrum of DIBs assigned to this molecule. Some differences in the appearance of the rotational contour of the bands of a molecule and the profiles of respective DIBs can result from different physical and chemical conditions that prevail in these environments. The next figure (21) shows the rotational contours of carbon chain radicals to visualise how the spectra change with the size of the molecule and with gas temperature. The profiles of two model molecules of the C_8, and C_{12} size are considered at two interstellar temperatures characteristic of polar and non-polar molecules (Fulara & Krełowski 2000). The rotational contour of a linear molecule composed of 12 heavy atoms is similar to the profiles of narrow DIBs. *A question arises as to which features of DIBs should be known (apart from their positions and their profiles) in order to make a reliable assignment of a given interstellar band to a specific molecule?*

The electronic spectrum of a molecule consists usually of the origin band and of several weaker vibronic bands. The relative intensities of the bands in the electronic spectrum of a molecule always remain constant. This is a very important point that can prove the hypothesis of whether a given molecule is the carrier of some DIBs or not. For this purpose, the relative intensities of DIBs in the spectra of a great number of reddened stars should be known with as high as possible precision. This is a very difficult task because the central depths of most DIBs do not exceed 10% of the continuum level.

A systematic search for the DIBs that would preserve the constant intensity ratios, irrespective of the direction of observation, has been undertaken by Weselak and co-workers (2000). We have found that the relative intensities of six DIBs are mutually very well correlated in the spectra of 41 reddened stars, the light of which penetrates the 'σ' or the 'ζ' type clouds. These are 5809, 6196, 6397, 6520, 6614 and 6660 Å DIBs. A very good correlation of the central depths of the 6614 and 6660 Å DIBs as well as of the 6614 and 6196 Å DIBs has been obtained. In the latter case DIBs have similar profiles with an asymmetric shape. However the half width of the 6196 Å DIB is a factor of two smaller than the half width of the 6614 Å DIB.

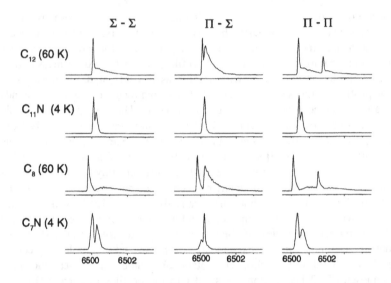

Figure 21. Rotationally - unresolved profiles of electronic absorption bands of model molecules at two interstellar temperatures characteristic of polar ('C$_7$N', 'C$_{11}$N') and non polar molecules ('C$_8$', 'C$_{12}$').

A question arises of whether DIBs that differ considerably in their half widths can originate from the same carrier? In order to answer this question simulations of the rotational contours of a molecule has been performed. The shape of the rotationally unresolved bands depends on several factors such as temperature, the rotational constant B in the ground and in the excited electronic state etc. For a given temperature the shape of the rotational contours depends on the difference in the value of B in the ground and the excited states. The simulations have been carried out for the Σ -Σ electronic transition of a linear gas - phase molecule built of 12 heavy atoms were for the temperature of 60 K. Three cases for which the rotational constant B' is equal to B" or is smaller than B" by 0.5 and 1 % respectively have been considered. The integral intensity of the bands was the same for all cases. The results of the simulations are depicted in fig. 22. A small change in geometry induces a large change of the width as well as in the intensities of the substructures that form the profile of the vibronic band. Results of these simulations show that bands, which differ by a factor of two in their half widths can originate from the same molecule.

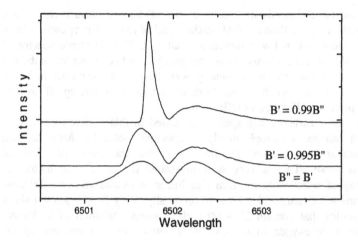

Figure 22. Simulated rotationally - unresolved contours of the electronic absorption bands of a gas - phase model molecule of C_{12} size at temperature of 60 K. Note that the bands, which differ by a factor of two in their half - widths may still originate from the same molecule.

3. CONCLUSIONS

1) The carbon chain radicals have strong electronic transitions in the visible or in the near infrared spectral region where DIBs are particularly numerous.

2) Electronic absorption spectra of the matrix - isolated molecules are not useful for direct comparison with the DIB spectrum. This is due to unpredictable value of the shift of absorption peaks in the matrix with respect to the gas - phase positions.

3) The profiles of absorption peaks of matrix-isolated molecules should not be compared with the profiles of DIBs. The profiles of absorption peaks in a matrix are not related to the size and to the structure of a molecule. They depend on the strength of interaction of the guest molecule with the host environment. They also depend on a quality of matrix, i.e. imperfections in the matrix crystal structure.

4) Electronic absorption spectra of the matrix - isolated species **provide accurate relative intensities of the peaks** that form the spectra. This is because the spectrum in the matrix experiment is measured in exactly the same conditions, in contrary to the gas phase measurement of the intensities of bands that may not be accurate. Spectroscopic studies of matrix-isolated species provide invaluable data that are the starting point for the gas - phase measurements.

5) At present the electronic spectra of several tens of linear carbon chain radicals, of more than ten PAH cations and of two fullerene cations have been measured in low temperature matrices. The electronic spectra of several transient molecules from this group have been recorded in the gas phase. In a few cases the wavelengths of the gas - phase electronic bands of such species match the positions of certain DIBs, but up till now no definitive identification of DIBs exists.

6) The linear carbon chain species discussed in this review form a very small sample of astrophysically - important molecules. Most of such molecular systems remain unknown from the point of view of spectroscopy. This is very well illustrated for the carbon molecules among of which only medium size linear molecules and two fullerene cations were characterized spectroscopically. Very little is known about molecules that contain 20 - 60 carbon atoms. Addition of hydrogen, nitrogen or oxygen to the carbon skeleton increases enormously the number of possible transient molecules that can be relevant to the origin of DIBs. All such molecular systems still await spectroscopic examination.

Acknowledgements

This work was supported by the Committee for Scietific Research grant No, 2 P03D 011 15.

REFERENCES

Ball, C.D., McCarthy M.C. and Thaddeus, P. (2000), *J. Chem. Phys.*, **112**, 10149.
Bell, N.B., Feldman, P.A., Travers, M.J., McCarthy, M.C., Gottlieb, C.A. and Thaddeus, P. (1997), *ApJ*, **483**, L61.
Bell, N.B., Watson, J.K.G., Feldman, P.A. and Travers, M.J. (1998), *ApJ*, **508**, 286.
Bernath, P.F., Hinkle, K.W. and Keady, J.J. (1989), *Science*, **244**, 562.
Birks, J. (1970), *Photophysics of aromatic molecules*, Wiley, New York.
Bondybey, V.E. and Miller, T.A. (1983), in Miller, T.A. and Bondybey, V.E. (eds.), *Molecular Ions: Spectroscopy, Structure and Chemistry*, North - Holland Publishing Company, p.125.
Bréchignac, P. and Pino, T. (1999), *A&A*, **343**, L49.
Callomon, J.H. (1956), *Can. J. Phys.*, **34**, 1046.
Cernicharo, J. and Guélin, M. (1996), *A&A*, **309**, L27.
Douglas, A.E. (1977), *Nature*, **269**, 130.
Forney, D., Freivogel, P., Fulara, J. and Maier, J.P. (1995a), *J. Chem. Phys.*, **102**, 1510.
Forney, D., Freivogel, P., Grutter, M. and Maier, J.P. (1996), *J. Chem. Phys.*, **104**, 4954.
Forney, D., Fulara, J., Freivogel, P., Jakobi, M., Lessen, D. and Maier, J.P. (1995b), *J. Chem. Phys.*, **103**, 48.

Forney, D., Grutter, M., Freivogel, P. and Maier, J.P. (1997), *J. Phys. Chem. A*, **101**, 5294.

Freivogel, P., Fulara, J., Jakobi, M., Forney, D. and Maier, J.P. (1995), *J. Chem. Phys.*, **103**, 54.

Freivogel, P., Fulara, J., Lessen, D., Forney, D. and Maier, J.P. (1994a), *Chem. Phys.*, **189**, 335.

Freivogel, P., Fulara, J., Maier, J.P. (1994b), *ApJ*, **431**, L151.

Freivogel, P., Grutter, M., Forney, D. and Maier, J.P. (1996), *Chem. Phys. Lett.*, **249**, 191.

Fulara, J., Freivogel, P., Forney, D., Maier, J.P. (1995), *J. Chem. Phys.*, **103**, 8805.

Fulara, J., Jakobi, M. and Maier, J.P. (1993), *Chem. Phys. Lett.*, **206**, 203.

Fulara, J. and Krełowski, J. (2000), *New Astron. Rev.*, **44**, 581.

Fulara, J., Lessen, D., Freivogel, P. and Maier, J.P. (1993), *Nature*, **366**, 439.

Galazutdinov, G.A., Krełowski, J. and Musaev, F.A. (1999), *MNRAS*, **310**, 1017.

Galazutdinov, G.A., Musaev, F.A., Krełowski, J. and Walker, G.A.H. (2000), *PASP*, **112**, 648.

Gasyna, Z., Schatz, P.N., Hare, J.P., Dennis, T.J., Kroto, H.W., Taylor, R. and Walton, D.R.M. (1991), *Chem. Phys. Lett.*, **183**, 283.

Guélin, M., Cernicharo, J., Kahane, C., Gomez-Gonzales, J. and Walmsley, C.M. (1987), *A&A*, **175**, L5.

Guélin, M., Cernicharo, J., Travers, M.J., McCarthy, M.C., Gottlieb, C.A, Thaddeus, P., Ohishi, M., Saito, S. and Yamamoto, S. (1997), *A&A*, **317**, L1.

Heger, M.L. (1922), *Lick Observatory Bull. N. 10*, 146.

Herbig, G.H. (1975), *ApJ*, **196**, 129.

Hinkle, K.W., Keady, J.J. and Bernath, P.F. (1988), *Science*, **241**, 1319.

Hollas, J.M. (1998), *High Resolution Spectroscopy*, Wiley.

Jacox, M.E. (1994), *Vibrational and Electronic Energy Levels of Polyatomic Transient Molecules*, Phys. Chem. Ref. Data.

Jenniskens, P., Désert, F.-X. (1994), *A&A Supp. Ser.*, **106**, 39.

Kerr, T.H., Hibbins, R.E., Fossey, S.J., Miles, J.R. and Sarre, P.J. (1998), *ApJ*, **495**, 941.

Kirkwood, D.A., Linnartz, H., Grutter, M., Dopfer, O., Motylewski, T., Pachkov, M., Tulej, M., Wyss, M. and Maier, J.P. (1998), *Faraday Discussions*, **109**, 109.

Kotterer, M. and Maier, J.P. (1997), *Chem. Phys. Lett.*, **266**, 342.

Kranze, R.H. and Graham, W.R.M. (1993), *J. Chem. Phys.*, **98**, 71.

Krełowski, J., Galazutdinov, G.A., Musaev, F.A. and Nirski, J. (2001), *MNRAS*, **328**, 810.

Linnartz, H., Motylewski, T. and Maier, J.P. (1998), *J. Chem. Phys.*, **109**, 3819.

Linnartz, H., Vaizert, O., Motylewski, T. and Maier, J.P. (2000), *J. Chem. Phys.*, **112**, 9777.

Maier, J.P. (1982), *Acc. Chem. Research*, **15**, 18.

McCarthy, M.C., Travers, M.J., Kovács, A., Gottlieb, C.A. and Thaddeus, P. (1996), *A&A*, **309**, L31.

Merrill, P.W. (1936), *PASP*, **48**, 179.

Motylewski, T., Linnartz, H., Vaizert, O., Maier, J.P., Galazutdinov, G.A., Musaev, F.A., Krełowski, J., Walker, G.A.H. and Bohlender, D.A. (2000), *ApJ*, **531**, 312.

Motylewski, T., Vaizert, O., Giesen, T.F., Linnartz, H. and Maier, J.P. (1999), *J. Chem. Phys.*, **111**, 6161.

Ohara, M., Shiromaru, H., Achiba, Y., Aoki, K., Hashimoto, K. and Ikuta, S. (1995), *J. Chem. Phys.*, **103**, 10393.

O'Keefe, A. and Deacon, D.A.G. (1988), *Rev. Sci. Instrum.*, **59**, 2544.

Pfluger, D., Motylewski, T., Linnartz, H., Sinclair, W.E. and Maier, J.P. (2000), *Chem. Phys. Lett.*, **329**, 29.

Romanici, D., Biennier, L., Salama, F., Kachanov, A., Allamandola, L.J. and Stoeckel, F. (1999), *Chem. Phys. Lett.*, **303**, 165.

Sarre, P.J., Miles, J.R., Kerr, T.H., Hibbins, R.E., Fossey, S.J. and Somerville, W.B. (1995), *MNRAS*, **277**, L41.

Sarre, P.J., Miles, J.R. and Scarrott, S.M. (1995), *Science*, **269**, 674.

Schnaiter, M., Mutschke, H., Dorschner, J., Henning, Th. and Salama, F. (1998), *ApJ*, **498**, 486.

Sinclair, W.E., Pfluger, D., Linnartz, H. and Maier, J.P. (1999), *J. Chem. Phys.*, **110**, 296.

Sinclair, W.E., Pfluger, D., Verdes, D. and Maier, J.P. (2000), *J. Chem. Phys.*, **112**, 8899.

Suzuki, H., Ohishi, M., Kaifu, N., Ishikawa, S., Kasuga, T., Saito, S. and Kawagushi, K. (1986), *PASJ*, **38**, 911.

Travers, M.J., McCarthy, M.C., Gottlieb, C.A. and Thaddeus, P. (1996), *ApJ*, **465**, L77.

Tulej, M., Fulara, J., Sobolewski, A., Jungen, M. and Maier, J.P. (2000), *J. Chem. Phys.*, **112**, 3747.

Tulej, M., Kirkwood, D.A., Maccaferri, G., Dopfer, O. and Maier, J.P. (1998a), *Chem. Phys.*, **228**, 293.

Tulej, M., Kirkwood, D.A., Pachkov, M. and Maier, J.P. (1998b), *ApJ*, **506**, L69.

Warren-Smith, W.F., Scarrott, S.M. and Murdin, P. (1981), *Nature*, **292**, 317.

Weselak, T., Fulara, J., Schmidt, M. and Krełowski, J. (2001), *A&A*, **377**, 677.

Zhao, Y., de Beer, E. and Neumark, D.M. (1996), *J. Chem. Phys.*, **105**, 2575.

Zhao, Y., de Beer, E., Xu, C., Travis, T. and Neumark, D.M. (1996), *J. Chem. Phys.*, **105**, 4905.

CHEMICAL REACTIONS ON SOLID SURFACES OF ASTROPHYSICAL INTEREST

Ofer Biham
Racah Inst. of Physics, The Hebrew University, Jerusalem 91904, Israel

Valerio Pirronello
DMFCI, Università di Catania, 95125 Catania, Sicily, Italy

Gianfranco Vidali
Syracuse University, Syracuse, NY 13244-1130, USA

Keywords: ISM – dust – molecules – catalysis

Abstract   Observed abundances of chemical species in interstellar clouds can be explained in most cases by reaction schemes involving only species in the gas phase. There is however clear evidence that reactions occurring on the surface of dust grains, helping the formation of key molecules, play a fundamental role into shaping the universe as we see it today.

   In this chapter we focus our attention on surface reactions on solids and in conditions close to those encountered in interstellar clouds. We will describe how experimental techniques of surface science have been used to study the recombination reaction of hydrogen on interstellar dust grain analogues and the oxidation of carbon monoxide in the interaction of oxygen atoms in water ice layers. Using theoretical methods and computer simulations, we show that it is possible to relate experimental results obtained in the laboratory to actual physical and chemical processes occurring in the interstellar space.

Introduction

The presence in space of submicron-sized dust grains mixed with gas is of fundamental importance for the processes of structure formation

V. Pirronello et al. (eds.), Solid State Astrochemistry, 211–250.
© 2003 *Kluwer Academic Publishers. Printed in the Netherlands.*

in the universe and the emergence of chemical complexity. The shape and dynamics of galaxies, the rate of star formation inside them and the abundances of atoms and molecules in interstellar clouds, all depend on the presence of dust. This is in spite of the fact that interstellar dust represents only a small fraction (of about 1% by mass) of diffuse matter in space. Dust grains play an important role in catalyzing chemical reactions on their surfaces and in shielding UV radiation from the inner part of the clouds (Whittet 1992). The formation of molecular species observed in the Interstellar Medium (ISM), is the product of a rich chemistry taking place in the rather extreme conditions encountered in diffuse and dense clouds. Such a rich chemistry is due to reactions occurring both in the gas phase and on the surfaces of dust grains (that act as catalysts). In the dense regions where icy mantles can accrete on grains, reactions induced in the bulk of the ice layer by UV photons and fast particles may also take place.

The relevance of surface reactions in astrophysics is related to the fact that they enable the formation of species that cannot be produced easily in the gas phase, such as H_2 (Pirronello et al. 1997a; Pirronello et al. 1997b; Pirronello et al. 1999; Manicò et al. 2001) and CO_2 (Roser et al. 2001) and enhance the formation of species that are formed more slowly in the gas phase (see e.g. Watanabe and Kouchi 2002). The most important surface reaction is the formation of molecular hydrogen in situations in which H atoms are neutral and in the ground state. Hydrogen is the most abundant species in the Universe and for this reason it is the most frequent collisional partner of other molecules and radicals, hence once ionized it initiates and controls gas phase chemistry in space. Furthermore H_2 and its daughter molecules once excited by collisions de-excite radiatively, emitting photons in the infrared that are rarely absorbed inside the cloud. Thus, they tend to reduce the energetic budget of the cloud itself favoring its gravitational collapse, the formation of stars and the shaping of galaxies.

We will also describe a possible way to form carbon dioxide, that has been observed in a wide variety of environments but almost exclusively in the solid phase and that is therefore considered a very good probe of surface chemistry.

The plan of the paper is the following: In the first Section we will briefly describe the process of catalysis and its main mechanisms. In the second section we will describe the gas-surface interaction. In the third section we will describe a typical experimental apparatus and the state of the art of the experimental methods that are used to characterize the surfaces and to investigate the elementary processes that make catalysis work. In the fourth section we will present the results on the formation of molecular hydrogen. In the fifth section we will present the theoretical framework used for the analysis of the experimental results on molecular hydrogen formation, and their astrophysical implications. In the last section results on carbon dioxide formation will be presented.

1. HETEROGENEOUS CATALYSIS

A reaction between two partners to form a product is said to be catalyzed when it is mediated by the intervention of a third species that at the end of the process is not altered. The importance of catalyzed reactions is mainly due to the fact that often such reactions are much faster than non catalyzed ones.

When a catalyst is in an aggregation state different from that of the reactants, like the case when the catalyst is a solid and the reactants are gaseous, the process is called heterogeneous. This is the case in the interstellar medium, where dust grains act as catalysts. There are two ways in which a grain surface can play its catalytic role. In one case the surface remains chemically passive (only being a third body that can absorb the energy excess of the formation process). In the other case the molecules forming the surface are directly involved in the reaction chains and reduce some of the energy barriers that are encountered.

Two mechanisms have been envisaged to explain how catalytic formation of molecules takes place: the Eley-Rideal mechanism and the Langmuir-Hinshelwood mechanism. In the Eley-Rideal mechanism the formation of the products occurs when either an atom or a molecule, coming from the gas phase, hits directly an already adsorbed atom or molecule and reacts with it. Because the efficiency of this process is linear with the surface density (coverage) of absorbed species and such a coverage is usually very small on interstellar grains, this mechanism can be considered to be important only in very particular conditions encountered in the ISM.

In the Langmuir-Hinshelwood mechanism the reaction takes place between two partners that are already adsorbed on the surface of the solids. If at least one of the partners is mobile they may encounter and react. Several steps have to occur for the Langmuir-Hinshelwood mechanism to work: collision and adsorption of the chemical partners onto the surface, diffusion of at least one of them, reaction and (in some cases) desorption of the products.

2. GAS-SURFACE INTERACTION

The retention and subsequent chemical and physical evolution of atoms near/on the surface depends on the strength of the interaction between the atoms of the gas phase and the atoms in the solid. Schematically, physical adsorption forces are responsible for atom-surface energies of the order of 1 to 300 meV. A better way to characterize this interaction, called also physisorption, is by saying that no significant overlap in the electron densities of the adsorbed atoms and the solid occurs (Bruch et al. 1997). Conversely, the strong overlap and rearrangement of the atom's and solid's electron distributions characterize chemical adsorption, or chemisorption, with bonds between atoms of the order of 0.5 eV and higher. The residence time t of an atom/molecule on a surface depends strongly on its adsorption energy E, since $t = \nu^{-1} \exp(E/kT)$ where ν is a characteristic vibrational frequency of an atom on a solid surface. At a grain temperature of 15 K, adsorption energies of the order of a few tenths of an eV or higher practically confine the particle almost forever. In this case, the virtually vanishing diffusion makes this type of chemical bond between the atom/molecule and the solid of reduced interest as the production of new molecules that can be released to the gas phase is concerned. Of course, reactions can still occur if an incoming atom reacts with this immobile atom and the energy gained in the reaction can be used to eject the newly formed molecule into the gas phase. Alternatively, absorption of a photon or cosmic ray can produce enough heat to promote diffusion and chemical reactions.

When diffusion is necessary for atoms to react, the weak adsorption forces are the ones to consider. Therefore, this presentation is centered mostly on physical adsorption phenomena, as they have been studied more extensively and are more relevant to the processes occurring in diffuse and dense clouds.

The interaction of a rare gas atom with a solid surface is an example of physical adsorption; other types of atoms and molecules can display this type of interaction, depending on the type of surface considered and the conditions of the interaction, as will be clarified later on. The origin of this interaction resides in the long-range van der Waals force. Between two atoms, the instantaneous electric dipole p_A of one atom (A) induces a dipole moment on the other atom, B, of magnitude $\alpha_B p_A / r^3$, where p_A/r^3 is the dipolar field due to A and α_B is the polarizability of B. In turn the atom B that has so acquired the dipole moment $\alpha_B p_A / r^3$, generates a dipolar field of strength $\alpha_B p_A / r^6$. Therefore, the interaction between A and B has a term that is proportional to the product of the atomic polarizabilities of A and B, and is inversely proportional to the distance to the 6th power. For the interaction of an atom and a surface, the potential energy is inversely proportional to distance cubed. This can be understood by considering that one can pair-wise sum the van der Waals interaction between the gas atom and all the atoms of the solid. More realistically, one has to consider the collective response of the solid by using the dielectric function of the solid instead of atomic polarizabilities. In a continuum model of the solid, the interaction can be described as $V(z) \sim -C_3/(z - z_0)^3$, where z is the distance from the plane of the nuclei of the surface, z_0 is the position of the reference plane and

$$C_3 = \frac{h}{4\pi} \int \frac{\epsilon(i\omega) - 1}{\epsilon(i\omega) + 1} \alpha(i\omega) d\omega. \tag{1}$$

Here, ϵ is the dielectric function of the solid evaluated at imaginary frequencies $i\omega$ and $\alpha(i\omega)$ is the atomic polarizability. We also have: $\epsilon(i\omega) = \epsilon_1(\omega) + i\epsilon_2(\omega)$ and $p^2 = h/(2\pi) \int \alpha(i\omega) d\omega$. The real ϵ_1 and imaginary ϵ_2 parts of ϵ are related by the Kramers-Kronig relations. Thus, it is possible to obtain the C_3 coefficients for many types of solids from the knowledge of the dielectric properties (Vidali et al. 1991). Consider the case of H interacting with a graphite surface. Using $C_3 = 380$ meV Å^3 and $z_0 \sim 1.1$ Å (Vidali et al. 1991), we get an interaction energy of ~ 70 meV at a distance of 2.8 Å. In the case of H on graphite, the resulting potential, when the repulsive contribution is added, gives a well depth of 43 meV and a ground state energy of 32 meV, which is the typical strength in the weak adsorption case.

While the van der Waals interaction governs the attractive part of the physisorption potential, the steep repulsive part is due to the Pauli exclusion principle. Such short range interaction can be calculated in a variety of ways, depending on the type of solid. For example, the embedding energy of an atom in a sea of electrons (jellium) can be calculated for the interaction of an atom with a surface of a metal, $V(r) \sim \alpha n(r)$, where α is a coefficient that depends on the atom (not to be confused with the atomic polarizability) and $n(r)$ is the electron density of the solid near the surface. The disadvantage of such a method is that it doesn't reproduce the long range attraction correctly. For the most part, semi-empirical descriptions of the interaction have been used. Vidali et al. (1991) have shown that the interaction of light atoms with solid surfaces can be reduced to a common, semi-empirical "universal" form by rescaling the interaction potential. This rescaling is given in terms of the depth of the adsorption energy well and of a length parameter that depends on the polarizability of the atom and the dielectric function of the solid: $V_0(z) = Dg(z^*)$, where $z^* = (z - z_m)/l$, $l = (C_3/D)^{1/3}$ with D the well depth, z_m the distance between the surface and the minimum of the interaction potential, and $g(z^*)$ is a universal function. Such a semi-empirical law can be used to obtain estimates of interaction energies between light atoms and solids for which there are no experimental data or theoretical analyses.

A similar "universal law" has been proposed for the strong chemical interaction of an atom with a solid, mostly for metals. However, chemical adsorption is dominated by the local chemical environment and fewer generalizations can be made (Smith et al. 1982; Rose et al. 1983).

3.   EXPERIMENTAL

In this section, we present the experimental equipment and some of the experimental methods used in surface science to characterize the surface of solids and to study processes that allow catalysis to work.

3.1.   THE APPARATUS

A typical experimental apparatus used in surface science studies is shown in Fig. 1. It is the equipment in the Physics Department of Syracuse University that has been used by us to investigate the hydrogen recombination (Pirronello et al. 1997a; Pirronello et al. 1997b; Pirronello et al. 1999; Manicò et al. 2001) and CO_2 formation (Roser et al. 2001) on dust grain analogues. It consists of two atomic beam lines, an ultrahigh vacuum (UHV) scattering chamber where the sample is located and a time-of-flight section. Before the beginning of this collaboration it was mainly used to investigate adsorption of atoms on monocrystalline surfaces and to characterize the structure of well ordered surfaces.

The two atomic beam lines converge on a target in an ultra-high vacuum chamber and are 38° apart. Each line has triple differential pumping; with the beams on, the pressure in the third stages, just before the UHV chamber, is in the low 10^{-8} torr to mid 10^{-9} torr range. There is a mechanical chopper in each line for in-phase detection, time-of-flight measurements or to reduce the flux of H atoms, as the pressure in the source cannot be changed easily without adversely affecting the dissociation rate. Each line is fitted with a hydrogen/deuterium dissociation source (S_1 and S_2 in Fig. 1). The atomic source consists of a water cooled Pyrex tube surrounded by an inductor in a radiofrequency (RF) cavity. The sources are mounted on stainless steel bellows and x-y micrometer positioning stages. Radiofrequency from a 300 watt 13.6 MHz power supply is fed to the RF cavities via a power splitter and impedance matching networks. Typically 100 watts are fed into both sources with less than 5 watts of reflected power. Ultra-high pure molecular hydrogen or deuterium are inserted into the sources at pressure of 0.1-0.5 torr measured before dissociation. After dissociation, H or D pass an aluminum nozzle. The beams can be cooled to about ~150 K using copper braids

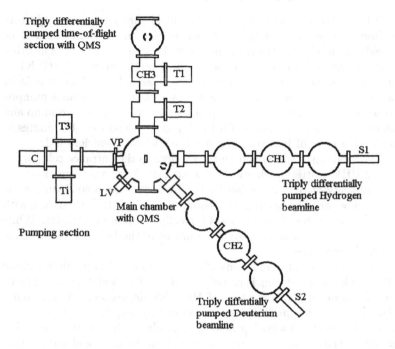

Figure 1 Apparatus at Syracuse University to study recombination reactions on surfaces of dust grain analogues (Roser et al., 2002).

connecting the nozzle to liquid nitrogen reservoirs. Peak dissociation rates over 90% have been measured downstream with the quadrupole mass spectrometer. Typically, the dissociation rate is between 70 and 85% and is stable during runs. These sources can be used to dissociate other molecules, such as O_2, in experiments of oxidation of CO ice by an oxygen beam (Roser et al. 2001).

The beams enter a UHV scattering chamber with operating pressure in the 10^{-10} torr range. The sample is attached to an OFHC (oxygen free high conductivity) copper block via a retaining copper ring; the copper block is attached to a liquid helium continuous flow cryostat via a silver disk. The cryostat is mounted on a rotatable platform and a bellows, so the sample can be rotated and translated vertically. The good thermal contact between sample and holder assures that a temperature of 5 K is routinely reached in the back of the sample, as measured by a calibrated silicon diode located in a well in the Cu block and surrounded by indium foil to assure good thermal conductivity. The sample can be heated to over 200 °C by a tungsten filament placed in the back of the sample. The

OFHC Cu block is shrouded by a Cu shield that has a small aperture in front so deposition of H and D occurs only on the sample, since the shield is at too high a temperature for sticking to occur on it. The other parts of the cold finger are at much higher temperatures ($\geq$ 160 K). A capillary stainless steel tubing can be positioned in front of the sample for depositing condensables, such as water vapor CO. The system is pumped by a 400 l/s turbomolecular pump, a 10" He closed-cycle cryopump and an ion pump. At a pressure of 5 10^{-10} torr, it would take 40 minutes to build up a layer of background gas if the sticking coefficient is 1.

The solid angle subtended by each source to the entrance collimators to the UHV scattering chamber is about 6 10^{-6} sr. Assuming a center-line intensity of 10^{17} atoms/sec/sr (Scoles 1988), and a 50 % duty cycle of the choppers, the beam at the sample is about $\sim 10^{12}$ atoms/sec with a $\sim$ 3mm diameter. The sample is typically $\sim$ 9 mm in diameter. While the apparatus is in air, a careful alignment of the beams is made using He-Ne laser beams.

The use of two atomic beams allows, as explained in detail elsewhere (Pirronello et al. 1997a; Pirronello et al. 1997b; Vidali et al. 1998a), the detection of the formation of HD with unprecedented sensitivity, thus bypassing the limitations of using one source, such limitations being: the presence of a background of molecular hydrogen in the reaction chamber; and the injection onto the sample of the undissociated fraction of molecular hydrogen from the beam source. With the sources in the line of site of the sample, one might wander whether the photons generated in the sources can influence the sample. For our experiments on amorphous ice, we calculated that, in the extreme case where the whole input power is converted into Lyman α photons, less than 1 photon per 1,000 hydrogen atoms is absorbed in the first few ice monolayers. The beams are detected by a rotatable, differentially pumped quadrupole mass spectrometer. The signal is sent to either a lock-in amplifier or a multichannel scaler. Thermal desorption measurements, described below, are carried out with the detector in front of the sample (entrance hole of detector: 6.3 mm at 6.2 cm from sample; the center of the detector is at 8.25 cm from the sample). If necessary it is possible to equip our apparatus with other surface science probes, such as a LEED (Low Energy Electron Diffraction) / Auger electron spectroscopy set-up to determine the structure and chemical compositions of surfaces, an Ar-ion gun for sample cleaning , a two-rotation axes, (XYZ) sample manipulator with temperature range from 140 K to 1200 K.

The temperature of the sample can be changed by using the heater or, in thermal desorption experiments when a high heating rate of the sample is required, by shutting off the flow of liquid helium in the continuous flow cryostat. The temperature rise of the sample vs. time obtained with this method is very reproducible, but it is not linear over the whole range of temperature (10 to 40 K) employed in these experiments.

Sample treatment varies according to the type of material. Graphite is peeled just prior insertion into the apparatus and then it is annealed at $\sim$ 200 °C for several hours. Silicates are cleaned with solvents in ultrasonic baths (acetone, methanol) and then placed in the UHV chamber. After UHV conditions are reached, the sample is flashed to $\sim$ 200 °C. The main background gas in a well-baked UHV chamber is hydrogen. Once cooled to a few K, samples can be cleaned of hydrogen by flashing them to 30 K and beyond.

An ice sample is prepared by depositing water vapor on the copper disk through a capillary aimed at the sample and placed about 1 cm from it. Before deposition the de-ionized water undergoes repeated cycles of freezing and thawing to remove trapped gases. A measured quantity of water vapor is admitted into the chamber through a heated stainless steel manifold and a UHV leak valve. The sample is held at 10 K or lower during deposition and the deposition rate to obtain high density amorphous ice is in the range of $\sim$8 layers/sec for a total thickness of the order of 1,200 layers.

The sample can be rotated, while maintaining 10^{-10} torr of background pressure and $T \sim 10 - 15K$, in front of a Xe flash lamp for irradiation with ultraviolet photons. The lamp is at focal point of a magnesium fluoride plano convex lens, which is positioned to focus a point source to a beam with a focal point at infinity. The light output can be changed by changing the repetition rate of the lamp and is measured at the sample to be of the order of $10^{15} photons/sec/cm^2$. The photon flux is of course much greater than the one experienced by a grain in the interstellar medium. However, it is scaled up, with respect to the ISM conditions, by roughly the same factor as the H flux is in our experiments.

Recently, we have begun the construction and testing of instrumentation to measure the kinetic energy and roto-vibrational energy of molecules leaving the surface as the result of recombination. To measure the translational energy of molecules leaving the surface we have built a time-of-flight (t-o-f) section located at 82 degrees from the hydrogen line (CH3 in Fig. 1). This line consists of three differentially pumped chambers; the middle one has a high speed motor with a pseudo-random wheel and the the final one has a quadrupole mass spectrometer. With this t-o-f section we hope to be able to make measurements of the velocity distribution of molecules leaving the surface after recombination.

To study the roto-vibrational states of the molecules ejected from the dust analogues following the recombination process we proceed in the following way. Since vibrational transitions in H_2 occurs in the ultraviolet where the use of tunable lasers is impractical, alternative methods have to be employed. We chose to use a (2+1) REMPI (Resonance Enhanced Multiphoton Ionization) spectroscopy scheme. A two photon absorption of $\sim 200nm$ light via a virtual state pushes the hydrogen molecule in an electronic singlet E,F Σ_g excited state. A third photon at $\sim 300nm$ ionizes the hydrogen molecule that had been excited; a multichannel plate detector records the arrival of the ion. For this scheme to work, a tunable and intense laser light is necessary. We use a Nd:YAG laser, frequency doubled, that pumps a dye laser. A tripling unit converts the 600 nm light from the dye laser into a 200 nm light with 1 mJ per pulse at 10 Hz.

3.2.   CHARACTERIZATIONS AND PROCESS INVESTIGATIONS BY SURFACE SCIENCE METHODS

Surface science methods are fundamental in gaining insight on the processes that allow the cathalitic mechanisms (in particular the Langmuir-Hinshelwood one) to work. Such processes after collision are sticking, mobility, diffusion and reaction upon encounter. Because all of them are strongly dependent on the structure and the composition of the surface, experimental methods need to be used to characterize them as much as possible. In this section we will describe briefly some of these methods.

3.2.1   Structural characterization of the surface.   A chemically inert He beam at thermal energy can be used to probe the surface condition, its dynamics or the structure and dynamics of atoms and molecules adsorbed on top of the substrate (Hulpke 1992). Modern He beams are supersonic, meaning that the flow velocity is much greater than the local sound velocity. Mach numbers of 80 and above are routinely achieved. Supersonic He beams are characterized by high forward intensity and narrow ($\frac{\delta v}{v} \sim 2\%$) velocity distributions; they are used to characterize the morphology, atomic structure and dynamics of the topmost layer of a solid surface. An application example of this method was given by Lin and Vidali (1996). The He beam is sent onto the graphite crystal and the specularly reflected beam is monitored under various conditions. From the shape of the specular reflection one can deduce the average size of the crystalline platelets and their relative orientation (Fig. 2). <u>Limitation</u>: only very well ordered surfaces can be characterized because only on them it is possible to obtain a well defined specularly reflected helium beam. For more details on the use of molecular beams in surface science studies see (Zangwill 1984; d'Evelyn & Madix 1984).

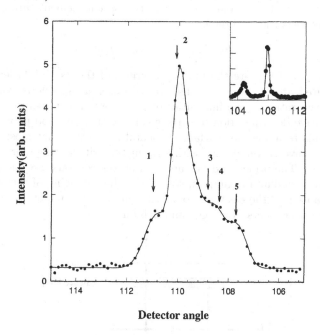

Figure 2 Reflected He beam from a graphite surface (Lin & Vidali, 1996).

Another powerful technique to characterize surfaces is Scanning Tunneling Microscopy; however, it has yet to be used on realistic analogues of interstellar dust grains.

3.2.2 Chemical characterization of the surface.

To find out what elements are present on the surface, Auger electron spectroscopy is often used. A 1.5-3 keV electron beam impinges on the surface and the emitted spectrum of electrons is detected. The precise energy position of the peaks which are produced by the emitted Auger electrons characterizes the chemical composition of the surface. To show this, let's look for example at the Auger decay shown in Fig. 3. The transfer of energy due to the sample bombardment ionize an atom in the solid, producing a hole in the core level $1s$. The excited atom can relax by emitting a X-ray or an Auger electron. This last one is the preferential path if the binding energy of the level, E_{1s}, is less than 2000 eV (Zangwill 1988). Then an electron of the $2s$ level drops into the hole and the transition energy ejects a second Auger electron from the $2p$ level. The emitted electron has a kinetic energy, E_{kin}, which is related to the characteristic energy of the parent element through the energy conservation expression

$$E_{kin} = E_{1s} - E_{2s} - E_{2p} \tag{2}$$

where E_{2s} and E_{2p} are the binding energies of the $2s$ and $2p$ levels of the excited atom. As each Auger transition is assignable to some element, the ratios of the intensities of Auger electron peaks provides a quantitative determination of surface composition. A good point in favor of this technique for surface elemental analysis is that most of the elements have Auger decays with E_{kin} in the critical range for surface sensitivity. The principal disadvantages, on the contrary, are the charging of non-conducting samples, sometimes the damaging of the surface (both caused by the electron bombardment) and the undetectability of the lightest elements, hydrogen and helium.

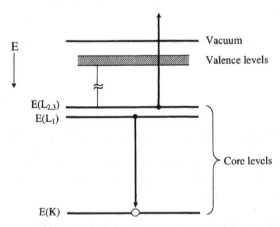

Figure 3   KLL Auger decay of a 1s core hole

Other methods to study the chemical composition of a surface include flash desorption, as discussed above, and SIMS, Static Secondary Ion Mass Spectrometry. In the latter case, an ion beam sputters (removes) a small portion of the surface, and the desorbing ions are studied with a quadrupole mass spectrometer and energy analyzer. Quantitative interpretation of the data is not always easy because the sputtering rate depends on specific properties of the surface+impurity system.

3.2.3   Sticking.   A method that can be applied when the sample surface is ordered enough on a 10-100 Å scale, so that there is a measurable amount of specular He beam reflection, is the one used for instance by Lin and Vidali (1996). They sent a H beam on the graphite surface while they monitored the reflection of a He beam with a quadrupole mass spectrometer. The attenuation in the reflection of the He beam due to the presence of the H atoms is proportional to the fraction of H covering the surface: $I(t)/I_0 = 1 - \Sigma JtS$ (Poelsema and Comsa 1989), where J is the flux, t the time, S the sticking coefficient, Σ the He-adsorbate scattering cross-section, and I_0 is the He reflected intensity before adsorption. From an estimate of the flux and of the He-adsorbed H scattering cross-section, the sticking coefficient of H on graphite at $\sim 16K$ was estimated to be between 0.06 and 0.1. The method is sensitive down to parts of 1% of one layer in coverage, chiefly because of the large He-adatom scattering cross section Σ.

On disordered surfaces another method can be used. A beam of atoms or molecules is sent onto the surface. The pressure in the scattering chamber is measured when the beam is on and with the sample outside of the beam path. A drop of the pressure when the sample is placed in the beam path is attributed to the sticking of atoms/molecules of the beam on the surface of the sample; in other words, the atoms/molecules in the beam that stick on the surface don't contribute to the pressure in the scattering chamber. Measurements can be difficult in cases in which the background pressure of the gas to be studied is already high, or when the gas reacts with the walls of the chamber; unfortunately this situation is encountered in the experiments involving hydrogen. Nonetheless, a measurement of H sticking on a carbonaceous surface, which is relevant to astrophysical conditions, has been attempted. It shows that sticking goes from a high value (0.8) at low temperature (below 8 K) to a low value $(0.2 - 0.4)$ in the astrophysically relevant range of $10 - 15$ K (Pirronello et al. 2000). One should also be aware that, in general, the sticking coefficient is coverage dependent (especially below one layer coverage). Thus, measurements of sticking probability should be taken at different coverages (Rettner et al. 1986).

3.2.4   Residence time.   If the beam is chopped by a mechanical selector, one can measure the scattered signal as a function of time. A time delay or change of shape of the scattered beam pulse yields information on residence times of atoms/molecules on the surface. Typical chopping frequencies go from a few Hertz to hundreds of Hertz (Comsa & David 1985), depending on the typical times of adsorption /reaction /desorption. As for the characterization of the surface discussed above, this method relies on the detection of a large enough signal scattered back into the gas phase. In practice, this method has been applied mostly to well-ordered surfaces.

A slightly different method is used to probe the residence time of atoms/molecules when this time is expected to be much longer than a few seconds. In this method, the surface is exposed to a certain flux of atoms/molecules for a given amount of time. Then, the amount that sticks is measured with progressively larger delay times between the end of the adsorption and the beginning of the desorption. The longer the delay time, the smaller the amount left on the surface, since some of the atoms and molecules evaporate from the surface during the interval between the end of the adsorption and the beginning of the desorption.

3.2.5   Diffusion.   Only in very few cases the diffusion of adatoms on surfaces can be directly probed, for example by using the technique of Field Ionization Microscopy (FIM) or Scanning Tunneling Microscopy (STM) (Zangwill 1984). In the first case, a sharp metal tip is placed at a high potential, so electric fields of the order of 10^9 volt/cm are reached at the tip. When a rare gas is admitted, it becomes ionized at the tip and the ions drift toward an imaging plate providing an image of the tip on the atomic scale. In the second case (STM) a sharp tip is placed near an electrically conductive sample and the tunneling current between tip and sample is used to regulate the distance of the tip from the sample. In both cases, by measuring the fluctuations of the electric current due to atoms adsorbed on the tip (in the FIM case) or on the sample just under the STM tip, the autocorrelation of the tip current can be evaluated. From it, one obtains the autocorrelation function of the atom's velocity and, thus, its mobility. To our knowledge, no experimental studies of diffusion on astrophysically relevant systems have been done, since FIM studies are confined to tips of refractory metals which can withstand the high electric fields without being stripped of their tip atoms. STM studies require a conductive sample; in principle, they could be applied to a variety of materials, but this technique for measuring atom mobility is still in its infancy.

Determination of the activation energy for diffusion, if present, or of other parameters relevant to the diffusion process can be obtained indirectly. In the case of formation of molecular hydrogen on dust grain analogues, an analysis of thermal desorption data can be helpful in sorting out the diffusion mechanisms at play, as described in Section 5.

3.2.6   Thermal Programmed and Flash Desorption.   In the technique of thermal programmed desorption and/or flash desorption

(TPD), the sample temperature is ramped quickly (and usually, but not necessarily, linearly) in order to desorb adatoms and molecules from a surface. A quadrupole mass spectrometer placed in front of the sample is used to measure, as a function of time, the reaction products desorbed form the surface.

This technique will be described in detail in the next section where the results of hydrogen recombination will be given.

4.   HYDROGEN RECOMBINATION UNDER ASTROPHYSICALLY RELEVANT CONDITIONS

A typical experiment to measure the recombination of atomic hydrogen on the surface of dust grain analogues is carried out in two phases. First, H and D are sent onto the surface for a given period of time (from tens of seconds to tens of minutes). The shorter runs are done to obtain a small coverage (i.e., a small number of atoms on the surface so as to approximate as much as possible conditions in the interstellar medium). The longer runs are done to find out the saturation condition, i.e., when the surface is nearly all covered with hydrogen. During the adsorption phase, HD molecules that are quickly formed and ejected from the surface are detected by a quadrupole mass spectrometer. In such a case, we probe reaction events that occur quickly, on the time scale of the length of the adsorption phase; such quick recombination occurs because either the diffusion of atoms on the surface is rapid or an atom from the gas phase reacts directly with an atom on the surface without becoming accommodated on the surface first (Eley-Rideal reaction). After the adsorption phase is over, the sample temperature is quickly (~ 0.06 K/sec) ramped and the HD signal is measured.

The so-called recombination efficiency, which is a number between 0 and 1 and is proportional to the ratio of the HD yield and the number of H and D atoms sent on the target, is computed by taking into account the fact that H and D yield also H_2 and D_2.

The experiment is performed at different adsorption temperature to investigate the important range of values that are encountered in the ISM. These results are shown in Fig. 4, for the three classes of solids that are particularly important in the interstellar medium: silicates and amorphous carbon in diffuse clouds and water ice in dense ones.

Interestingly (at the lowest sample temperature during adsorption) it is found that most of HD detected is formed due to diffusion of H and D following thermal activation during the heat pulse. Only a small fraction of HD is formed during the irradiation process, showing that, at

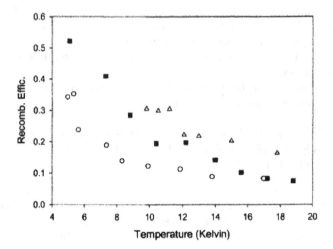

Figure 4 Recombination efficiency of hydrogen on olivine (circles), amorphous carbon (squares) and ice samples (triangles) as a function of the adsorption temperature.

least under our experimental conditions, prompt-reaction mechanisms (Duley 1996) or fast tunneling (Hollenbach & Salpeter 1971) are not that important.

This important result is obtained by analyzing the kinetics of desorption as a function of the coverage of H and D adatoms on the surface after irradiation performed at the lowest possible temperatures; the technique used is the already mentioned Thermal Programmed Desorption (TPD).

Typical desorption spectra obtained from an amorphous ice surface (Manicò et al. 2001) using the TPD technique are shown in Figs. 5 as a function of the coverage of the partners that are sent on the surface.

Hydrogen and deuterium adatoms, that at the increase of the ramp temperaure become able to overcome the energy barrier for diffusion, are set in motion; shortly thereafter, they start encountering one another, recombining, leaving the surface as HD and producing the peak signal in the quadrupole. The positions of the peak maxima that shift toward lower ramp temperatures (hence to shorter times from the beginning of the temperature ramp) as the coverage increases clearly show that H and D atoms are still present on the surface when the warm-up begins. In this case the kinetics of the process is said to be of the second order

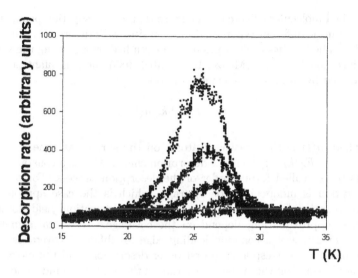

Figure 5 Hydrogen Recombination on amorphous high density ice. TPD spectra shown here were obtained after (bottom to top) 2, 4, 6, 8, 12 minutes exposure at 15 K.

because the production and desorption rates of the final products HD are proportional to the product of the surface concentrations of the reaction partners. The higher the coverage of H atoms, the shorter the average time before they encounter and form the HD molecule. This evidence is even clearer in the case of olivine (Pirronello et al. 1997b) where a lower coverage could be obtained. From this type of measurements we may safely conclude that tunneling by itself does not assure high mobility, as it has been assumed up to now following the influential papers of Hollenbach and Salpeter (1970, 1971), but that (at least at temperature lower than those encountered in interstellar space) thermal activation is necessary.

A reasonable possibility is that during the warm up phase thermally assisted tunneling takes place, a process in which thermal activation allows the H adatom to reach a higher energy level inside the adsorption well it is occupying, before it might tunnel into an adjacent site.

From TPD spectra one can also obtain information on the "effective" energy barriers that are encountered during the process. More exactly the location (on the ramp temperature scale) of the peak of the desorption curve, provides useful information about the desorption energy, while the area below the curve is proportional to the total amount of desorbed molecules. There can be more than one desorption peak, since there can be different types of adsorption sites with different binding energies. The kinetics of desorption can be studied by analyzing the shape of the desorption peak (Menzel 1975; Yates 1985) and it is quantitatively described by the Polanyi-Wigner equation:

$$-dn(t)/dt = k_m n(t)^m, \tag{3}$$

where $n(t)$ is the density of atoms on the surface at time t, $k_m = k^{(m)} \exp(-E_d/k_B T)$, E_d is the desorption energy, $k^{(m)}$ is a constant, and m is the so called formal order of the desorption process. For $m = 0$, desorption is independent of coverage, which is the case appropriate for desorption from several layers, since the desorption yield doesn't, in first approximation, depend on the coverage. A typical signature is the presence of a common leading edge for different coverages. the case of $m = 1$ corresponds to first order desorption, and the molecules already formed on the surface leave during the desorption independently from each other; a distinguishing trait is the common trailing edge and the same peak position as a function of coverage. The case of $m = 2$ corresponds to second order desorption, that is the reaction rate is proportional to the product of the concentrations of the reactants. In this case, the peak shifts towards higher temperatures as the coverage decreases.

Examples of thermal desorption of HD from a silicate and amorphous carbon are presented in the next Section. Other examples of actual desorption spectra displaying different desorption orders, but not for astrophysically relevant substrates, can be found in (Kay et al. 1989).

The position of the maximum in $dn(t)/dt$ (taking place at time t_{max} and temperature T_{max}) gives useful information about the desorption energy barrier. Fast temperature ramps are needed to obtain a picture of the state of the adsorbate at the time of desorption; otherwise, the adsorbate layer might go through changes, such as phase transitions or quasi-isothermal desorption. Using a linear temperature ramp, i.e. taking $T = T_0 + bt$, where T_0 is the initial temperature, t is the time and b is the heating rate, and setting $d^2 n(t)/dt^2 = 0$, we obtain

$$\frac{bE_d}{k_B T_{max}^2} = mk^{(m)} n(t_{max})^{m-1} \exp(-E_d/k_B T_{max}). \tag{4}$$

In the $m = 2$ case the peak position shifts continuously to lower temperature as the coverage increases, but the value of the desorption energy does not change. This energy E_d should be considered to be the energy for the whole recombination process, thus including, if applicable, activation energy for diffusion, recombination and desorption. Typical values of E_d for HD formation and desorption on surfaces of astrophysical interest (silicates, amorphous carbon and ices) are in the 20-60 meV range. An example of a more complete analysis of experimental data using rate equations is given in the next section; for more details, see (Biham et al. 1998; Katz et al. 1999; Biham et al. 2001).

5.   THEORETICAL MODELING OF HYDROGEN RECOMBINATION

In this section we will describe the analysis, based on a Rate Equation approach, that has been used to deduce from experimental results energy barriers for mobility and desorption of H adatoms on analogues of interstellar grains and a new model, based on a Master Equation approach, that allows to apply experimental results to the formation of molecular hydrogen on the small interstellar dust grains.

5.1.   ANALYSIS OF EXPERIMENTAL RESULTS

5.1.1   Rate Equations.   Consider an experiment in which a flux of H atoms is irradiated on the surface. H atoms that stick to the surface, once the surface temperature is raised, perform hops as random walkers with increased frequency and recombine when they encounter one another. Let $n_H(t)$ [in monolayers (ML)] be the coverage of H atoms on the surface and $n_{H_2}(t)$ (also in ML) the coverage of H_2 molecules. We obtain the following set of rate equations:

$$\dot{n}_H = f(1 - n_H - n_{H_2}) - W_H n_H - 2a n_H^2 \qquad (5)$$
$$\dot{n}_{H_2} = \mu a n_H^2 - W_{H_2} n_{H_2}. \qquad (6)$$

The first term on the right hand side of Eq. (5) represents the incoming flux in the Langmuir-Hinshelwood kinetics. In this scheme H atoms deposited on top of H atoms or H_2 molecules already on the surface are rejected. f represents an *effective* flux (in units of MLs^{-1}), namely it already includes the possibility of a temperature dependent sticking coefficient. The second term in Eq. (5) represents the desorption of H atoms from the surface. The desorption coefficient is

$$W_H = \nu \cdot \exp(-E_1/k_B T) \qquad (7)$$

where ν is the attempt rate (standardly taken to be $10^{12}\ s^{-1}$), E_1 is the activation energy barrier for desorption of an H atom and T is the temperature. The third term in Eq. (5) accounts for the depletion of the H population on the surface due to recombination into H_2 molecules, where

$$a = \nu \cdot \exp(-E_0/k_B T) \tag{8}$$

is the hopping rate of H atoms on the surface and E_0 is the activation energy barrier for H diffusion. Here we assume that there is no barrier for recombination. If such a barrier is considered, it can be introduced as discussed in Pirronello et al. (1997b, 1999). The first term on the right hand side of Eq. (6) represents the creation of H_2 molecules. The factor 2 in the third term of Eq. (5) does not appear here since it takes two H atoms to form one molecule. The parameter μ represents the fraction of H_2 molecules that remains on the surface upon formation, while a fraction of $(1 - \mu)$ is spontaneously desorbed due to the excess energy released in the recombination process. The second term in Eq. (6) describes the desorption of H_2 molecules. The desorption coefficient is

$$W_{H_2} = \nu \cdot \exp(-E_2/k_B T), \tag{9}$$

where E_2 is the activation energy barrier for H_2 desorption. The H_2 production rate r is given by:

$$r = (1 - \mu) \cdot a n_H^2 + W_{H_2} n_{H_2}. \tag{10}$$

This model can be considered as a generalization of the Polanyi-Wigner model. It gives rise to a wider range of simultaneous applications, compared to the Polanyi-Wigner equation. In particular, it describes both first order and second order desorption kinetics (or a combination) for different regimes of temperature and flux.

In the experiments analyzed here, both the temperature and the flux were controlled and monitored throughout. Each experiment consists of two phases. In the first phase the sample temperature is constant up to time t_0, under a constant irradiation rate f_0. In the second phase, the irradiation is turned off and linear heating of the sample is applied at the rate b $(K s^{-1})$:

$$\begin{aligned} f(t) &= f_0; & T(t) = T_i : & & 0 \le t < t_0 & \quad (11) \\ f(t) &= 0; & T(t) = T_i + b(t - t_0) : & & t \ge t_0. & \quad (12) \end{aligned}$$

Here T_i is the constant temperature of the sample during irradiation.

In the case that the rejection terms in $f(1-n_H-n_{H_2})$ are neglected and the effective flux becomes simply f (a valid assumption at low coverages), the rate equations can be solved analytically. However, the solution is expressed in terms of intractable nested integral expressions, and is of little use to us. In the case we study here, in which the rejection terms are taken into account, no such solution exists and the equations are integrated numerically.

5.1.2   Analysis.

We will now examine to what extent the rate equation model can describe the experimental results. To this end we performed numerical integration of Eqs. (5)-(6) with the aid of a Bulirsch-Stoer stepper algorithm (Press et al. 1992). The result of the integration is a set of TPD curves that are a function of the chosen set of parameters. A standard TPD experimental run includes the time dependence of the flux $f(t)$ and temperature $T(t)$ as well as the four parameters E_0, E_1, E_2 and μ. An approximate value for $f(t)$, in the required units of MLs^{-1}, is obtained by integrating the TPD spectra, generating the total *yield* of the various experiments. The flux is then obtained from the exponential fit indicated by Langmuir-Hinshelwood kinetics. It is important to stress that this is a lower bound value for the flux, and this value is reached only if there is no H desorption at all. We are now left with the four parameters E_0, E_1, E_2 and μ which are assumed to be independent of the flux or temperature. These parameters form a four dimensional space that has to be explored in order to find the values for which the calculated TPD curves provide the best fit to the experimental TPD ones.

The merit function to be minimized in the fitting procedure is the standard χ^2 function, which is the sum over the squares of the differences between the experimental points and the calculated ones. To obtain the parameters that best fit the experimental data one needs to probe the parameter space and find the set of values of the parameters that give rise to the global minimum of the merit function. The probing procedure we used is based on random search. A point in the four dimensional parameter space is randomly picked. A numerical integration of the model's equations is performed with the chosen parameters. The merit function obtained from the comparison of the resulting curve with the experimental curve is then evaluated. If this set represents an improvement over the current minimum, the new point is accepted as the new minimum. If, however, the new set did not score lower (i.e. better) than the known minimum, it was promptly discarded. The probability distribution for picking the next random set of parameters was taken to be a Lorentzian centered around the current set of parameters and thus favored nearby points. The relatively slow drop-off of the Lorentzian function also allowed the occasional taking of longer steps and thus prevented the process from getting stuck in a local minimum.

5.1.3   Results.

The experimental TPD curves and the fits obtained by the rate equations are shown in Figs. 6 and 7 for the olivine sample and in Figs. 8 and 9 for the amorphous carbon sample.

The values of the parameters E_0, E_1, E_2, and μ, that best fit the experimental results were obtained. For the olivine sample it was found that $E_0 = 24.7$ meV, $E_1 = 32.1$ meV, $E_2 = 27.1$ meV and $\mu = 0.33$, while for the amorphous carbon sample $E_0 = 44.0$ meV, $E_1 = 56.7$ meV, $E_2 = 46.7$ meV and $\mu = 0.413$. The parameter set generated for each sample represents a simultaneous best fit for all the six TPD curves. Fitting each curve separately typically produce better fits, but at the expense of an increased range in the values of the parameters. These variations allow us to generate approximate error estimates. It is found that the energy barriers E_0 and E_2 are very well determined by this process (to within several tenths of a meV). The barrier E_1 is not as well determined, and its values given above, for both samples, are to be taken as lower bounds to the correct value (within 3 meV). The parameter μ is determined to within ± 0.1, and thus justifies our assumption, within this model, that not all H_2 molecules immediately desorb upon recombination. Attempting to artificially force $\mu = 0$ and do the fits with the remaining three parameters, degrades the fit substantially and cannot recreate the entire range of behavior of the data simultaneously.

Although Eq. (3) can be used to fit the entire range of experimental TPD results, this equation, which includes a single activation energy, does not provide as much insight as our model [Eqs. (5)-(6)]. For example, applying Eq. (3) in the case of Fig. 6 (olivine, 1st order desorption kinetics), we use $m = 1$ and arrive at $E_d = 26.8$ meV. This is equivalent to the E_2 found for olivine using this model. However, as we shift to Fig. 7 (olivine, 2nd order desorption kinetics), we must now take $m = 2$, which results in $E_d = 24$ meV. Here the rate limiting process is the diffusion on the surface, and therefore E_d is closer to E_0. Unlike Eq. (3), our model does not require setting a parameter (such as m). Furthermore, it provides the best-fit values of each of the relevant activation energies for both the first and second order kinetics and using the same framework.

The model [Eqs. (5)-(6)]. can also describe the steady state conditions which are reached when both the flux and the temperature are fixed. The steady state solution is then easily obtained by setting $\dot{n}_H$ and $\dot{n}_{H_2}$ to 0 and solving the quadratic equation for n_H.

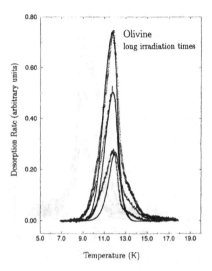

Figure 6 TPD curves for higher coverage experiments on an olivine slab. Irradiation times are (in minutes) 8.0 (•), 5.5 (×) and 2.0 (+). Fits are in solid lines (Katz et al., 1999).

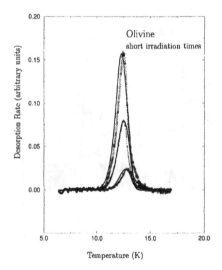

Figure 7 TPD curves for lower coverage experiments on an olivine slab. Irradiation times are (in minutes) 0.55 (•), 0.2 (×) and 0.07 (+). Fits are in solid lines (Katz et al., 1999).

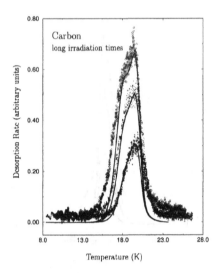

Figure 8 TPD curves for higher coverage experiments on amorphous carbon. Irradiation times are (in minutes) 32.0 (•), 16.0 (×) and 8.0 (+). Fits are in solid lines (Katz et al., 1999).

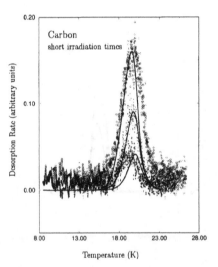

Figure 9 TPD curves for lower coverage experiments on amorphous carbon. Irradiation times are (in minutes) 4.0 (•), 2.0 (×) and 1.0 (+). Fits are in solid lines (Katz et al., 1999).

In this work, we present a model that captures the kinetics of the diffusion-recombination-desorption process. The parameters thus obtained can then be used to study the astrophysically relevant cases. For example, by assuming steady state conditions, we can obtain the recombination efficiency as a function of flux f and temperature T for a range of parameters that goes from the astrophysically relevant (extremely low flux, $10 - 15K$) to the ones used in the laboratory (low flux, $5 - 30K$). The recombination efficiency η is defined as the ratio between the production rate r [Eq. (10)] and the deposition rate $f/2$ (in molecules). The recombination efficiency for the olivine sample under steady state conditions as a function of T is shown in Fig. 10 for an astrophysically relevant value of the flux f.

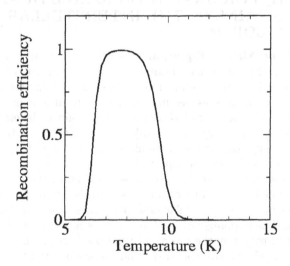

Figure 10 The recombination efficiency η, obtained from the rate equations, for olivine under steady state conditions at a constant flux, as a function of the grain temperature. The flux is $f = 0.18 \times 10^{-8}\ MLs^{-1}$. The system exhibits a window of high efficiency between 7 - 9 K and a tail of decreasing efficiency above 9 K (Biham et al., 2001).

The main conclusion from this Figure is that the recombination efficiency is highly temperature dependent. There is a window of high efficiency along the temperature axis. Outside this window the efficiency drops sharply. It is found that under astrophysically relevant irradiation rates the high efficiency temperature range for olivine is between 7 - 9 K while for amorphous carbon it is between 12 - 16 K.

We find that, for both samples, the parameter μ (the probability of an H_2 molecule to remain on the surface upon recombination) has little effect on the production rate r of molecular hydrogen under steady state conditions. This is easy to understand, since under steady state conditions the production rate R must be equal to the recombination rate on the surface, and, thus, must be independent of μ. The coverage of hydrogen molecules on the surface is adjusted accordingly. Similarly, the energy barrier for desorption of molecular hydrogen, E_2, has little effect on the recombination efficiency, under steady state conditions, as long as it remains significantly smaller than the barrier for atomic desorption, E_1.

5.2. H_2 FORMATION ON SMALL DUST GRAINS IN THE INTERSTELLAR MEDIUM

5.2.1   The Master Equation Model.

We will now consider the formation of H_2 molecules on small dust grains in interstellar clouds. In this case it is more convenient to rescale our parameters such that instead of using quantities per unit area - the total amount per grain will be used. The number of H atoms on the grain is denoted by N_H. Its expectation value is given by $\langle N_H \rangle = S \cdot n_H$ where S is the number of adsorption sites on the grain. Similarly, the number of H_2 molecules on the grain is N_{H_2} and its expectation value is $\langle N_{H_2} \rangle = S \cdot n_{H_2}$ (we assume that each adsorption site can adsorb either an H atoms or an H_2 molecule). The incoming flux of H atoms onto the grain surface is given by $F_H = S \cdot f_H$ (atoms s^{-1}). The desorption rates W_H and W_{H_2} remain unchanged. The hopping rate a_H (hops s^{-1}) is replaced by $A_H = a_H/S$ which is approximately the inverse of the time t_s required for an atom to visit nearly all the adsorption sites on the grain surface. This is due to the fact that in two dimensions the number of distinct sites visited by a random walker is linearly proportional to the number of steps, up to a logarithmic correction (Montroll & Weiss 1965). The H_2 production rate of the single grain is given by $R = S \cdot r$ (molecules s^{-1}). The rate equations (neglecting the Langmuir-Hinshelwood rejection term) will thus take the form

$$\frac{d\langle N_H \rangle}{dt} = F_H - W_H \langle N_H \rangle - 2A_H \langle N_H \rangle^2 \tag{13}$$

$$\frac{d\langle N_{H_2} \rangle}{dt} = \mu A_H \langle N_H \rangle^2 - W_{H_2} \langle N_{H_2} \rangle, \tag{14}$$

where the first term in (14) accounts for the flux of hydrogen molecules from the gas phase that are adsorbed on the grain surface. While for large grains Eqs. (13) and (14) provide a good description of the recombination processes, in the limit in which the number of atoms on the grain becomes small they may not be suitable anymore (Tielens 1995; Charnley et al. 1997; Caselli et al.1998; Shalabiea et al.1998). This is due to the fact that these equations take into account only average concentrations and ignore fluctuations as well as the discrete nature of the H atoms. These properties become significant in the limit of very small grains and low incoming flux of H atoms, exactly the conditions encountered in diffuse interstellar clouds where hydrogen recombination on silicate and carbon surfaces is expected to be relevant. As the number of H atoms on a grain fluctuates in the range of 0, 1 or 2, the recombination rate cannot be obtained from the average number alone. This can be easily understood, since the recombination process requires at least two H atoms simultaneously on the surface. Comparisons with Monte Carlo simulations have shown that the rate equations tend to over-estimate the recombination rate. A modified set of rate equations which exhibits better agreement with Monte Carlo simulations was introduced by Caselli et al. (1998) and applied by Shalabiea et al. (1998) to a variety of chemical reactions. In these equations the rate coefficients are modified in a semi-empirical way to take into account the effect of the finite grain size on the recombination process.

In order to to resolve this problem we will now introduce a different approach based on a master equation that is suitable for the study of H_2 formation on small grains. Each grain is exposed to a flux F_H of H atoms. At any given time the number of H atoms adsorbed on the grain may be $N_H = 0, 1, 2, \ldots$. The probability that there are N_H hydrogen atoms on the grain is given by $P_H(N_H)$, where

$$\sum_{N_H=0}^{\infty} P_H(N_H) = 1. \tag{15}$$

The time derivatives of these probabilities, $\dot{P}_H(N_H)$, are given by

$$
\begin{aligned}
\dot{P}_H(0) &= -F_H P_H(0) + W_H P_H(1) + 2 \cdot 1 \cdot A_H P_H(2) \\
\dot{P}_H(1) &= F_H \left[P_H(0) - P_H(1) \right] + W_H \left[2P_H(2) - P_H(1) \right] \\
&+ 3 \cdot 2 \cdot A_H P_H(3) \\
\dot{P}_H(2) &= F_H \left[P_H(1) - P_H(2) \right] + W_H \left[3P_H(3) - 2P_H(2) \right] \\
&+ A_H \left[4 \cdot 3 \cdot P_H(4) - 2 \cdot 1 \cdot P_H(2) \right]
\end{aligned}
$$

$$\vdots$$

$$
\begin{aligned}
\dot{P}_H(N_H) = \ & F_H \left[P_H(N_H - 1) - P_H(N_H) \right] \\
+ \ & W_H \left[(N_H + 1)P_H(N_H + 1) - N_H P_H(N_H) \right] \\
+ \ & A_H \left[(N_H + 2)(N_H + 1)P_H(N_H + 2) - N_H(N_H - 1)P_H(N_H) \right].
\end{aligned}
$$

$$\vdots$$

$$\tag{16}$$

Each of these equations includes three terms. The first term describes the effect of the incoming flux F_H on the probabilities. The probability $P_H(N_H)$ increases when an H atom is adsorbed on a grain that already has $N_H - 1$ adsorbed atoms [at a rate of $F_H P_H(N_H - 1)$], and decreases when a new atom is adsorbed on a grain with N_H atoms on it [at a rate of $F_H P_H(N_H)$]. The second term includes the effect of desorption. An H atom desorbed from a grain with N_H adsorbed atoms decreases the probability $P_H(N_H)$ [at a rate of $N_H W_H P_H(N_H)$, where the factor N_H is due to the fact that each of the N_H atoms can desorb] and increases the probability $P_H(N_H - 1)$ at the same rate. The third term describes the effect of recombination on the number of adsorbed H atoms. The production of one molecule reduces this number from N_H to $N_H - 2$. For one pair of H atoms the recombination rate is proportional to the sweeping rate A_H multiplied by 2 since both atoms are mobile simultaneously. This rate is multiplied by the number of possible pairs of atoms, namely $N_H(N_H - 1)/2$. Note that the equations for $\dot{P}_H(0)$ and $\dot{P}_H(1)$ do not include all the terms, because at least one H atom is required for desorption to occur and at least two for recombination. The rate of formation of H_2 molecules on the surface (in units of molecules s^{-1}) is thus given by

$$
A_H \sum_{N_H=2}^{\infty} N_H(N_H - 1)P_H(N_H). \tag{17}
$$

For simplicity, we assume here that $\mu = 0$, namely that all the molecules are desorbed upon formation. The expectation value for the number of H atoms on the grain is

$$
\langle N_H \rangle = \sum_{N_H=0}^{\infty} N_H P_H(N_H). \tag{18}
$$

The time dependence of the expectation value, obtained from Eq. (16) is given by

$$
\frac{d\langle N_H \rangle}{dt} = F_H - W_H \langle N_H \rangle - 2A_H \langle N_H(N_H - 1) \rangle. \tag{19}
$$

The recombination rate R (molecules s^{-1}), namely the net rate in which H_2 molecules desorb into the gas phase is

$$R = A_H \langle N_H(N_H - 1) \rangle. \tag{20}$$

The recombination efficiency is given by

$$\eta = \frac{R}{(F_H/2)}. \tag{21}$$

Eq. (19) resembles the rate equation (13) apart from one important difference: the recombination term $\langle N_H \rangle^2$ is replaced by $\langle N_H^2 \rangle - \langle N_H \rangle$. On a macroscopically large grain it is expected that the difference between these two terms will be small and Eq. (13) would provide a good description of the system. However, on a small grain, where $\langle N_H \rangle$ is small these two terms are significantly different and it is necessary to use the master equation rather than the rate equations.

In principle the master equation consists of infinitely many equations for each atomic or molecular specie. In practice, for each specie such as atomic hydrogen we simulate a finite number of equations for $P_H(N_H)$, $N_H = 1, \ldots, N_{max}$, where $P_H(N_H) = 0$ for $N_H > N_{max}$. Obviously, N_{max} cannot exceed the number of adsorption sites, S, on the grain. In the equations for $P_H(N_{max} - 1)$ and for $P_H(N_{max})$, the terms that couple them to $P_H(N_{max} + 1)$ and $P_H(N_{max} + 2)$ are removed (these terms describe the flow of probability from $P_H(N_H)$, $N_H > N_{max}$ to $P_H(N_H)$, $N_H \leq N_{max}$). Terms such as $F_H P_H(N_H)$ that describe probability flow in the opposite direction are also removed. The latter terms are evaluated separately and frequently during the integration of the master equation in order to examine whether N_{max} should be increased. The condition for adding more equations is typically $\dot{P}_H(N_{max} + 1) > \epsilon$ at a certain time t, where ϵ is a small parameter, suitably chosen according to the desired precision.

Note that the master equation is typically needed when $\langle N_H \rangle$ is of order unity. Under such conditions most of the probability $P_H(N_H)$ is concentrated at small values of N_H and therefore N_{max} is expected to be small. In a time dependent simulation, when $\langle N_H \rangle$ increases reaching the limit $\langle N_H \rangle \gg 1$ (thus requiring $N_{max} \gg 1$) the master equation can be easily replaced by the rate equations, during the run. One simply has to evaluate $\langle N_H \rangle$ at a certain time t and from that point to continue the run using the rate equations. The opposite move of switching from the rate equations to the master equation (when $\langle N_H \rangle$ decreases towards $\langle N_H \rangle \approx 1$) is nearly as simple. One has to pick as an initial condition for the master equation a narrow distribution $P_H(N_H)$ that satisfies the average $\langle N_H \rangle$ given by the rate equations, and after some relaxation time it will converge to the proper distribution. In simulations of more complex reactions involving multiple species, the coupling between different species typically involves only averages such as $\langle N_H \rangle$ (this is an approximation that will be discussed below). Therefore, one can simultaneously use the rate equations for some species and the master equation for others, according to the criteria mentioned above.

5.2.2   Simulations and Results.   To examine the effect of the finite grain size on the recombination rate of hydrogen in the interstellar medium we performed simulations of the recombination process using the master equation and compared the results to those obtained from the rate equation (13). The parameters we have used are given below. Assuming, for simplicity, a spherical grain of diameter d we obtain a cross section of $\sigma = \pi d^2/4$. The estimate of the number of adsorption sites on the grain was based on the experimental data for the olivine and amorphous carbon surfaces (Pirronello et al. 1997a, 1997b, 1999) using the following procedure. The flux of the H and D beams was estimated as $b \cong 10^{12}$ (atoms cm^{-2} s^{-1}). The beams passed through a chopper that reduced their flux by a factor of $c = 20$. A measurement of the flux in units of ML per second was done using the data for the total HD yield vs. exposure time [Fig. 3 in Pirronello et al. (1997a)]. The theoretical Langmuir-Hinshelwood mechanism provides a prediction for the coverage of adsorbed atoms after irradiation time t, which is

$$n_H(t) = 1 - \exp(-f_H \cdot t). \tag{22}$$

Fitting the total HD yield to this expression we obtained good fits that provide the flux values $f_H = 2.7 \cdot 10^{-4}$ (in ML s^{-1}) for the olivine experiment and $f_H = 9.87 \cdot 10^{-4}$ for the amorphous carbon experiment. From these two measurements we obtain the density of adsorption sites (sites cm^{-2})

$$s = \frac{b}{c \cdot f_H}. \tag{23}$$

For the olivine surface it is found that $s \cong 2 \cdot 10^{14}$ and for the amorphous carbon surface $s \cong 5 \cdot 10^{13}$ (sites cm^{-2}).

Observations indicate that there is a broad distribution of grain sizes, that roughly resembles power-law behavior, in the range of 10^{-6}cm $< d < 10^{-4}$cm (Mathis et al. 1977; Mathis 1996; O'Donnell & Mathis 1997). The number of adsorption sites on a (spherical) grain is given by $S = \pi d^2 s$. In the simulations we focus on diffuse clouds and use as a typical value for the density of H atoms $\rho_H = 10$ (atoms cm^{-3}). The temperature of the H gas is taken as $T = 100K$. The typical velocity of H atoms in the gas phase is given by (see e.g. Landau & Lifshitz 1980)

$$v_{\mathrm{H}} = \sqrt{\frac{8}{\pi} \cdot \frac{k_B T}{m}} \qquad (24)$$

where $m = 1.67 \cdot 10^{-24}$ (gram) is the mass of an H atom. We thus obtain $v_H = 1.45 \cdot 10^5$ (cm s^{-1}). The density of grains is typically taken as $\rho_g = 10^{-12}\rho_{\mathrm{H}}$ and hence in our case $\rho_g = 10^{-11}$ (grains cm^{-3}). The sticking probability of H atoms onto the grain surface is taken as $\xi = 1$. Experimental results indicate that the sticking probability is close to 1 for temperatures below about 10K and possibly somewhat lower at higher temperatures. Since there is no high quality experimental data for the temperature dependence $\xi(T)$, and in order to simplify the analysis we chose $\xi = 1$.

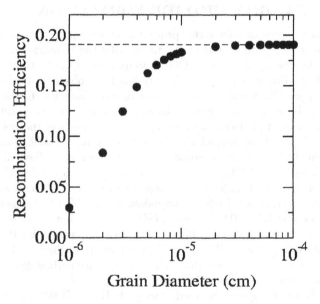

Figure 11 The recombination efficiency η for an olivine grain as a function of the grain size at $T = 10K$ and $f = 0.18 \times 10^{-8}$ MLs^{-1}. The results predicted by the rate equations under similar conditions, which are independent of the grain size, are also shown (dashed line). It is observed that as the grain size decreases below $d = 0.1\mu$m the recombination efficiency quickly decreases (Biham et al., 2001).

We will now analyze the processes that take place on a single grain using numerical integration of the master equation and comparison to the rate equations. The flux of H atoms onto the grain surface is given by $f_H = \rho_H v_H/(4s)$ (ML s^{-1}), where the factor of 4 in the denominator is the ratio between the surface area and the cross section for a spherical grain. Using the parameters above we obtain that $f_H = 0.18 \cdot 10^{-8}$ for olivine and $f_H = 0.73 \cdot 10^{-8}$ for amorphous carbon. The total flux on a grain of diameter d is given by $F_H = f_H \cdot S$ (atoms s^{-1}).

The recombination efficiency η for an olivine grain as a function of the grain size is shown in Fig. 11. The dashed line shows the recombination efficiency obtained from the rate equations, which is independent of the grain size. It is observed that for grain diameter smaller than about 10^{-5} (cm) the recombination efficiency sharply drops below the rate-equation value. This is due to the fact that in this range $\langle N_H \rangle < 2$, hence most often an H atom resides alone on the grain and no recombination is possible.

6.   CARBON DIOXIDE FORMATION

As an example of a catalytic process in which the surface is actively involved in the process it is worth to describe the formation of carbon dioxide by the association of CO molecules and O atoms on a water ice surface (Roser et al. 2001). CO_2 has been observed by ISO (Infrared Space Observatory) in a variety of environments (Gerakines et al. 1999; Boogert et al. 2000). Its high abundance in the solid phase (van Dishoeck et al. 1996), and its very low abundance in the gas-phase, suggest that CO_2 is formed almost exclusively in the solid phase and that it might be a sensitive probe of gas-grain chemistry. Several experiments showed that CO_2 formation occurs in ices containing CO pure or mixed with H_2O under UV or particle irradiation (d'Hendecourt et al. 1986; Grim et al. 1989; Allamandola et al. 1997; Gerakines et al. 1996; Moore et al. 1991). However, ISO observations indicate that CO_2 is abundant also in quiescent dark clouds, such as toward Elias 16 (Whittet et al. 1998), where UV and particle fluxes are negligible. Therefore there must be a mechanism that is able to produce carbon dioxide even in the absence of ionizing radiation.

Well before CO_2 was observed, Tielens & Hagen (1982) made extensive Monte Carlo modeling of surface reactions on interstellar grains. Assuming an activationless CO + O reaction, in agreement with experiments of Fournier et al. (1979), they suggested that carbon dioxide in ices is ubiquitous and abundant, a view that has been confirmed by ISO. In their experiments Fournier et al. (1979) irradiated a mixture of $N_2O/CO/Ar$ (with a mole ratio of 1:1:330) with 8.4 eV photons at 7 K. They interpreted their data as if $O(^1S)$ atoms, photolytically produced with enough kinetic energy, diffuse and react with CO to yield CO_2. In a similar experiment Grim & d'Hendecourt (1986) found no appreciable quantities of CO_2 and concluded that there is an activation energy barrier for the reaction to occur. In both experiments of Fournier et al. (1979) and of Grim & d'Hendecourt(1986) O atoms are obtained by dissociating O_2 with UV photons, a mechanism that might yield excited radicals and render the simulation somewhat more distant from interstellar conditions.

Roser et al. (2001) carried out experiments on carbon dioxide formation in the same ultra-high vacuum (UHV) apparatus in which molecular hydrogen formation was investigated. In this experiment only one of the two available beam lines was used, to send a beam of oxygen atoms on a CO-ice surface. The products of the reaction between oxygen atoms and CO molecules on/in CO-ice were detected by means of quadrupole mass spectrometer placed in the UHV scattering chamber. The O beam is produced by radio-frequency (RF) dissociation of O_2 in a Pyrex source. The O_2 dissociation efficiency ranged between 30 and 40%. Because no UV irradiation of the ice was performed to produce oxygen atoms, this experiment should be considered as a closer simulation of surface processes occurring in quiescent dark clouds than it had been attempted so far. Isotopically labelled $^{13}C^{16}O$ (99.99% nominal chemical purity and 88.1% nominal isotopic purity) was introduced via a leak valve and a capillary into the UHV chamber concurrently with O atom irradiation of the sample in order to obtain an intimate mixture in the solid phase of O atoms and CO molecules. The stainless steel CO delivery line was held about 1 cm from the sample. The amount of CO molecules delivered on the sample was obtained by the pressure change in the known volume of the supply line.

CO_2 formation was investigated over a range of CO/O abundance ratio from 5.6 to 21 and with a CO-ice thickness of ~ 100 layers. The dilution of O in CO has been chosen to minimize the probability of undesired interaction between two oxygen atoms or between an oxygen atom and the undissociated O_2 leading to the formation of ozone, a fact that could have played a role in the negative result obtained by Grim & d'Hendecourt (1986). To improve the signal to noise ratio, it was used a beam of ^{18}O atoms, and tuned the detector to mass 47 to measure the amount of $^{13}C^{16}O^{18}O$ formed by the interaction of $^{13}C^{16}O$-ice and ^{18}O atoms. The amount of CO_2 produced was measured during deposition as well as during a thermal programmed desorption (TPD) experiment. In a first set of experiments, after each deposition at about 5 K, it was performed a TPD experiment. In a second set of experiments it was placed on top of the already deposited CO and O mixture a water-ice cap of ~ 100 layers. In this way it was possible to maintain CO and O together for a longer time during the warm-up. An amorphous layer of water-ice was deposited by putting the sample directly in front of the gas line (Jenniskens et al. 1995). In all runs, the amount of CO deposited was kept constant while the dose of oxygen atoms was changed. In the experiments done without the water-ice cap ("first set") no detectable amount of CO_2 was produced. CO and O sublimed quickly (the maximum in CO desorption is reached at ~ 40 K) making it unlikely for them to receive enough thermal energy to overcome a reaction energy barrier even if it is only slightly higher than the desorption one. This result indicates also that the Eley-Rideal mechanism is not efficient.

In the second set of experiments, done with the addition of the water-ice cap, the signal of carbon dioxide ($^{13}C^{16}O^{18}O$ - mass 47) was obtained, as shown in Fig. 12, when the substrate temperature was high enough for the sublimation of the water-ice layer (above 150 K). We believe that this procedure may have increased the probability for partners such as CO and O to interact while they were diffusing through the layers of porous amorphous water-ice.

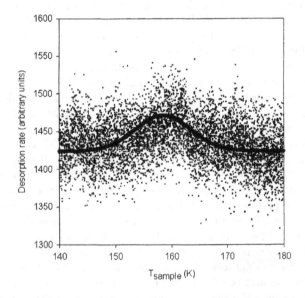

Figure 12 Mass 47 signal during thermal desorption. Representative thermal desorption curve of $^{13}C^{16}O^{18}O$ (mass 47) from a O/CO=0.1 ice mixture with a water ice cap (Roser et al., 2001).

The formation of CO_2 most probably occurs during the migration of CO and O through the water-ice layer, but CO_2 molecules (larger than CO and O), once formed, remain trapped in the water-ice until the whole ice sublimes, an effect that had already been observed previously (Pirronello et al. 1982). Fig.13 shows the yield of $^{13}C^{16}O^{18}O$ as a function of the mixing ratio of ^{18}O to $^{13}C^{16}O$. In the range of the relative abundance investigated, the yield of carbon dioxide is roughly proportional to the oxygen dose. The fact that in these experiments the O abundance constitutes the rate limiting factor in the formation of CO_2 is in agreement with observations (Gerakines et al. 1999; Chiar 1997).

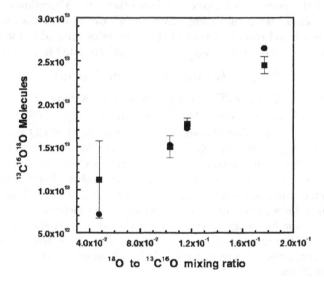

Figure 13 $^{13}C^{16}O^{18}O$ Yield versus $^{18}O/^{13}C^{16}O$ ratio. Squares represent the measured $^{13}C^{16}O^{18}O$ yields. The error bars indicate the range in the value of the signal due to different methods that have been employed to subtract the background. Circles represent the results of the fit (Roser et al., 2001).

To gain insights in the kinetics of CO_2 formation it was built a simple model that takes into account the migration and escape of CO and O from the water ice during warm-up. To model the process it was assumed that the reaction $CO + O \longrightarrow CO_2$ takes place during the diffusion of both CO and O in the amorphous water-ice cap and that such a diffusion can be regarded as a sequence of desorptions from and adsorptions on sites in the inner walls of cavities in the amorphous ice cap. In this model the barriers to overcome to obtain carbon dioxide molecules are simply the desorption energy barrier and the formation barrier, if any.

For CO the desorption barrier from an adsorption site in a cavity is roughly equal to 1740 K, which is the measured surface binding energy for CO on water-ice (Sandford & Allamandola 1988). The desorption barrier for O atoms has been evaluated to be around 800 K (Tielens & Hagen 1982; Tielens & Allamandola 1987).

The porous structure of the amorphous water ice is represented as made of interconnected voids having diameters of an average size equal to 20 Å (Langel et al. 1994; Mayer & Pletzer 1986). Inside this voids CO and O bind to the water-ice surface. Their migration in such a structure is a succession of: 1) desorption from one site; 2) free flight inside the cavity of the pore; 3) adsorption onto another site. Migration of O or CO stops either when they escape from the ice cap or when they occupy the same site and react to form CO_2 (that remains trapped in the water ice structure) with a rate, R_{CO_2}), that can be estimated as follows:

$$R_{CO_2} = k_r \left(\alpha_{CO} P_O N_{CO} + \alpha_O P_{CO} N_O \right) \tag{25}$$

where $k_r = \exp(-E_a/K_bT)$ is the probability to overcome the reaction barrier "E_a", $\alpha_{CO} = \nu \exp(-\frac{E_d(CO)}{K_bT})$ and $\alpha_O = \nu \exp(-\frac{E_d(O)}{K_bT})$ are the migrating rates (number of steps in the unit of time) of CO molecules and O atoms, respectively; E_d is the energy barrier for such a process; $P_O = N_O/N_{site}$ is the probability that a migrating CO molecule occupies the same site of an O atom (being N_{site} the number of sites available on the inner walls of the interconnected pores); $P_{CO} = N_{CO}/N_{site}$ is the probability for a migrating O atom to meet a CO molecule at a site; ν is the attempt frequency to escape from the adsorption site, and N_{CO} and N_O are the number of CO molecules and O atoms trapped in the water ice cap, respectively, at time t. Using the expressions given above, equation 25 reads:

$$R_{CO_2} = \nu \{ \exp\{-[E_d(O) + E_a]/K_bT\} + \exp\{-[E_d(CO) + E_a]/K_bT\} \} \frac{N_{CO} N_O}{N_{site}} \tag{26}$$

This equation is coupled with those describing CO and O loss from the water ice layer, i.e.:

$$\frac{dN_{CO}}{dt} = -\frac{\nu \exp[-E_d(CO)/K_bT]}{steps} N_{CO} \tag{27}$$

$$\frac{dN_O}{dt} = -\frac{\nu \exp[-E_d(O)/K_bT]}{steps} N_O \tag{28}$$

where *steps* is the average number of the three-step migration process described above, that in a sort of density gradient driven diffusion has been estimated of the order of one hundred.

Integrating this set of differential equations (using the measured time dependence of the sample temperature during warm-up) we obtained the value of the activation energy barrier E_a for CO_2 formation that provides the best fit to the data (Fig. 13). The result is $E_a = 290$ K.

In this experiment the presence of water-ice can have a two-fold effect: it allows to maintain CO and O in the solid at higher temperatures (that makes easier to overcome the not too high formation barrier) and, as already noticed by Fournier et al. (1979), the icy matrix allows the otherwise forbidden transition

$$CO(X\ ^1\Sigma^+) + O(^3P) \longrightarrow CO_2(X\ ^1\Sigma_g^+) \tag{29}$$

to occur.

Results obtained in the second set of experiments, in which a water ice-cap has been deposited on a O+CO mixture, can be considered a definitive piece of experimental evidence that reactions on grains can form CO_2 without the intervention of energetic agents in agreement with Gerakines et al. (1999).

The astrophysical relevance of this experimental result is shown by the timescales to produce CO_2 on interstellar grains. Using the activation energy barrier, $E_a=290$ K, it is possible to estimate the timescale to produce a carbon dioxide molecule when an O atom from the gas phase lands on a site of the ice manle occupied by a CO molecule or viceversa. Such a timescale is of the order of a second at 10 K and is already negligible at a grain temperatures of 15 K - that is, close to the temperature values of the ice mixture in quiescent clouds toward Elias 16 (Whittet et al. 1998) and in Sgr A (Gerakines et al. 1999). Such timescales are short enough to allow a large fraction of solid CO to be converted into CO_2 in mantles on grains of quiescent clouds.

7.   SUMMARY

In this chapter we have described the importance of surface reactions to form key species that are difficult to be produced by gas phase reactions in interstellar conditions.

We focused on the formation of molecular hydrogen and carbon dioxide and presented both experimental surface science methods to gain insight in the processes that actually occur and theoretical methods to apply the results obtained in the laboratory in astrophysically relevant conditions.

Acknowledgments

We would like to acknowledge the important contribution of our collaborators to the work reviewed here. The experimental research at Syracuse University could not have been done without the careful and thoughtful work of graduate students J. Roser, C.Liu, L.Shen and J.-S. Lin, and undergraduate students R. D'Agostino and R. Conde. The contributions of the Catania group to both experimental and theoretical work have been possible thanks to the dedication and ability of the postdoc collaborator G. Manicò and the graduate student G. Ragunì. Giulio Manicò is also gratefully acknowledged for providing the description of the Auger method of analysis. The graduate students I. Furman and A. Lipshtat, and undergraduate student N. Katz have contributed greatly to the theoretical studies done at the Hebrew University. G.V. was supported by NASA through grants NAG5-4998 NAG5-6822 and NAG5-9093. V.P. was supported by the Italian Ministry for University and Scientific Research through grant 21043088. O.B. was supported by the Adler Foundation for Space Research, of the Israel Science Foundation.

REFERENCES

Allamandola, L.J., Bernstein, M.P. and Sanford, S.A. (1997), in Cosmovici, C.B., Bowyer, S. and Wertheimer, D. (eds.), *Astronomical and Biochemical Origins and the Search for Life in the Universe*, Editrici Compositori, Bologna, p.23.

Biham, O., Furman, I., Katz, N., Pirronello, V. and Vidali, G. (1998), *MNRAS*, **296**, 869.

Biham, O., Furman, I., Pirronello, V. and Vidali, G. (2001), *ApJ*, **553**, 595.

Boogert, A.C.A. et al. (2000), *A&A*, **353**, 349.

Bruch, L.W., Cole, M.W., Zaremba, E. (1997), *Physical Adsorption: Forces and Phenomena*, Oxford Science Publications.

Caselli, P., Hasegawa, T.I. and Herbst, E. (1998), *ApJ*, **495**, 309.

Charnley, S.B., Tielens, A.G.G.M. and Rodgers, S.D. (1997), *ApJ*, **482**, L203.

Chiar, J.E. (1997), *Origins Life Evol. Biosphere*, **27**, 79.

Comsa, G. and David, R. (1985), *Surf. Sci. Rep.*, **5**, 145.

d'Evelyn, M.P. and Madix, R.J. (1984), *Surf. Sci. Rep.*, **3**, 413.

d'Hendecourt, L.B., Allamandola, L.J., Grim, R.J.A. and Greenberg, J.M. (1986), *A&A*, **158**, 119.

Duley, W.W. (1996), *MNRAS*, **279**, 591.

Duley, W.W. and Williams, D.A. (1986), *MNRAS*, **223**, 177.

Fournier, J., Deson, J., Vermeil, C., Pimentel, G.C. (1979), *Journal of Chem. Phys.*, **70**, 5726.

Gerakines, P.A. et al. (1999), *ApJ*, **522**, 357.

Gerakines, P.A., Schutte, W.A., Ehrenfreund, P. (1996), *A&A*, **312**, 289.

Grim, R.J.A., d'Hendecourt, L.B. (1986), *A&A*, **167**, 161.

Grim, R.J.A., Greenberg, J.M., de Groot, M.S., Baas, F., Schutte, W.A., Schmitt, B. (1989), *A&A Supp. Ser.*, **78**, 161.

Hollenbach, D. and Salpeter, E.E. (1970), *J. Chem. Phys.*, **53**, 79.

Hollenbach, D. and Salpeter, E.E. (1971), *ApJ*, **163**, 155.

Hollenbach, D., Werner M.W. and Salpeter, E.E. (1971), *ApJ*, **163**, 165.

Chemical Reactions on Solid Surfaces of Astrophysical Interest   249

Hulpke, E. (ed.) (1992), *Helium Beam Scattering from Surfaces*, Springer Series in Surface Science, v.22, Springer-Verlag.
Jenniskens, P., Blake, D.F., Wilson, M.A., Pohorille, A. (1995), *ApJ*, **455**, 389.
Katz, N., Furman, I., Biham, O., Pirronello, V. and Vidali, G. (1999), *ApJ*, **522**, 305.
Kay, B.D., Lykke, K.R., Creighton, J.R. and Ward, S.J. (1989), *J. Chem. Phys.*, **91**, 5120.
Landau, L.D. and Lifshitz E.M. (1980), Statistical Physics Part I, Third Edition, Pergamon Press, Oxford, p.86.
Langel, W., Fleger, H.-W. and Knözinger, E. (1994), *Ber. Bunsenges. Phys. Chem.*, **98**, 81.
Lin, J. and Vidali, G. (1996), in Greenberg, M. (ed.), *The Cosmic Dust Connection*, Kluwer, p.323.
Manicò, G., Ragunì, G., Pirronello, V., Roser, J.E. and Vidali, G. (2001), *ApJ*, **548**, L253.
Mathis, J.S. (1996), *ApJ*, **472**, 643.
Mathis, J.S., Rumpl, W. and Nordsieck, K.H. (1977), *ApJ*, **217**, 425.
Mayer, E. and Pletzer, R. (1986), *Nature*, **319**, 298.
Menzel, D. (1975), in Gomer, R. (ed.), *Interaction on Metal Surfaces*, Springer Verlag, NY, p.102.
Montroll, E.W. and Weiss, G.H. (1965), *J. Math. Phys.*, **6**, 167.
Moore, M.H., Khanna, R., Donn, B. (1991), *J. Geophys. Res.*, **96**, 17541.
O'Donnell, J.E. and Mathis, J.S. (1997), *ApJ*, **479**, 806.
Pirronello, V., Biham, O., Liu, C., Shen L. and Vidali, G. (1997b), *ApJ*, **483**, L131.
Pirronello, V., Biham, O., Manicò, G., Roser, J. and Vidali, G. (2000), in Combes, F., Pineau de Forets, G. (eds.), *Molecular Hydrogen in Space*, Cambridge University Press, Cambridge, p.71.
Pirronello, V., Brown, W.L., Lanzerotti, L.J., Marcantonio, K.J., Simmons, E.H. (1982), *ApJ*, **262**, 636.
Pirronello, V., Liu, C., Roser J.E. and Vidali, G. (1999), *A&A*, **344**, 681.
Pirronello, V., Liu, C., Shen L. and Vidali, G. (1997a), *ApJ*, **475**, L69.
Poelsema, B. and Comsa, G. (1989), *Scattering of thermal energy atoms from disordered surfaces*, Springer Tracts in Modern Physics, Springer Verlag, New York.
Press, W.H., Teukolsky, S.A., Vetterling, W.T. and Flannery, B.P. (1992), *Numerical Recipes in C, The Art of Scientific Computing*, Cambridge University Press.
Rettner, C.T., DeLouise, L.A. and Auerbach, D.J. (1986), *J. Chem. Phys.*, **85**, 1131.
Rettner, C.T., Michelsen, H.A. and Auerbach, D.J. (1995), *J. Chem. Phys.*, **102**, 4625. Rettner, C.T. (1994), *J. Chem. Phys.*, **101**, 1529.
Rose, J.H., Smith, J.R. and Ferrante, J. (1983), *Phys. Rev. B*, **28**, 1835.
Roser, J.E., Manico', G., Pirronello, V. and Vidali, G. (2002), *ApJ*, **581**, 276.
Roser, J.E., Vidali, G., Manicò, G. and Pirronello, V. (2001), *ApJ*, **555**, L61.
Sandford, S.A., Allamandola, L.J. (1988), *Icarus*, **76**, 201.
Scoles, G. (ed.) (1988), *Atomic and Molecular Beam Methods*, Oxford University press, v.1.
Shalabiea, O.M, Caselli, P. and Herbst, E. (1998), *ApJ*, **502**, 652.
Smith, J.R., Ferrante, J. and Rose, J.H. (1982), *Phys. Rev. B*, **25**, 1419.
Tielens, A.G.G.M. (1995), unpublished.
Tielens, A.G.G.M. and Allamandola, L.J. (1987), in Hollenbach, D.J. and Thronson, H.A. Jr. (eds.), *Interstellar Processes*, Kluwer, Dordrecht, p.397.
Tielens, A.G.G.M. and Hagen, W. (1982), *A&A*, **114**, 245.
van Dishoeck, E.F. et al. (1996), *A&A*, **315**, L349.

Vidali, G., Ihm, G., Kim, Y.-S. and Cole, M.W. (1991), *Surf. Sci. Rep.*, **12**, 133.

Vidali, G., Pirronello, V., Liu, C., Shen, L.Y. (1998a), *Astrophys. Lett. Comm.*, **35**, 423.

Vidali, G., Roser, J.E., Manicò, G. and Pirronello, V. (2002), in Salama, F. (ed.), *Proceedings of NASA Laboratory Astrophysics Workshop*, in press.

Vidali, G. and Zeng, H. (1996), *Appl. Surf. Sci.*, **92**, 11.

Watanabe, N. and Kouchi, A. (2002), *ApJ*, **567**, 651.

Whittet, D.C.B. (1992), *Dust in the Galactic Environment*, IOP Publishing.

Whittet, D.C.B. et al. (1998), *ApJ*, **498**, L159.

Yates, J.T. (1985), in *Methods of Experimental Physics: Solid State Physics: Surfaces*, v.22, Academic Press, p.425.

Zangwill, A. (1984), *Physics at Surfaces*, Cambridge University Press.

FROM INTERSTELLAR POLYCYCLIC AROMATIC HYDROCARBONS AND ICE TO ASTROBIOLOGY

Louis J. Allamandola and Douglas M. Hudgins
Astrochemistry Laboratory, NASA Ames Research Center, MS 245-6, Mountain View, CA 94035-1000, USA

Keywords: PAHs - Ices - Laboratory simulations - astrobiology

Abstract: Tremendous strides have been made in our understanding of interstellar material over the past twenty years thanks to significant, parallel developments in observational astronomy and laboratory astrophysics. Twenty years ago the composition of interstellar dust was largely guessed at, the concept of ices in dense molecular clouds ignored, and the notion of large, abundant, gas phase, carbon rich molecules widespread throughout the interstellar medium (ISM) considered impossible. Today the composition of dust in the diffuse ISM is reasonably well constrained to micron-sized cold refractory materials comprised of amorphous and crystalline silicates mixed with an amorphous carbonaceous material containing aromatic structural units and short, branched aliphatic chains. In dense molecular clouds, the birthplace of stars and planets, these cold dust particles are coated with mixed molecular ices whose major components are very well constrained. Lastly, the signature of carbon-rich polycyclic aromatic hydrocarbons (PAHs), shockingly large molecules by earlier interstellar chemistry standards, is widespread throughout the Universe. This paper presents a detailed summary of these disparate interstellar components and ends by considering both of them as important feedstock to the chemical inventory of the primordial Earth. Particular attention is paid to their possible role in the chemistry that led to the origin of life. An extensive reference list is given to allow the student entry into the full depth of the literature.

The first part of this paper focuses on interstellar PAHs. The laboratory and theoretical underpinning which supports the PAH model is reviewed in some detail. This is followed by a few specific examples which demonstrate how these data can be used to analyze the interstellar spectra and probe local conditions in different emission zones. These examples include tracing the

V. Pirronello et al. (eds.), Solid State Astrochemistry, 251–316.
© 2003 *Kluwer Academic Publishers. Printed in the Netherlands.*

evolution of carbon as it passes from its birthsite in circumstellar shells through the ISM, determining specifics about the cosmic PAH population in many different environments including PAH size and structure, and probing local conditions in the different emission zones. The second part of this paper summarizes the laboratory and observational background leading to our current understanding of interstellar/precometary ices. Although the most abundant interstellar ice components are the very simple molecules such as H_2O, CH_3OH, CO, CO_2, and NH_3, more complex species including accreted PAHs and those formed by UV and cosmic ray processing within the ice must also be present. Here we give a detailed summary of the photochemical evolution on those ices found in the densest regions of molecular clouds, the regions where stars and planetary systems are formed. Ultraviolet photolysis of these ices produces a host of new compounds, some of which show intriguing prebiotic behavior. The last part of this paper draws all this information together and considers the possible roles these compounds might have played in early Earth chemistry. As interstellar ices are the building blocks of comets and comets are thought to be an important source of the species which fell on the primitive Earth, their composition may be related to the origin of life. Three potential roles are considered, ranging from these interstellar materials simply providing the raw materials used for a completely endoginous origin to the opposite extreme that they delivered species poised to take part in the life process, or perhaps even at the earliest stages of what would be perceived of as a living system.

INTRODUCTION

The origin of life on earth is intimately tied to the nonbiological formation and chemical evolution of the compounds of carbon with five elements: H, N, O, S, and P. Since the Sun and planets formed some 4.6 billion years ago in a Universe perhaps 15 -20 billion years of age, these elements have had a long and complex history before incorporation into the pre-solar nebula. Although it was once believed that the chemical compounds in space are rather simple, with chemical complexity limited by harsh radiation fields and extremely low densities, recent research has begun to call this belief into question. Today, compelling evidence is mounting that a substantial fraction of the carbon incorporated into planets, their satellites, asteroids, and comets in developing planetary systems is in the form of complex organic molecules. Thus, it is entirely possible that the extraterrestrial evolution of chemical complexity may play a crucial - perhaps even a determinant - role in defining the early, prebiotic chemical state of these planetary systems. Moreover, the relative cosmic abundances of O, C, and N illustrated in Figure 1 underscore the fact that (ignoring the chemically inert helium) these elements are *by far* the most abundant chemically reactive elements after hydrogen, dwarfing the amounts of the next tier of elements including that of silicon. This implies that if life exists anywhere else in the cosmos, it is most likely composed of these most abundant elements and therefore evinces a chemistry similar to our own.

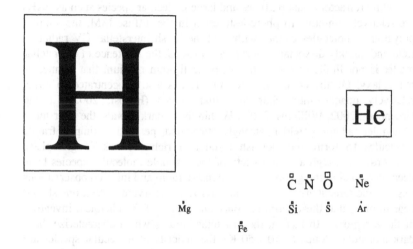

Figure 1. Astronomer's periodic table of the elements with the areas of the boxes illustrating the relative cosmic abundances of the most plentiful elements. Figure reproduced courtesy of Ben McCall.

The history of the biogenic elements (e.g. C, N, and O) begins with their nucleosynthesis deep within the interiors of late-type stars. These elements are dredged up and thrown off into the surrounding interstellar medium (ISM) during the periods of intense mass-loss that punctuate the late asymptotic giant branch (AGB) phase. For carbon-rich AGB stars, extensive observational studies have shown that a wide array of organic materials are formed during these episodes of late-stellar mass-loss. Included in these outflows are simple molecules (<20 atoms) such as acetylene, carbon monoxide, and the polyacetylenic and cyanopolyacetylenic chains; large, robust molecules (tens to hundreds of atoms) such as polycyclic aromatic hydrocarbons (PAHs); and small (100-1000 Å, several thousands of atoms or more) amorphous carbon and silicate dust particles. As this ejecta gradually disperses through the protoplanetary and planetary nebula (PPN and PNe) stages, the surrounding diffuse ISM is gradually enriched with these carbonaceous materials. In the diffuse ISM, these compounds and particles are further modified through a variety of physical and chemical processes including: UV irradiation; cosmic ray bombardment; gas-phase chemistry; accretion and reaction upon grain surfaces; and destruction by shock waves generated by supernova explosions. Numerous reviews relevant to this wide range of phenomena can be found elsewhere in this volume and in the scientific literature (Allamandola and Tielens, 1989; Pirronello, Krelowski and Manicò, 2002).

While refractory dust particles and large molecular species such as PAHs are relatively immune to photodestruction in the diffuse ISM, the simpler polyatomic molecules cannot withstand the harsh interstellar UV radiation field and quickly dissociate. Fortuitously for us, the existence of interstellar matter is not limited to the tenuous, optically-thin medium that permeates our galaxy. On the contrary, much material is also concentrated in large, relatively opaque interstellar molecular clouds (Herbst, 2002; Li and Greenberg, 2002; Williams, 2002). Within these 'dark' clouds, the interstellar ultraviolet radiation field is strongly attenuated, permitting simple, fragile molecules to form and flourish through a rich network of gas phase reactions. Although a wide variety of these simple molecular species have been identified in the gas-phase by extensive radio and infrared observations (e.g. Li and Greenberg, 2002; Winnewisser, 1997; Irvine, 1998), one should bear in mind that these represent only one aspect of the chemical inventory of these regions. In fact, at the low temperatures which characterize these dark molecular clouds ($\approx$ 10 - 50 K), the majority of molecular species are expected to be frozen out upon the surfaces of cold, refractory grains (e.g. Sandford and Allamandola, 1993, and references therein). Thus, the molecular inventory of cold, dark interstellar clouds must be shaped by a *combination* of ion-molecule reactions in the gas phase and gas-grain surface reactions (e.g. Charnley et al., 1992; Charnley, 1997). Moreover, the attenuated diffuse ISM UV which penetrates a dense cloud, as well as UV from internal sources and penetrating cosmic ray particles are sufficient to drive *in-situ*, solid-state reactions within the icy grain mantles leading to a variety of even more complex species.

It is within cold, dark molecular clouds such as this that new stars and planetary systems are born. Once part of a molecular cloud becomes unstable under its own gravitational field, it will begin to collapse, forming a protostar. As this collapse proceeds, the angular momentum possessed by the infalling material draws it into a disk. Planetary systems are thought to coalesce from the remnants of this protostellar accretion disk after the young star springs to life (the "Solar Nebula"). Thus, the raw material from which planetary systems form contains the biogenic elements in the same diverse states of molecular complexity found in the parent molecular cloud. Naturally, this biogenic material may also be modified to some extent by chemistry taking place during the collapse as well as in the accretion shock. Chemical processes at work during this epoch include equilibrium gas-phase reactions in the warmer regions of the nebula and non-equilibrium processes in colder regions and on coalescing planetessimals. Ultimately, the biogenic compounds present, whether produced in the nebula or accepted unchanged from the ISM, are incorporated into the condensed matter that became the planets, satellites, asteroids, and comets. Thus, the compounds that emerge from the interstellar/protostellar crucible provide the seed from which life must spring, and the study of these organic compounds is crucial to our understanding of the origin and early evolution of life.

Once a terrestrial planet is sufficiently cool to retain volatile materials, comets and meteoritic materials continuously pepper it with copious quantities of their complex organic inventory (e.g. Oro and Cosmovici, 1997). Consequently, from the standpoint of Astrobiology, it is important to understand both the nature of the large reservoir of complex, carbon-rich materials in the ISM, as well as the composition and chemistry encountered in cold interstellar ice grains - the building blocks of comets. Both of these sources can be much more chemically complex than the gaseous interstellar material. Furthermore, since icy grain mantles in all likelihood represent the largest repository of interstellar molecules in dense clouds, they tie up a large fraction of the biogenic elements available in these clouds. Given the likelihood that a substantial amount of material was delivered to the early earth through the influx of cometary matter, and in light of the molecular complexity now known to characterize such ices, it is reasonable to postulate that such materials may have played more than just a "spectators" role in the origin of life.

In this paper we review the foundations and summarize the current knowledge of the most complex organic molecules that comprise this extraterrestrial primordial soup. In section 1, the basis for the presence of PAHs in the ISM will be presented, followed by an overview of the laboratory studies that have been carried out to verify and refine the model. The salient astrophysical implications of this laboratory work are presented and insights into the size distribution, structure, and ionization state as a function of interstellar object type are drawn based on the cutting-edge astronomical data returned by the Infrared Space Observatory (ISO). The following section (§2) focuses on the evidence for interstellar/precometary ices and presents a summary of the laboratory work that has led to our current understanding of interstellar ice composition. The photochemical evolution of these ices in the absence and in the presence of PAHs is then considered with an eye toward the abiotic production of biologically active compounds. Finally, the possible contribution of this chemistry to the chemistry of the primordial Earth and the origin of life is considered.

1.   INTERSTELLAR PAHS: THE LARGEST INTERSTELLAR MOLECULES

1.1   INTERSTELLAR PAHS: THE OBSERVATIONAL FOUNDATION

A discrete emission feature at 11.3 μm discovered by Gillett, Forrest, and Merrill in 1973 (Gillett et al., 1973) was the first member of a now-well-known family of interstellar emission bands to be reported. Subsequent observations by a host of observers (c.f. Russell et al., 1977; Aitken, 1981; Willner, 1984; Phillips et al., 1984; and references therein) revealed that this was just one element in a recurring set of prominent emission bands at 3.3, 6.2, 7.7, and 8.6 μm as well as a complex array of minor bands, plateaus, and underlying continua. This spectrum is now known to be an integral part of the IR emission from many different astronomical objects including H II regions (Figure 2), planetary nebulae, reflection nebulae, and the ISM of our galaxy as well as that of other galaxies (A&A Special Edition, 1996). Moreover, these features carry 20-40% of the total IR luminosity from many of these galactic objects (Puget, 1989), indicating that the carrier represents an abundant and ubiquitous component of the ISM. Calculations show that the carriers of these features are as abundant as the most abundant simple polyatomic molecule, NH_3 (10^{-6}-10^{-7} with respect to hydrogen; e.g. Leger and Puget, 1984; Allamandola et al., 1985). Remarkably, recent ISO observations have even indicated that these features are sufficiently prevalent and intense in distant galaxies that they might be useful as red-shift indicators (e.g. Genzel et al., 1998).

It was first proposed in the mid-1980's that this widespread emission spectrum might be diagnostic of gas phase polycyclic aromatic hydrocarbon (PAH) molecules and closely related species (Leger and Puget, 1984; Allamandola et al., 1985). The rationale underlying this suggestion is straightforward. First, the emission bands are non-thermal in nature - that is, they are observed even in regions where the dust temperature is too low for the material to be emitting thermally (Leger and Puget, 1984; Allamandola et al., 1985; Greenberg, 1971; Purcell, 1976; Sellgren, 1984). Thus, the emission must be excited by the absorption of individual UV/visible photons, implying that the carriers are free, gas phase molecules rather than a solid state material. Second, there is a direct correlation between carbon abundance and the intensity of the emission features (Cohen et al., 1986; Cohen et al., 1989), implying that the gas phase carriers are carbon-based molecules. Third, the emission features are observed even from extremely harsh environments, indicating that the gaseous, carbon-rich molecules are exceptionally stable. Finally, the positions of the interstellar emission features provide insight into the chemical nature of the material from which they originate. Significantly, prominent bands in the interstellar emission spectrum fall at all the positions that would be expected to arise from the vibrational transitions of aromatic molecules. Taken together, these elements provide strong evidence that polycyclic aromatic hydrocarbons are prevalent in the ISM.

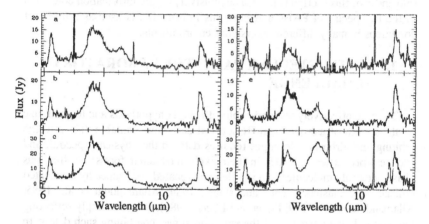

Figure 2. The 6 - 12 μm spectra of six compact HII regions illustrating the appearance of the common interstellar emission features at 6.2, 7.7, 8.6, and 11.2 μm. The sources are IRAS: (a) 8116-1646; (b) 18162-2048; (c) 19442-2427; (d) 21190+5140; (e) 22308+5812; and (f)18434-0242. Figure adapted from Roelfsma et al. (1996).

Obviously, the existence of a previously-unsuspected, yet very abundant interstellar material such as this could have important ramifications in many areas of astrophysics, and a number of new perspectives on long-standing problems were born in the years following the introduction of the PAH model. For example, since the ubiquity and complexity of the PAH population reflect a chemical history very different from the traditional ion-molecule interstellar chemistry, they were posited as the molecular "missing link" to larger carbonaceous dust grains (Allamandola et al., 1985). Indeed, observations are beginning to support this picture (Tielens et al., 1993). Others suggested that PAHs could fundamentally influence temperature (d'Hendecourt and Leger, 1987; Lepp and Dalgarno, 1988), and ionization level, chemistry, and radiative transfer (Omont, 1986) within molecular clouds, thereby influencing star formation; that interstellar PAHs may have been the source of much of the carbon in meteorites and interplanetary dust particles (Allamandola et al., 1987a; Allamandola et al., 1989; DeVries et al., 1993; Clemett et al., 1993); and that PAHs in one form or another might account for some of the enigmatic visible/near-IR diffuse interstellar bands (DIBs) (van der Zwet and Allamandola, 1985; Leger and d'Hendecourt, 1985; Crawford et al., 1985), a very long-standing interstellar puzzle. Comprehensive discussions of the PAH model as it stood at the end of the 1980's can be found in the literature (Allamandola et al., 1989; Puget and Leger, 1989) and a more thorough discussion of the widespread evidence for PAHs can be found in reference (Allamandola, 1996). The purpose of the remainder of this section is to focus exclusively on the information contained in the IR spectra of PAHs and how this information is used as a probe of conditions in many different interstellar environments.

1.2   INTERSTELLAR PAHS: THE LABORATORY CHALLENGE

To more effectively test and exploit the PAH hypothesis and to capitalize on the wealth of astronomical IR spectral data now available thanks to the orbiting and airborne telescopes requires data on the physical, spectral, and chemical properties of PAHs in their likely interstellar forms: i.e. free, gas phase neutral molecules, ions, dehydrogenated or super-hydrogenated radicals, and clusters (Leger and Puget, 1984; Allamandola et al., 1985; Allamandola et al., 1989; Puget and Leger, 1989). Beyond simply reflecting the physical characteristics of the emission zones, combining such data with the latest observations provides a unique window on the chemical evolution of carbonaceous material throughout the ISM. For example, significant spectral variations have been found between objects of a particular class (e.g. Figure 2) and between the various classes of IR emitting objects (protoplanetary nebulae, planetary nebulae, HII regions, reflection nebulae, etc), implying that different PAH populations are favored in different regions (Sloan et al., 1993; Bregman et al., 1993a,b; Bernstein et al., 1996; Sloan et al., 1996; Sloan et al., 1999). Further, comprehensive astrophysical models based on such observations hold the promise to provide insight into the chemical make-up of different objects; to trace the chemical evolution of those objects as they change from one stage to another; and to probe the ionization balance and other conditions within the emission zones over the wide range of objects which emit the features. Nevertheless, to gain this valuable insight, such models require a thorough understanding of the spectroscopic properties of PAHs, and the interplay between PAH vibronic states and their associated radiative photoprocesses with the interstellar radiation field - fundamental molecular information which can only be obtained through appropriate laboratory experiments.

Unfortunate in this regard, early testing and exploitation of the PAH hypothesis was severely hampered by a general lack of knowledge of the spectroscopic properties of PAHs under astrophysically relevant conditions. At the time of its inception, the laboratory data available to the PAH model were limited primarily to spectra measured from pure PAH crystals, from PAHs dispersed in salt pellets or organic solvents, or from PAHs embedded in glassy melts. Under such conditions, the individual PAH molecules are not effectively isolated and interact strongly with each other and/or with the surrounding medium. These conditions strongly influence the measured spectra and are far from the cold, isolated conditions encountered in the ISM. Furthermore - and in retrospect, most importantly - all of the then-available spectroscopic data were for *neutral* PAHs, while the I/S PAHs were, instead, believed to be *ionized*. Consequently, all that could be said by the end of the 1980's was that the interstellar spectra "resembled" the spectra of some neutral PAHs.

Be that as it may, it is one thing to say that one needs astrophysically relevant spectroscopic data, but it is quite another to actually realize that goal in the laboratory. Within the framework of the PAH model, the interstellar infrared emission arises from transiently-heated, gas-phase PAHs in both neutral and ionized forms. The goal of laboratory work in this area has been to measure the spectroscopic properties of these species under conditions which approximate those encountered in the ISM. At first glance, it might seem that spectral measurements of gas phase PAHs might easily be undertaken through the use of a simple single- or multipass gas cell, suitably modified for the study of these compounds. Specifically, since all but the smallest PAHs have vanishingly low vapor pressures at room temperature, the gas cell would have to be heated to facilitate warming the sample within to achieve a measurable vapor pressure. Indeed, in this manner it has been possible to measure the absorption and emission spectra of a few species and these studies have produced important insights into the astrophysical problem (Flickinger and Wdowiak, 1990; Flickinger et al., 1991; Brenner and Barker, 1992; Kurtz, 1992; Joblin et al., 1995). Nevertheless, this technique faces a number of serious practical difficulties for all but the very smallest PAHs. Moreover, these difficulties become increasingly problematic as molecular size increases and the volatility of the sample decreases. Since the optical surfaces of the cell cannot be heated as effectively as the rest of the cell, for PAHs larger than about 14 carbon atoms the pure material slowly but inexorably crystallizes out on these cooler surfaces over the course of an experiment. As this happens, the overall throughput of the cell gradually drops and the spectrum of the solid PAH becomes increasingly prominent until it completely dominates the measured spectrum.

Another problem one encounters in laboratory studies of PAHs centers on the overall *efficiency* of the experimental technique - that is, the amount of spectroscopic data one can collect per mass of PAH sample consumed. This is of concern because many PAHs - particularly the large PAHs thought to be important in the ISM - are often obtainable only in milligram quantities and/or at great expense. Furthermore, clean-up after an experiment will entail exposure to the often carcinogenic/mutagenic PAH material and disposal of the resulting potentially hazardous waste. The standard gas-cell technique is particularly *inefficient* because it requires a relatively large sample of the solid PAH to fill a cell with PAH vapor and maintain it against crystallization, and because recovery of the spent material after the experiment is usually not practical. What's more, even if one should overcome all of these difficulties, one would only have succeeded in studying *neutral* PAHs. To study *ionized* PAHs one must overcome all of these difficulties plus wrestle with the problem of generating and maintaining a measurable population of PAH ions under these conditions.

Finally, experimental challenges aside, perhaps the most insidious problem associated with undertaking astrophysically relevant spectroscopic studies of PAHs in the laboratory stems from the counterintuitive photophysics of the interstellar emission process. In a conventional laboratory sample of PAH vapor, generated by appropriate warming of a solid PAH, the individual PAH molecules are fully thermalized at the vaporization temperature, T_{vap}, i.e. $T_{vap} \approx T_{vib} \approx T_{rot}$, where T_{rot} and T_{vib} are the rotational and vibrational temperatures of the molecules which reflect the internal energy content of their respective modes. Typically, even for modestly sized PAHs, T_{vap} is on the order of a few hundred Kelvin or higher and the observed infrared bands are heavily broadened as the spectral intensity for each vibrational transition is distributed over an extensive manifold of rotational substates. This is *not*, however, the case for the interstellar PAHs. In the ISM, the average time between excitation events is typically far longer than that required for radiative relaxation. Thus, the PAH molecules are not in thermal equilibrium with their surroundings, instead spending the majority of their time in a fully rotationally and vibrationally relaxed state at an effective temperature of perhaps only ~100 K. When a fully relaxed species such as this absorbs a UV/visible photon, the initial electronic excitation energy is almost immediately converted to vibrational excitation, and the effective vibrational temperature of the molecule skyrockets. However, under the collisionless conditions of the ISM, angular momentum conservation *prevents* the redistribution of any internal energy into rotations. Thus, *the effective rotational temperature of the molecule remains low.* As a result, in contrast to the broad, double-humped rovibrational envelopes observed in the laboratory spectrum, the interstellar emission bands display a much narrower profile resulting from essentially pure vibrational transitions (Allamandola et al., 1989; Barker and Cherchneff, 1989; Leger et al., 1989). Thus, while the early gas-phase measurements discussed above provided an important milestone and test of the PAH hypothesis, further testing and exploitation of the model necessitated that other tools be brought to bear on the problem.

Faced with these limitations, verification and advancement of the PAH model has catalyzed the development of other creative experimental procedures that reduce or eliminate the difficulties associated with the straightforward, heated gas-cell method described above. These include both gas phase and solid-state, astrophysically relevant spectroscopic techniques. In the Astrochemistry Laboratory at NASA Ames a major effort has been underway for the last decade directed towards addressing the need for astrophysically-relevant, laboratory spectroscopic data on PAHs from the ultraviolet (190 nm) through the far-infrared (200 μm). Here we will focus on the results of our IR studies of PAHs as this is where the astronomical and laboratory databases are most extensive and most directly relevant to the astrobiological issues discussed in §3. Recent excellent reviews of the current research into the spectroscopy and astrophysical impact of PAHs in the ultraviolet/visible region of the spectrum can be found elsewhere (Salama et al., 1996; Salama et al., 1999).

To overcome the practical experimental difficulties outlined above, we have employed the matrix isolation technique to generate a database of infrared spectra of PAHs and PAH ions. In this technique, PAH vapor is generated by warming a sample of the solid in a small test tube mounted on a high vacuum chamber. The vapor effuses from the tube and is co-condensed with an overabundance of an inert gas (typically argon for infrared studies) onto a cryogenically cooled (10 K) infrared window suspended within the vacuum chamber. In this highly diluted sample, each PAH molecule is isolated from its neighbors and interacts only very weakly with the inert matrix, resulting in a quasi-gas phase condition. Furthermore, this condition can be maintained indefinitely as long as the window temperature is maintained. Cold PAH ions are generated by subsequent *in-situ* UV photolysis of the matrix-isolated neutral species and their spectral features distinguished from that of the neutral by comparison of spectra measured before and after photolysis. More complete discussions of the matrix isolation technique and the various experimental methods that have been employed to generate and study the IR spectral characteristics of neutral PAHs, PAH cations, and PAH anions can be found elsewhere (Hudgins et al., 1994; Hudgins and Allamandola, 1995a; Halasinski et al., 2000). Among its advantages, this technique is extremely efficient. Essentially all of the vaporized material that exits the reservoir tube is incorporated into the matrix-isolated sample. Since only ~100 - 200 μg of matrix-isolated PAH are required for an experiment, PAH samples of only a few milligrams are sufficient for many experiments. Clean-up is also greatly simplified, entailing minimal waste and exposure hazard, since the PAH residue remaining after the experiment is small and confined to the sample window. Additionally, although the matrix-isolation technique is limited to the measurement of absorption spectra, it faithfully reproduces the low rotational temperatures of interstellar PAHs, and careful modeling together with the latest experimental studies of jet-cooled, gas phase PAHs have shown that a simple thermal model is adequate for calculating of the astrophysical emission spectrum of PAHs based on their absorption spectra (Leger et al., 1989; Schutte et al., 1993b; Cook and Saykally, 1998). Finally, regarding the fidelity of argon matrix-isolated vibrational spectra relative to their corresponding gas phase spectra, over 40 years of matrix isolation research has shown that for most species, infrared band positions typically fall within 5 to 10 cm^{-1} ($\leq$ 1%) of their corresponding gas phase values. Theoretical and gas phase experimental studies have specifically corroborated this accuracy for PAHs and their ions.

Using the matrix-isolation technique, we have generated a spectral database which includes the infrared spectra of over 100 neutral, cationic, and anionic PAHs ranging in size from $C_{10}H_8$ to $C_{48}H_{20}$ (Hudgins et al., 1994; Hudgins and Allamandola, 1995a,b; Hudgins and Allamandola, 1997; Hudgins and Sandford, 1998a,b,c; Hudgins et al., 2000; Bauschlicher et al., 1997; Langhoff et al., 1998; Hudgins et al., 2001; Mattioda et al., 2002). The species currently included in the database are listed in Table 1. Amongst the species currently represented in the dataset are: (1) the thermodynamically most stable PAHs through coronene, $C_{24}H_{12}$, the molecules most likely to be amongst the smallest interstellar PAHs; (2) a representative sampling of species from the fluoranthene family, aromatic hydrocarbons which incorporate a five-membered ring in their carbon skeleton; (3) dicoronylene, $C_{48}H_{20}$, the largest PAH studied to date; and (4) a variety of "aza-PAHs", polycyclic aromatic compounds with a nitrogen atom incorporated into their carbon skeleton. We have also begun the process of making these data available to the scientific community on the internet at <http://web99.arc.nasa.gov/~astrochm/pahdata/index.html>. As discussed in detail in the next section, this data, together with that deriving from similar experimental studies conducted by Vala and coworkers at the University of Florida (Szczepanski et al., 1992; Szczepanski and Vala, 1993a,b; Szczepanski et al., 1993a,b; Vala et al., 1994; Szczepanski et al., 1995a,b) and extensive theoretical studies (Talbi et al., 1993; de Frees et al., 1993; Pauzat et al., 1995; Pauzat et al., 1997; Pauzat et al., 1999; Pauzat and Ellinger, 2001; Langhoff, 1996; Bauschlicher and Langhoff, 1997;

Bauschlicher, 1998a,b; Bauschlicher and Langhoff, 1998; Bauschlicher and Bakes, 2000) has established that *mixtures of free molecular PAHs, dominated by PAH ions, can account for the global appearance of the interstellar emission spectra and the variations of those spectra.*

Table 1. The inventory of the PAH species studied to date in the Astrochemistry Lab.

Formula	Name	Ion State
C_9H_7N	quinoline	0
	isoquinoline	0
$C_{10}H_8$	naphthalene	0,+
$C_{12}H_8$	acenaphthylene	0
$C_{13}H_9N$	acridine	0,+
	7,8-benzoquinoline	0,+
$C_{14}H_{10}$	phenanthridine	0,+
	anthracene	0,+
	phenanthrene	0,+
$C_{14}H_{11}N$	2-aminoanthracene	0,+
$C_{15}H_9N$	9-cyanoanthracene	0,+
$C_{15}H_{12}$	1-methylanthracene	0,+
	9-methylanthracene	0,+
$C_{16}H_{10}$	fluoranthene	0,+
	pyrene	0,+
$C_{16}H_{12}$	4,5-dihydropyrene	0
$C_{16}H_{16}$	1,2,3,6,7,8-hexahydropyrene	0
$C_{18}H_{10}$	benzo[ghi]fluoranthene	0
$C_{17}H_{11}N$	1-azabenz[a]anthracene	0,+
	2-azabenz[a]anthracene	0,+
	1-azachrysene	0,+
	2-azachrysene	0,+
	4-azachrysene	0,+
$C_{18}H_{12}$	1,2-benzanthracene	0,+
	chrysene	0,+
	tetracene	0,+
	triphenylene	0,+
$C_{19}H_{14}$	7,8-dihydro-9H-cyclopenta[a]pyrene	0
$C_{20}H_{12}$	benzo[a]fluoranthene	-,0,+
	benzo[b]fluoranthene	0,+

Formula	Name	Ion State
$C_{20}H_{12}$	benzo[j]fluoranthene	-,0,+
	benzo[k]fluoranthene	0,+
	benzo[a]pyrene	0,+
	benzo[e]pyrene	0,+
	perylene	0
$C_{20}H_{14}$	7,8-dihydrobenzo[a]pyrene	0
$C_{21}H_{13}N$	9,10-dihydrobenzo[e]pyrene	0,+
	dibenz[a,h]acridine	0,+
	dibenz[a,j]acridine	0,+
$C_{22}H_{12}$	benzo[ghi]perylene	0,+
$C_{22}H_{14}$	pentacene	-,0,+
$C_{22}H_{16}$	7,14-dihydrodibenz[a,h]anthracene	0
$C_{24}H_{12}$	coronene	0,+
$C_{24}H_{14}$	dibenz[a,e]pyrene	0,+
	dibenz[a,l]pyrene	0,+
	naphtho[2,3;a]pyrene	0,+
$C_{36}H_{16}$	3,4;5;6;7;8,12,13-tetrabenzoperopyrene	0,+
	3,4;5;6;10;11;12,13-tetrabenzoperopyrene	
$C_{40}H_{18}$	dipyreno-(1';3':10,2),(1'',3'':5,7)-pyrene	0,+
$C_{40}H_{22}$	dianthraceno-(2;3':3,4),(2'';3'':9,10)-pyrene	0,+
$C_{42}H_{18}$	1,12;2;3;4,5;6;7;8,9,10,11-hexabenzocoronene	0,+
$C_{42}H_{22}$	1,18;4,5,9,10,13,14-tetrabenzoheptacene	0,+
$C_{44}H_{20}$	2,3;12,13;15,16-tribenzoterrylene	0,+
$C_{48}H_{20}$	1;2;3;4;5;6;7;8;9,10;12,13-hexabenzoperopyrene	-,0,+
	dicoronylene	

The culmination of all this work is the exploitation of PAHs as probes of local conditions in objects that span all stages of the lifecycle of interstellar matter. This entails quantitative comparisons of experimentally-measured infrared spectra of PAHs and PAH-related materials with the observed interstellar infrared emission spectra with the goal of gaining insight into the physical and chemical conditions within those emitting regions. Through comprehensive astrophysical modeling based on sound experimentally-measured spectroscopic properties, it will be possible to understand circumstellar dust formation and the entire evolutionary cycle of interstellar carbon-rich compounds from their production in late-stellar outflows to their incorporation into interstellar dust and new stellar systems. This information will further our knowledge of the role PAHs play in the chemical and dynamic evolution of interstellar clouds and the process of star formation.

Finally, we would be remiss if we were to fail to point out the emergence of two elegant new experimental techniques which permit direct spectroscopic measurements of gas-phase PAHs under conditions which are directly relevant to the astrophysical problem. These studies provide an important assessment of the astrophysical utility of the matrix-isolation spectral database and provide direct insight into the photophysics of the interstellar emission process.

First, working at the Netherlands National Free Electron Laser Facility, Meijer and coworkers have developed a technique for directly measuring the infrared absorption spectrum of cold, gas phase PAH cations. In this technique, a jet cooled beam of PAH cation - inert gas atom van der Waals clusters is probed with the tunable infrared beam of a free electron laser. Infrared absorptions by the PAH cation precipitate dissociation of the fragile van der Waals clusters and are detected as a reduction in total cluster flux. Using this technique, Piest et al. have succeeded in measuring the IR spectrum of the naphthalene and phenanthrene cations in the gas phase (Piest et al., 1999; Piest et al., 2001). Of direct interest here is the notable agreement between the infrared spectra obtained using this technique and those measured for the matrix isolated cations. This agreement is illustrated in Table 2 which compares the results of the two techniques for the naphthalene cation, $C_{10}H_8^+$. Inspection of the table reveals that the majority of the argon matrix isolated vibrational band frequencies fall within 4 cm^{-1} (~0.4%) of the gas phase values determined using the cluster dissociation method (only one band differs by as much 8 cm^{-1}). Such a modest matrix shift is negligible in the context of the astrophysical problem, where the natural linewidth of a vibrationally excited PAH emitting under interstellar conditions is ~30 cm^{-1}. This result reinforces the validity of the use of matrix isolation data for the analysis of the interstellar infrared emission spectra.

Table 2. The frequencies and relative intensities for the cation of the PAH naphthalene ($C_{10}H_8$) recently measured in the gas phase compared to the values measured using matrix isolation. The excellent agreement between the matrix and gas phase data for this PAH cation validates the matrix isolation approach to this problem.

Gas Phase[a] Freq. (cm^{-1})	Relative Intensity	Matrix[b] Freq. (cm^{-1})	Relative Intensity	Matrix[c] Freq. (cm^{-1})	Relative Intensity
589	0.04	---		---	
759	0.05	758.7	0.27	---	
1019	0.11	1016/1023.2	0.05	1016	0.2
1121	0.01	---		---	
1168	0.08	---		---	
1215	1.0	1214.9	0.2	1215/1218	1.0
		1218.0	1.0		
1284	0.01	---		---	
1393	0.23	1400.9	0.04	1401	0.04
1523	0.4	1518.8/1525.7	0.39	1519/1525	0.16
1539	0.14				

[a] data taken from Piest et al. (1999); [b] data taken from Hudgins et al. (1994); [c] data taken from Szczepanski et al. (1992)

Second, working at the University of California, Berkeley, Saykally and coworkers have directly measured the infrared emission from cold, vibrationally excited gas phase PAHs, shedding light on the photophysics of the interstellar UV/vis-to-IR conversion process (Cook and Saykally, 1998; Schlemmer et al., 1994; Wagner et al., 2000; Kim et al., 2001). This work has been critically needed both to provide a demonstration of the infrared fluorescent emission mechanism and to test the validity of using the *absorption* properties of matrix-isolated PAHs to calculate the *emission* from vibrationally excited PAHs. Although these too are extremely challenging experiments, the infrared fluorescent emission from a number of neutral PAHs and the pyrene cation has successfully been measured under conditions identical to that of the I/S emitters. These studies have revealed heretofore unknown details of the emission process and support the quantum-/statistical-mechanical, molecular excitation-emission process put forward to calculate the emission of vibrationally excited PAHs and the use of matrix-isolation spectra. This approach has been used in many models to describe how PAH molecules convert interstellar UV-visible radiation into IR (Allamandola et al., 1989; Puget and Leger, 1989; Schutte et al., 1993b). In summary, gas-phase and matrix-isolation laboratory studies of PAH spectra represent complementary rather than redundant aspects of the experimental work needed, each addressing different, critical aspects of the astrophysical problem.

1.3   INTERSTELLAR PAHS: MID-INFRARED SPECTRAL PROPERTIES

One of the early important results of all the laboratory and theoretical studies on neutral and ionized PAHs is the remarkably dramatic effect ionization has on their infrared spectra (Hudgins et al., 1994; Hudgins and Allamandola, 1995a,b; Hudgins and Allamandola, 1997; Hudgins and Sandford, 1998a,b,c; Hudgins et al., 2000; Bauschlicher et al., 1997; Langhoff et al., 1998; Hudgins et al., 2001; Mattioda et al., 2002; Szczepanski et al., 1992; Szczepanski and Vala, 1993a,b; Szczepanski et al., 1993a,b; Vala et al., 1994; Szczepanski et al., 1995a,b; Talbi et al., 1993; de Frees et al., 1993; Pauzat et al., 1995; Pauzat et al., 1997; Pauzat et al., 1999; Pauzat and Ellinger, 2001; Langhoff, 1996; Bauschlicher and Langhoff, 1997; Bauschlicher, 1998a,b; Bauschlicher and Langhoff, 1998; Bauschlicher and Bakes, 2000; Allamandola et al., 1999). This effect is illustrated in Figure 3.

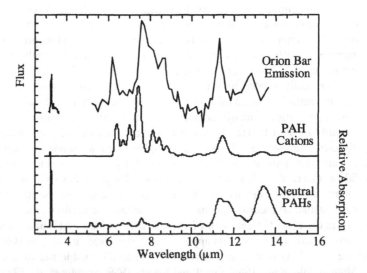

Figure 3. The absorption spectrum produced by coadding the spectra of the PAHs anthracene, tetracene, 1,2-benzanthracene, chrysene, pyrene, and coronene in their (a) neutral and (b) ionized forms compared to the emission spectrum of the ionization ridge in the Orion Nebula. This comparison shows that, for PAHs, ionization has a much greater influence on relative intensities than on peak frequencies, and that ionized PAHs are better candidates for the emission band carriers.

The infrared spectra of neutral PAHs are dominated by strong features arising from aromatic CH stretching vibrations near 3.3 μm and CH out-of-plane bending vibrations between 11 and 14 μm. Weaker features arising from aromatic CC stretching and CH in-plane bending vibrations are observed in the 6 to 9 μm range. In ionized PAHs, on the other hand, the situation is completely reversed. Enhanced by an order of magnitude relative to their neutral counterparts, the 6 to 9 μm CC stretching and CH in-plane bending modes now dominate the spectra of PAH cations. Conversely, *suppressed* by an order of magnitude, the CH stretching features have all but disappeared from the cation spectra. The CH out-of-plane bending modes are also suppressed in the cations, but much more modestly so (≈ 2x). Thus, as illustrated in Figure 3, ionization produces a global pattern of band intensities that is in much better agreement with the pattern of intensities observed in the interstellar emission spectrum than is the case for neutral PAHs. This relieves what has, since the time of its inception, been one of the most troubling qualitative difficulties of the interstellar PAH model - the disparity between the global appearance of the interstellar emission spectra and that of the initially available laboratory spectra of neutral PAHs (Allamandola et al., 1999).

Within the framework of the PAH model, the interstellar spectrum arises from the combined emission of a complex *mixture* of PAHs. Therefore, to have any hope of reproducing the appearance of this spectrum, one must consider *not the spectrum of any one PAH, but the composite spectrum of a variety of different PAHs*. Previous comparisons between the interstellar features and laboratory spectra have been forced to rely on the spectra of a single PAH or on limited combinations of a very few PAHs. With the advent of an extensive database of astrophysically relevant PAH spectra, comes the potential for much more comprehensive modeling of the emitting PAH population. This is illustrated in Figure 4 which shows a comparison between the *emission* from the ionized ridge in the Orion Nebula and the composite *absorption* spectrum generated by coaddition of 11 PAH spectra, judiciously chosen from the laboratory database (See Allamandola et al., 1999, for a discussion of the fitting procedure). Note that since the expected matrix shift (5-10 cm⁻¹) and vibrational excitation red-shift between absorption and emission band positions (5-15 cm⁻¹, Joblin et al., 1995) are significantly smaller than the 30 cm⁻¹ natural linewidth expected from the interstellar emitters, these effects have not been compensated for in the figure.

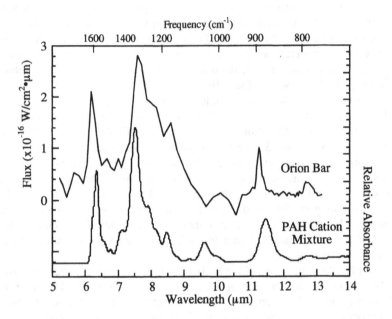

Figure 4. Comparison of the *emission* spectrum from the Orion ionization ridge to the composite *absorption* spectrum of a mixture composed only of *ionized* PAHs. The mixture consists of: 20% each benzo(k)fluoranthene[+]; 20% dicoronylene[+]; 10% coronene[+]; 10% benzo(b)fluoranthene[+]; 10% 9,10-dihydrobenzo(e)pyrene[+]; 10% phenanthrene[+]; 5% benzo(ghi)perylene[+]; 5% tetracene[+]; 5% benz(a)anthracene[+]; 3% chrysene[+]; and 2% fluoranthene[+]. The Orion spectrum is reproduced from Bregman et al. (1989).

1.4   INTERSTELLAR PAHS: PROBING THE EVOLUTION OF INTERSTELLAR CARBON

The following examples, adapted from Allamandola et al. (1999), serve to illustrate how an analysis of the interstellar emission spectrum can yield important insight into the nature and properties of the PAH population, and how this information reflects the physical and chemical conditions within the emission regions themselves.

The protoplanetary nebula phase likely represents the earliest stage in the lifecycle of cosmic PAHs (Allamandola et al., 1985; Allamandola et al., 1989; Gauger et al., 1990; Cherchneff et al., 1992; Cadwell et al., 1994). During the epoch of copious mass loss that punctuates the last stages of a star's life, C, N, and O produced during the final fitful stages of nucleosynthesis deep within the star are dredged up and cast off together with the majority of the dying star's atmosphere. If the abundance of carbon exceeds that of oxygen in this shell, a rich variety of carbon-rich compounds are formed. IRAS 22272+5435, whose spectrum is shown in Figure 5, is a carbon-rich object undergoing just such a transformation (Buss et al., 1993; Kwok et al., 1995). The observed infrared emission is excited by the remaining, relatively cool (T~ 5,300 K, Kwok et al., 1995), central giant star. Eventually, the outer layers of the star will be thrown off, exposing the ejecta to the harsh ionizing radiation of the still-extremely-hot (T ~ 50,000 to 150,000 K), stellar core and ushering in the planetary nebular phase. Thus, this transition phase, which lasts on the order of 10^3 years, is sometimes referred to as the *proto*-planetary nebula phase. The best fit we have found to the spectrum of IRAS 22272+5435 using the ca. 1999 database is also shown in Figure 5. Inspection of the composition of this mixture (given in the figure caption) reveals that it is dominated by neutral PAHs (~60%) and that it includes species with a broad range of stabilities, from large, condensed PAHs (e.g. dicoronylene) to naphthalene, the smallest PAH. Note also that the mixture is internally consistent in that the neutral and cationic forms of the same PAHs have been used to construct the fit (i.e there are no PAHs present in ionized form, but not neutral form, and vice-versa). A mixture such as this is certainly reasonable when one takes into consideration the nature of the object. Here, in the region where aromatic compounds are beginning to appear and before they have been exposed to the ferocious radiation field of the coming planetary nebula phase, it is logical to expect that the emitting material would contain a diverse mixture of species, representing a wide range of thermodynamic stabilities. Furthermore, given the relatively benign radiation field produced by a 5,300 K star, it is also expected that both neutral and ionized species should contribute to the emission. The PAH population which provides the fit shown in Figure 5 reflects exactly these characteristics.

Next, consider the very different environment represented by the ionization ridge in the Orion Nebula. The spectrum of that region, together with that of the best-fit mixture of species drawn from the database, was shown previously in Figure 4. The Orion ionization ridge represents the interface between a cold, dense molecular cloud and an H II region (Bregman et al., 1989). Here the material originally produced in late stellar outflows has been "aged" for perhaps a billion years. During this time the material has been passed back and forth many times between diffuse and dense cloud phases of the ISM, alternately bathed in the harsh galactic interstellar radiation field and frozen into icy grain mantles. Now this material is being exposed to the ionizing radiation from the adjacent hot young O stars which make up the Trapezium.

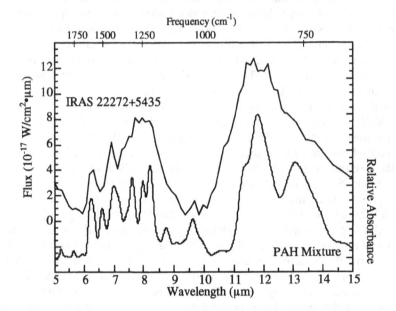

Figure 5. The *emission* spectrum from the protoplanetary nebula IRAS 22272+5435 compared to the *absorption* spectrum produced from a mixture of neutral and cationic PAHs. The mixture is comprised of ("°" indicates a neutral species; "+" indicates a cation): 18% dicoronylene°, 14% each naphthalene+ and 9,10-dihydrobenzo(e)pyrene+, 11% 9,10-dihydrobenzo[e]pyrene°, 10% each benzo[j]fluoranthene° and coronene°, and 3% each benzo[a]fluoranthene+, benzo[j]fluoranthene+, coronene+, hexabenzocoronene+, dicoronylene+, benzo[a]fluoranthene°, naphthalene°, and hexabenzocoronene°. The spectrum of IRAS 22272 +5435 is reproduced from Buss et al. (1993).

The composition of the mixture that provides the best fit to the Orion spectrum is quite revealing about the nature of the PAH population there. Unlike the proto-planetary nebula, where a substantial contribution from neutral PAHs was required to achieve a satisfactory fit, a mixture composed entirely of cationic PAHs is indicated for this region. Furthermore, the role of less stable PAH structures (i.e. less condensed) in the mix is substantially reduced compared to the protoplanetary nebula case. Instead, PAHs having more highly condensed (and therefore more thermodynamically favored) structures dominate the emission. In fact, four of the thermodynamically most favored PAHs in the mixture contribute 60% of the match to Orion shown in Figure 4. Closer inspection of the figure also shows that, although the model spectrum reproduces all the major peaks and relative intensities of the Orion spectrum reasonably well, the 1300 cm^{-1} feature is significantly narrower. This reflects the lack of the prominant 1280 cm^{-1} component (Bregman, 1989) and the broad emission plateau underlying the 1613, 1310 and 1160 cm^{-1} bands observed in the astronomical emission (Cohen et al., 1986; Cohen et al., 1989). These deficiencies are likely attributable to one or more classes of PAHs that are, as yet, not represented in the database and to the lack of any contribution from carbonaceous grains. The PAH population reflected in the model spectrum from Figure 4 is again entirely consistent with what one would expect given the nature and history of this object. The molecules found in this region are those which have survived the interstellar gauntlet and the fierce radiation from the nearby O/B stellar association. Lesser stable components of the carbon-rich material initially ejected into the ISM have long since been 'weeded out' - either destroyed or isomerized into more stable structures by energetic processing. In addition, in the presence of the intense ultraviolet radiation from the nearby, young O/B stars, it is expected that a substantial portion of the molecular population is likely to be ionized. Thus, it is entirely reasonable that we find the best-fit PAH mixture for the Orion ionization ridge reflects a disproportionately large contribution from the hardiest species and from ionized species.

While the PAH mixtures used to provide the spectral fits in the above examples are not unique, there is not a lot of variation possible in the choice of the dominant PAHs in each. Since IRAS 22272+5435 and the Orion ionization ridge represent very different epochs in the evolution of cosmic carbon, the spectral differences reveal how carbonaceous material evolves as it passes from its circumstellar birth site into the general ISM. While there can be great variability in the appearance of the UIR spectrum between objects or from one region to another within one object (A&A Special Edition, 1996), these differences can readily and naturally be accommodated by different PAH populations. The differences in the astronomical spectra are a direct consequence of differences in the composition of the emitting PAH population. The PAH population, in turn reflects a variety of physical and chemical conditions such as radiation field flux and energy, ionization states, carbon abundance, etc., in the emitting regions. Thus, given the ubiquity and intensity of the interstellar infrared emission features, PAHs hold the potential to provide a powerful probe of interstellar environments which span all the stages in the lifecycle of cosmic carbon.

Lastly, it should be emphasized that within the framework of the PAH model, the overall interstellar emission spectrum reflects contributions from both free, individual PAH molecules *and* from amorphous carbon grains comprised of a complex network of aromatic moieties. Earlier work has suggested that the particles can account for the broad, low contrast plateaus which underlie and vary with respect to the prominent sharp features (Cohen et al., 1986; Flickinger and Wdowiak, 1990; Blanco et al., 1988; Allamandola et al., 1987b). Figures 3, 4, and 5 provide a compelling proof of concept that mixtures of free molecular PAHs and PAH ions can account for the discrete emission features. Indeed, considering the fact that while PAHs containing fewer than 50 carbon atoms likely account for 95% of the observed interstellar emission at 3.3 μm, they contribute only ≈ 30% of the emission at 6.2 μm, 20% at 7.7 μm, and no more than a few percent at 11.2 μm (Schutte et al., 1993b), it is remarkable that such good matches can be obtained using a database that samples only the very smallest end of the interstellar PAH population.

1.5   INTERSTELLAR PAHS: PROBING THE INFRARED EMISSION ZONES

As illustrated in the previous section, the ability of the laboratory spectra to accommodate the global appearance and variations of the interstellar spectrum supports the PAH hypothesis and provides a powerful new probe of a wide variety of interstellar conditions and histories. The most recent such applications have been aimed at achieving a fuller understanding of the PAH model and the nature of the interstellar infrared emitters in light of the latest spectroscopic observations provided by the ISO satellite. These applications are providing significantly deeper insight into the nature of interstellar PAHs and carbon's interstellar lifecycle. Two examples of this work will be presented and discussed briefly here. The first deals with the recent discovery of a discrete 16.4 μm emission band and associated 15 to 20 μm emission plateau and their interpretation within the context of the PAH model (Van Kerckhoven et al., 2000; Moutou et al., 2000). The second deals with a detailed analysis of the complex substructure of the interstellar emission spectrum from 10.5 - 14 μm, and the implications of this analysis regarding the size and structure of the emitting PAH population (Hony et al., 2001).

1.5.1 The 15 to 20 μm Emission Plateau and a New Emission Band at 16.4 μm: Probing the presence of extremely large PAHs and PAH clusters.

Within the framework of the PAH model, spectral features observed in the 15 to 30 μm region are attributed to skeletal distortion modes of PAHs (i.e. in-plane and out-of-plane C-C-C bending). Given that these weak modes characteristically tend to congregate in this wavelength region, a broad, low-level continuum emission arising from the composite of many such bands was expected from PAH sources (Allamandola et al., 1989). As shown in Figure 6, this expectation, initially based on a very limited set of PAH spectra but now supported by the full weight of our infrared database, is borne out by the latest ISO observations.

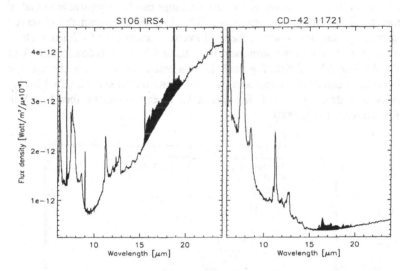

Figure 6. Two clear examples of the 15 - 20 μm plateau emission discovered by ISO and attributed to the emission from a mixture of PAHs (shaded). Figure adapted from Van Kerckhoven et al. (2000).

This identification adds further credence to the PAH model and can be used to deduce specific characteristics of the interstellar PAH population in a given environment. Interestingly, the available PAH spectral data indicate that, despite the dramatic effect observed in other regions of the mid-infrared spectrum, ionization has little or no impact on the absolute intensities of the PAH bands in the 15 - 30 μm range. Overall, the integrated intensities of the individual PAH bands observed in this region typically range from 2 to 30 km/mol independent of ionization state - roughly 5-10% that of the most intense bands in the spectrum (usually found between 11 and 14 μm in the neutrals and between 7 and 8 μm in the cations) (Allamandola et al., 1989; Hudgins et al., 1994; Hudgins and Allamandola, 1995a,b; Cook and Saykally, 1998; Hudgins and Allamandola, 1997; Hudgins and Sandford, 1998a,b,c; Hudgins et al., 2000; Bauschlicher et al., 1997; Langhoff et al., 1998; Hudgins et al., 2001; Mattioda et al., 2002; Szczepanski et al., 1992; Szczepanski and Vala, 1993a,b; Szczepanski et al., 1993a,b; Vala et al., 1994; Szczepanski et al., 1995a,b; Talbi et al., 1993; de Frees et al., 1993; Pauzat et al., 1995; Pauzat et al., 1997; Pauzat et al., 1999; Pauzat and

Ellinger, 2001; Langhoff, 1996; Bauschlicher and Langhoff, 1997; Bauschlicher, 1998a,b; Bauschlicher and Langhoff, 1998; Bauschlicher and Bakes, 2000; Piest et al., 1999; Piest et al., 2001; Schlemmer et al., 1994; Wagner et al., 2000; Kim et al., 2001). That notwithstanding, the *sum* of the intensities of the bands in the 15 - 30 μm range can be comparable to that of the strongest band in the spectrum. This is consistent with the observed interstellar emission in which the total intensity under the 15 - 20μm plateau often exceeds that of the prominent interstellar 6.2 μm emission feature (Van Kerckhoven et al., 2000). Figure 7 shows a comparison between the average interstellar emission plateau and two composite model spectra based on the laboratory data. Details of the interstellar spectrum can be found in Van Kerckhoven et al. (2000).

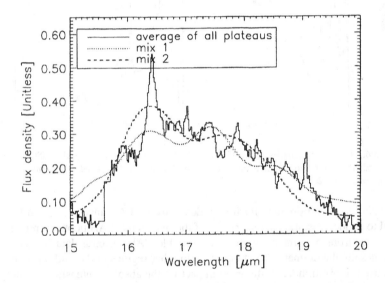

Figure 7. Comparison between the average 15 - 20 μm interstellar emission plateau and laboratory spectra. The average emission was obtained by coadding all the interstellar plateaus normalized to the integrated intensity. The first laboratory spectrum (mix 1) is produced by the coadded spectra of anthracene (33%), 1,2-benz[a]anthracene (33%), and pentacene (33%). The second laboratory spectrum (mix 2) is a spectrum produced by the coadded spectra of benzo(k)fluoranthene (20%), pentacene (40%), anthracene (20%), and 1,2 benzanthracene (20%). Figure adapted from Van Kerckhoven et al. (2000).

The model spectra were prepared by direct coaddition of astrophysically realistic simulations of the individual PAH spectra. Those simulations were prepared using the experimentally-measured band positions and intensities and assigning each band a nominal 30 cm^{-1} Gaussian profile consistent with the natural linewidth of the interstellar emitters. The figure clearly illustrates the broad feature produced by coadding the absorption spectra of just 7 matrix isolated PAHs. Note that not only is there good agreement between the interstellar and the laboratory data with respect to the intensity of the emission plateau compared to the rest of the spectrum, but also that there is good agreement as to the extent of the plateau (i.e. the wavelength range). This figure demonstrates that a complex of interwoven PAH bands in the 15-20 μm region can produce a structured quasi-continuum reminiscent of the observed plateau.

While there is some modest variation in the profile and extent of the emission plateau in different sources, the most pronounced distinction involves a relatively narrow, discrete band superimposed on the plateau and centered near 16.4 μm (see Figure 7). While this feature is present in many of the regions displaying 15 - 20 μm plateau emission, its intensity relative to that of the plateau is highly variable and there are regions where it is completely absent, suggesting a related but distinct origin.

Another singular aspect of this band is its FWHH - $\leq$ 10 cm^{-1} - substantially narrower then any other member of the emission band family and that expected based on our understanding of the photophysics of the interstellar emission process discussed above. While this latter characteristic remains enigmatic, we are not prevented from seeking an interpretation of a discrete feature at this position within the context of the PAH model.

Infrared modes of PAHs over roughly the 15 - 17 μm spectral range are characterized by *in-plane* C-C-C bending motions, whereas those falling longward of about 17 μm are dominated by *out-of-plane* warping of the carbon skeleton (Bauschlicher, private communication.). Interestingly, one class of PAHs that consistently show a band between 16 and 17 μm which could overlap to generate a striking feature at this location are the fluoranthenes - PAHs which incorporate a five-membered ring in their skeletal structure (Hudgins and Sandford, 1998c; Hudgins et al., 2000; Moutou et al., 2000). Based on this trend, it is reasonable to posit that a vibrational motion peculiar to the five-membered ring might be responsible for the feature observed at this position. This hypothesis, however, is not borne out by a detailed analysis of the specific atomic displacements associated with these modes. Such an analysis indicates that the fluoranthene modes between 16 and 17 μm arise *not* from a vibration of the five-membered ring, *per se*, but instead from an in-plane distortion of a "pendant" hexagonal ring in their structures (i.e. a benzenoid ring fused to the remaining molecule along a single face). This distortion is illustrated in Figure 8. It involves an in-phase oscillation of the two opposing C atoms immediately adjacent to the fused face of the pendant ring along a line parallel to that face. This characterization is confirmed by a similar analysis of other species from our database which carry a pendant benzenoid ring but no five-membered ring in their structures, and which exhibit a distinct feature near 16.4 μm. Thus, while it does not appear that the 16.4 μm band specifically traces PAHs which carry five membered rings, it is nonetheless reasonable to suggest that the distinctive symmetry of the pentagonal ring represents one way to enhance the IR activity in C-C-C bending modes in the 16 to 17 μm range which otherwise would be weak or forbidden. This effect may also become particularly important for asymmetric, non-condensed PAHs.

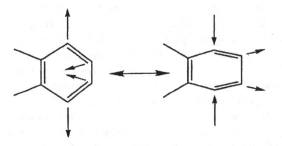

Figure 8. The in-plane C-C-C bending of the pendant ring that characterizes the vibrational modes between 16 and 17 μm in the molecules in our database. It involves an in-phase oscillation of the two opposing C atoms immediately adjacent to the fused face of the pendant ring along a line parallel to that face. Figure adapted from Van Kerckhoven et al. (2000).

While this analysis does reveal the origin of the vibrational motion that produces a band in the 16 to 17 μm region of the laboratory spectra, it should be noted that it is unlikely that the interstellar emission feature can be explained by this class of vibration alone. A pendant benzenoid ring will also necessarily contribute a quadruply-adjacent set of CH groups to the molecule. The out-of-plane bending mode of such a group typically falls near 14 μm and is in all cases substantially stronger than the apical distortion band associated with that ring. Consequently, if the latter mode were at the heart of the interstellar 16.4 μm band, there should be a significantly stronger emission feature near 14 μm. While the latest observations do reveal a band near this position, it is too weak to account for the implied quartet CH out-of-plane bending feature. Thus, while the pendant ring apical distortion identified here may contribute to the discrete 16.4 μm interstellar emission band, it cannot, in and of itself, completely account for the feature.

Finally, the total integrated intensity of the 15 - 20 μm emission plateau has been found to be as much as an order-of-magnitude higher in regions of massive star formation than in those regions where intermediate- and low-mass stars form. It has been suggested that this variation is reflective of an increase in the relative contribution from extremely large molecular PAHs and/or PAH clusters ($\gtrsim$ 500 C atoms (Van Kerckhoven et al., 2000)). Possessing large heat capacities (from a molecular standpoint), these species never attain the high ($\geq$1000 K) peak temperatures achieved by the species which dominate the 3 - 15 μm emission (i.e. $\leq$ 50 to ~200 C atoms). Consequently, they emit a disproportionate fraction of their energy at long wavelengths. Perhaps the prominence of this feature in regions of massive star formation reflects the effects of more extensive PAH coagulation compared to those regions which spawn intermediate and low-mass stars. Exploration of this possibility and its implications will require further observation and laboratory study.

1.5.2 PAH emission in the 10 to 15 μm region: Probing the size and structure of the emitters.

High resolution ISO/SWS observations in the 10 - 15 μm spectral region for a wide variety of objects which exhibit the interstellar infrared emission bands has revealed a wealth of new structure and detail. This structure is illustrated in Figure 9 which shows the 10 - 15 μm spectra of three representative sources: NGC 7027, CD-42 11721, and IRAS 18317-0757. Figure 9 reveals discrete features at 10.6, 11.0, 11.23, 12.0, 12.7, 13.5, and 14.2 μm with intensities relative to one another that are highly variable. A detailed discussion of the observations and spectral characteristics of the interstellar features in this region are provided in Hony et al. (2001), and is not repeated here. Our purpose will be to consider what the spectral structure in this region reveals about the nature of the interstellar PAH population.

Chemists have long recognized the diagnostic value of the aromatic CH out-of-plane bending features in the 11 to 15 μm spectral region for the classification of the aromatic ring edge structures present in a particular sample (Bellamy, 1958). Specifically, the positions of the bands in this spectral region reflect the number of adjacent CH groups on the peripheral rings of the PAH structure (Allamandola et al., 1985; Leger et al., 1989; Allamandola et al., 1999; Bellamy, 1958; Hudgins and Allamandola, 1999; Cohen et al., 1985; Roche et al., 1989; Witteborn et al., 1989). Traditionally, aromatic rings carrying CH groups which have no neighboring CH groups (termed "non-adjacent" or "solo" CH groups) show IR activity between 11.1 and 11.6 μm. Likewise, activity between 11.6 and 12.5 μm is indicative of two adjacent CH groups ("doubly-adjacent" or "duet" CH's) on the periphery of the PAH. Three adjacent CH groups ("triply-adjacent" or "trio" CH's) are indicated by activity in the 12.4 to 13.3 μm region, and four adjacent CH groups ("quadruply-adjacent" or "quartet" CH's) by activity between 13 and

13.6 µm. Five adjacent CH groups ("quintuply-adjacent" or quintet CH's) are indicated by features falling in the 13 to 13.7 µm range. Trios and quintets also show a weak CCC bending mode in the 14 - 14.5 range. Other such CCC bending modes were discussed above. Over the years the reliability of this region to yield insight into the molecular structure and ring sidegroup placement on aromatic samples has been verified again and again (see Hudgins and Allamandola, 1999, and references therein). However, most of these chemist's guidelines were based on studies of small PAHs where varying patterns of sidegroup substitution were employed to achieve different degrees of CH adjacency. Furthermore, these chemist's 'rules-of-thumb' are based on spectroscopic studies of aromatic molecules in solution or solid mixtures which, as discussed in §2.2, are of questionable astrophysical applicability.

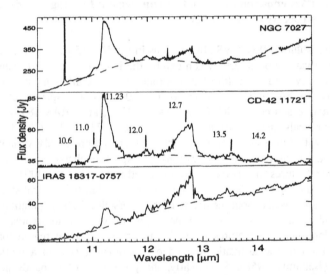

Figure 9. The 10 - 15 µm spectra of three typical sources revealing an extremely rich collection of variable emission features with bands at 10.6, 11.0, 11.23, 12.0, 12.7, 13.5, and 14.2 µm. These features are perched on top of an emission plateau of variable strength, which extends across the entire region (dotted curves underlying spectra). Figure adapted from Hony et al. (2001).

From the standpoint of the astrophysical problem, there are two particularly pertinent issues regarding the characteristic PAH modes in the 10 - 15 µm range that can be addressed by an analysis of the laboratory database. The first is the effect of ionization on the characteristic wavelength regions of the various CH adjacency classes. The second is the intrinsic integrated absorption strengths (A values) of the various adjacency classes and the effect of ionization on those strengths. Together, this information provides the tools not only to *qualitatively* infer the sorts of PAH edge structures present in the interstellar population, but also to *quantitatively* determine their relative amounts.

The first such analysis of the laboratory database was presented by Hudgins and Allamandola (1999) and subsequently refined and extended in Hony et al. (2001). The key points of those reports are summarized here in Figure 10 which schematically compares the average interstellar emission spectrum with the wavelength regions associated with different CH adjacency classes for neutral and ionized, isolated PAHs. Perusal of this figure shows that, while the ranges for matrix-isolated neutral PAHs do not differ substantially from those reported in the literature (e.g. Bellamy, 1958), ionization produces some notable changes in region boundaries.

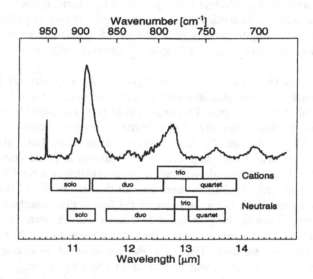

Figure 10. A comparison of the average interstellar spectrum (top) with the ranges characteristic of the out-of-plane bending modes of neutral and cationic PAHs (bottom). Details of the average interstellar spectrum are given in Hony et al. (2001). The boxes indicate the wavelength regions associated with the out-of-plane bending vibrations of the various adjacency classes of the peripheral CH groups as determined from matrix isolation spectroscopy. Figure adapted from Hony et al. (2001).

Considering these modified domains and taking into account the roughly 0.1 μm red- shift in the peak position for PAHs emitting at temperatures of 500 - 1000 K (Flickinger et al., 1991; Brenner and Barker, 1992; Joblin et al., 1995; Schutte et al., 1993b) we are left with the following conclusions:

• The broad, weak interstellar emission feature between 10.6 and 10.7 μm and the stronger, distinct interstellar band peaking near 11.0 μm fall in the region unambiguously attributable to PAH cations.

• The bulk of the 11.2 μm interstellar band falls squarely within the region for the solo-CH modes of *neutral* PAHs and at the extreme long wavelength end of the range for solo cationic modes (note, however, that the neutral solo modes may extend out to 11.7).

• The domains indicated in Figure 10 show that regardless of the region definitions used, there can be little doubt that the weak interstellar 12 μm band arises from duo modes.

• The blue shaded profile of the moderately strong band at 12.7 μm corresponds most closely to the regions characteristic of the trio-CH modes of PAH ions and the duo-CH modes of neutral PAHs. The trio modes of neutral species do not provide a good match.

• The weak 13.5 μm feature falls squarely in the quartet domain for both the neutral and cationic species. However, this region also overlaps the lower of the two wavelength domains characteristic of quintet-CH groups, so a contribution from this class of CH group - however unlikely - cannot be ruled out.

Taken together, these conclusions regarding the edge-structural units make it possible to draw plausible average molecular structures for the PAHs which dominate the emission. This important insight is obtained through an analysis of the band *intensities* in the interstellar spectra rather than their positions. To this end, the absolute intensities of the solo-, duet-, trio-, and quartet-CH out-of-plane bending features as measured in the laboratory are presented in Table 3. Inspection of these data reveals that on a per-CH basis, the solo modes are 2 to 6 times more intense than the modes of the other adjacency classes. This is certainly one of the factors that contribute to the relative prominence of the interstellar 11.2 μm band. *The data also show that, on average, ionization has little effect on the intensities of the CH out-of-plane bending features.* Thus, for the purposes of analyzing band intensities in this region, it will be satisfactory to adopt a single average intensity for each of the adjacency classes, independent of ionization state.

To demonstrate the variability of the interstellar PAH population, we will focus on the extreme spectra of NGC 7027 and IRAS 18317-0757 - two objects whose emission spectra effectively bracket the range of observed relative band intensities. Furthermore, we will assume that the interstellar 11.2 μm emission band arises from solo-CH modes, the 12.0 μm band from duo modes, the 12.7 μm from trios, and the 13.5 μm from quartets. The salient data for the aforementioned objects is taken from Hony et al. (2001) and summarized in Table 4.

Table 3. Average integrated intensities per CH group for the various classes of aromatic CH out-of-plane bending modes arranged according to adjacency class.

Adjacency	$\langle A \rangle_{neutr.}$ (km/mol)		$\langle A \rangle_{cation}$ (km/mol)		$\langle A \rangle_{avg.}$ (km/mol)	
	per Ch^a	per group[a]	per CH^a	per group[a]	per CH^a	per group[a]
Solo	25.7	25.7	24.1	24.1	24.8	24.8
Duet	4.4[b]	8.8	3.7[b]	7.4	4[b]	8
Trio	10.1	30.1	9.0	27.0	9.6	28.8
quartet	11.5	46.0	12.6	50.4	12.0	48.0

[a] 1 solo group = 1 CH; 1 duet group = 2 CH's; 1 trio group = 3 CH's; 1 quartet group = 4 CH's.

[b] The average A value per CH group for the duet modes decreases rapidly with increasing molecular size settling down to a value near 4 km/mol for PAHs larger than about 24 C atoms. Therefore, this value is most appropriate to use for the purposes of constraining the edge structures of the large PAHs that dominate the emission at these wavelengths.

Table 4. Intensities of the principal interstellar emission features in the 10 - 15 µm region for two example objects.

Object	$I_{11.2}{}^a$	$I_{12.0}{}^a$	$I_{12.7}{}^a$	$I_{13.5}{}^a$	$I_{11.2}/I_{12.0}$	$I_{11.2}/I_{12.7}$	$I_{11.2}/I_{13.5}$
NGC 7027	142.7	6	35.9	9.7	23.8	4.0	14.7
IRAS 18317-0757	10.5	1	15.6	2.0	10.5	0.67	5.3

[a] intensities given in units of 10^{-14} W/m^2.

Taking the data from the last column of Table 3, we can calculate the ratio of the average intensity of the solo-CH out-of-plane bending features to that of each of the other adjacency classes. For example, the solo-to-duet intensity ratio is:

$$\frac{\langle A \rangle_{avg}^{solo}}{\langle A \rangle_{avg}^{duet}} = \frac{24.8 km/mol}{8 km/mol} = 3.1 \qquad (1)$$

Similarly we find $\langle A \rangle_{avg}^{solo}/\langle A \rangle_{avg}^{trio} = 0.86$ and $\langle A \rangle_{avg}^{solo}/\langle A \rangle_{avg}^{quartet} = 0.52$. Now, these ratios can be used to calibrate the relative abundances of the various adjacency classes in the interstellar PAH population. This is accomplished by dividing the actual observed interstellar emission band ratios given in the right hand columns of Table 4, by the corresponding intrinsic band intensity ratio:

$$\frac{N_{solo}}{N_x} = \frac{\left(I_{11.2}/I_y \right)}{\left(\langle A \rangle_{solo}/\langle A \rangle_x \right)} \qquad (2)$$

where the subscript "x" indicates one of the multiply adjacent classes (duet, trio, or quartet) and the subscript "y" indicates the position of the interstellar feature attributed to that class. For example, the ratio of the number of solo- to duet-CH groups in the emitting PAH population of NGC 7027 is:

$$\frac{N_{solo}}{N_{duet}} = \frac{\left(I_{11.2}/I_{12.0}\right)}{\left(\langle A \rangle_{solo}/\langle A \rangle_{duet}\right)} = \frac{23.8}{3.1} = 7.7 \qquad (3)$$

The ratios for the remaining adjacency classes for NGC 7027 and those of IRAS 18317-0757 have been calculated in similar fashion and are tabulated in Table 5.

Table 5. The relative abundances of solo-, duet-, trio-, and quartet-CH groups in the emitting PAH population of two representative objects.

	NGC 7027	IRAS 18317-0757
N_{solo}/N_{duet}	7.7	3.4
N_{solo}/N_{trio}	4.7	0.78
$N_{solo}/N_{quartet}$	28.2	10.2

Beyond simply quantifying the relative abundances of the various adjacency classes of peripheral CH bonds in the interstellar PAH population, these abundance ratios can further be used to construct example molecular structures that reflect the implied characteristics of the PAHs that dominate the 10 - 15 μm interstellar emission in each object. Indeed, although the CH groups directly reflect only the edge structures of the emitting PAHs, these ratios indirectly constrain the minimum size of the dominant emitters. This is illustrated in Figure 11 which shows a series of example molecular structures that satisfy the abundance ratios of Table 5. In the figure, each structure is given along with its overall C/H ratio, its solo-/duet-CH group ratio, and its solo-/trio-CH group ratio. In general, solo-CH groups are indicative of long straight molecular edges, while duets and trios reflect corners. Quartets likely indicate pendant rings attached to such structures and the addition of one such ring for every few molecules in the interstellar population would be sufficient to account for their implied low relative abundance. In examining these structures, one should bear in mind that they represent just a few of many structures one might draw to satisfy the interstellar intensity ratios of Table 5. However, significantly, it is *not* possible to satisfy the interstellar intensity ratios with much smaller PAHs. Thus, to reproduce the dominance of the 11.2 μm feature and the relative intensities of the bands at 12.0, 12.7, and 13.5 μm, one is naturally driven towards rather large molecules with at least 100-200 carbon atoms. This is entirely in keeping with previous theoretical calculations of the molecular sizes of the PAH species which account for most of the emission in these features (Schutte et al., 1993b).

Structure 1 in Figure 11 reflects the abundance ratios derived for NGC 7027 and illustrates the type of PAH structure which must dominate the 10 - 15 μm emission in that object. Of course, as mentioned above, many other structures are possible which satisfy the observational constraints equally well. Nevertheless, all such structures would share a number of important characteristics with this one - namely, they would all possess reasonably condensed structures with long, straight edges and a minimum of corners, and only the occasional pendant ring thrown in for color. *Thus, all of these structures would point to the same conclusion - that the PAH family in NGC 7027 is dominated by large compact PAHs.*

The situation for IRAS 18317, on the other hand, is very different. The observed ratios in that object imply a proportionally greater number of corners and/or uneven edges and a somewhat higher occurrence of pendant rings in the emitting population. Structure 4 in Figure 11 illustrates one way of achieving this. This effect could also be achieved by going to smaller, compact structures of the type shown in structure 1 (with shorter "flat" edges) or by breaking up structure 4 in two or more fragments. Structures 2 and 3 shown in Figure 11 represent intermediate cases that fall between the extremes of structures 1 and 4.

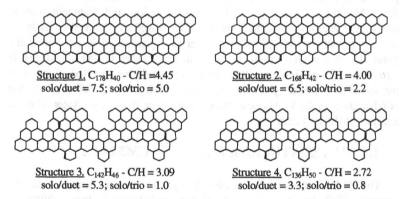

Structure 1. $C_{178}H_{40}$ - C/H = 4.45
solo/duet = 7.5; solo/trio = 5.0

Structure 2. $C_{168}H_{42}$ - C/H = 4.00
solo/duet = 6.5; solo/trio = 2.2

Structure 3. $C_{142}H_{46}$ - C/H = 3.09
solo/duet = 5.3; solo/trio = 1.0

Structure 4. $C_{136}H_{50}$ - C/H = 2.72
solo/duet = 3.3; solo/trio = 0.8

Figure 11. A series of example molecular structures that satisfy the structural constraints implied by the solo/duet and solo/trio abundance ratios given in Table 6. The ratios of the number of solo-CH groups to the number of duet- and trio-CH groups, as well as the overall C/H ratio, is given below each structure.

These have solo/duo and solo/trio ratios consistent with the relative interstellar band intensities of other objects which lie between the extremes of NGC 7027 and IRAS 18317-0757 (Hony et al., 2001). We surmise that the spectral evolution shown in Figure 9 belies a structural signature of the PAH population from large, highly condensed structures in some emission regions to open, uneven structures in others.

The nature of the molecular structures which dominate an emission region necessarily reflects the integrated history of the PAH family in that region. It is therefore interesting to note that the 10 - 15 µm spectra of PNe typically exhibit the relatively strong 11.2 µm (solo-CH) band associated with large, highly condensed PAHs like structure 1 in Figure 11. As the principle site of PAH formation, the species found in late stellar ejecta have formed within the last few thousand years - but a moment by interstellar standards. Because "bay regions" - cavities in the edge structure of a PAH (c.f. Figure 11, structures 2 - 4) - are particularly susceptible to the addition of carbon atoms, it is not surprising that such edge structures should quickly be filled-in in regions of active PAH growth. In addition, the ferocious radiation field and dynamic processes characteristic of PNe would tend to weed out all but the most stable PAH structures. Both of these factors would be expected to drive the population toward more highly condensed structures, consistent with our analysis of the ISO spectra of these objects.

In contrast, HII regions (as exemplified by IRAS 18317-0757) tend to exhibit relatively strong 12.7 µm (trio-CH) bands. These are regions where hot, young O/B stars illuminate material which has been processed for perhaps millions of years through both dense- and diffuse-cloud phases of the ISM. This processing involves both ablation and fragmentation by shocks and energetic particle bombardment, as well as chemical degradation such as oxidation and reduction by energetic processing in icy interstellar grain mantles as discussed in the following sections. As illustrated in structures 2 - 4 in Figure 11, these processes would be expected to gradually but inexorably eat away at the structure, reducing the overall level of ring condensation in the PAH population. Once again, this is consistent with our analysis of the ISO data.

1.6   INTERSTELLAR PAHS: A SUMMARY

The I/S PAH model has now moved beyond merely seeking to verify the presence of PAHs in space to the active exploitation of these ubiquitous species as probes of the ISM. The latest high-quality astronomical observations, supported by an ever-increasing database of astrophysically relevant laboratory spectra of PAHs and PAH ions, is now providing insight into the conditions in IR-emitting regions at an unprecedented level of detail. Clearly, PAHs hold the potential to be an invaluable probe of the ISM, much as CO has been for the last quarter-century.

Given their demonstrated widespread abundance in molecular clouds, it is natural to wonder how PAHs might interact the other ubiquitous dust component of these regions - icy grains. In addition, since the members of the extended PAH family (small and large molecules, clusters, and particles) represent the single largest repository of organic material in our galaxy, it is natural to wonder what role these materials might play in the development of organically-based life. These questions and their implications are considered in the following sections.

2. INTERSTELLAR/PRECOMETARY ICES: THE BIRTHPLACE OF COMPLEX, MULTIFUNCTIONAL MOLECULES

As with the PAHs discussed in §2, over the past 25 years infrared observations, combined with realistic laboratory simulations, have revolutionized our understanding of interstellar ices and dust, the feedstock of the Solar Nebula and the building blocks of comets. In this section we discuss the evidence for these ices. Although the dominant ice species are revealed by the IR spectra of dense molecular clouds, other species can also be important. These include species accreted directly from the gas phase and those produced within the ice itself by UV photon and high energy particle bombardment.

It is likely that the Solar Nebula incorporated material with a heritage in both the dense and diffuse ISM since there are many mixing cycles between these phases prior to star and planet formation (Allamandola et al., 1993; Chiar et al., 1998a). It is thought that roughly 10% of dense cloud material becomes incorporated into stars and planets through each cycle, with the remainder dispersed back into the diffuse medium. While it is well beyond the scope of this paper to discuss the properties of each interstellar phase, detailed discussions of the general ISM can be found elsewhere (Li and Greenberg, 2002; Williams, 2002; Winnewisser, 1997; Irvine, 1998; Ehrenfreund and Fraser, 2002; Sandford, 1996; Dorschner and Henning, 1995; van Dishoeck et al., 1993). In view of the astrobiological theme of this paper, we focus on the composition of IS ices, the birthplace and main reservoir of the extraterrestrial, complex, molecular material which ultimately seeded the planet with species that may have played a role in the evolution of life on earth.

2.1 THE BUILDING BLOCKS OF INTERSTELLAR ICES

Once a dense cloud is formed, thanks to the attenuation of the general IS radiation field, gas-phase and gas-grain chemistry leads to the production and sustaining of more complex species in the gas than possible in the diffuse ISM. At the same time, since the dust in a dense molecular cloud is so cold, (~20K), any polyatomic molecule striking a dust grain should condense (Sandford and Allamandola, 1993), one might expect grain composition to reflect gas composition. However, as shown in Table 6, the simplest and most abundant gas-phase *polyatomic* molecules known from radio observations are much lower in abundance than their frozen counterparts. Thus, at this epoch, direct accretion of IS gas-phase species plays a very minor role in determining IS ice composition since the ice grains are chemical factories in their own right, generally harboring far greater amounts of material than the gas.

Table 6. Comparison between the gas-phase and solid-state abundances for several molecular species normalized to hydrogen. This comparison shows that the interstellar ices contain the bulk of the simple interstellar polyatomic molecules.

Molecule	GAS PHASE[a]		ICE[b]	Ice/Gas Ratio
	TMC-1	OMC-1	NGC 7538 IR9	
CO	8×10^{-5}	5×10^{-5}	6×10^{-6}	0.12
H_2O	-	-	6×10^{-5}	?
CH_3OH	2×10^{-9}	1×10^{-7}	6×10^{-6}	60-3000
NH_3	2×10^{-8}	-	6×10^{-6}	300
CO_2	5×10^{-8}	-	8×10^{-6}	160
CH_4	-	-	6×10^{-7}	?

[a] Gas phase composition: TMC-1, (Ohishi et al., 1992); OMC-1 (Blake et al., 1987); CO_2, (van Dishoek et al., 1996). [b] Ice mantle composition: (Allamandola et al., 1992); CO_2, (DeGrauw et al., 1996); CH_4, (Boogert et al., 2000).

Throughout the cloud's lifetime, processes such as accretion of gas phase species, simultaneous reactions on the surfaces involving atoms, ions, and radicals, as well as energetic processing within the body of the ice by ultraviolet photons and cosmic rays all combine to determine the mantle composition (Herbst, 2002; Li and Greenberg, 2002; Ehrenfreund and Fraser, 2002; Greenberg, 1978; Tielens and Hagen, 1982; d'Hendecourt et al., 1985; Greenberg, 1989; Charnley et al., 2001; Moore et al., 1983; Moore et al., 1996; Bernstein et al., 1995; Pirronello and Averna, 1988; Pirronello, 1997; Pirronello et al., 1997; Gerakines et al., 1995a; Palumbo et al., 2000; Hiraoka et al., 1998; Schutte, 1999; Charnley, 2001; Chiar, 1994). These proceses are illustrated in Figure 12. Since hydrogen is 3 to 4 of orders of magnitude more abundant than the next most abundant reactive elements such as C, N, and O (Figure 1), overall grain surface chemistry is strongly moderated by the H/H_2 ratio in the gas. In regions where this ratio is large, H atom addition (hydrogenation) dominates and species such as CH_4, NH_3 and H_2O are expected to be prominent (d'Hendecourt et al., 1985). If the H/H_2 ratio is substantially less than one, however, reactive heavy atoms such as O and N are free to interact with one another forming molecules such as CO, CO_2, O_2 and N_2. Thus, two qualitatively different types of ice mantle may be produced by grain surface reactions, one dominated by hydrogen-rich, polar molecules, capable of hydrogen-bonding and the other dominated by hydrogen-poor, non-polar (or only slightly polar), highly unsaturated molecules. Figure 12 also shows the first generation of products one might expect upon mantle photolysis (UV irradiation).

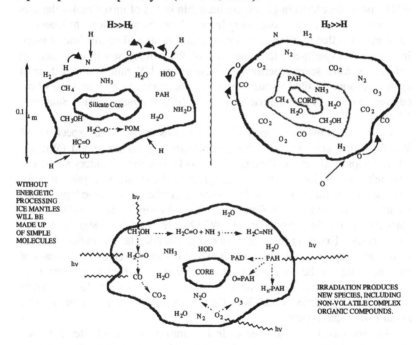

Figure 12. Schematic drawings of the types of ice mantles expected in dense molecular clouds. In regions where the H/H_2 ratio is much greater than one (top left), surface reactions tend to be reducing and favor the production of simple hydrides of the cosmically abundant O, C, and N. In contrast, oxidized forms of these species are favored in regions where this ratio is much less than one (top right). Irradiation and thermal processing (bottom) of these ice mantles creates considerably more complex species and, ultimately, non-volatile residues.

Interstellar ice compositions are revealed through their IR spectra. A star, fortuitously situated in or behind a molecular cloud, can provide a reasonably featureless, continuous mid-IR spectrum. As this radiation passes through the cold cloud, the intervening molecules in the gas and dust *absorb* at their characteristic vibrational frequencies. (This is quite distinct from the situation described earlier for the PAHs where it is the IR emission of vibrationally excited molecules that is measured.) Since the ISM between the "outside boundary" of the cloud and the Earth is far less dense, the IR absorption spectrum of objects obscured by molecular cloud material is mainly that of the dust in the dense molecular cloud. Since the ice features tend to dominates these spectra, interstellar ice composition can be analyzed by directly comparing the astronomical data with the spectra of ices prepared in a laboratory which duplicate the salient interstellar conditions. This can be accomplished using a cryogenic sample chamber similar to that used in the PAH studies described in §2, except here thin layers of mixed molecular ices are deposited on the cold sample window. Sample ice thickness is comparable to that of all the interstellar grain mantles along the line of sight. In a typical experiment, the spectrum of the sample would be measured before and after several periods of exposure to UV radiation and thermal cycles. A more detailed discussion of this approach can be found elsewhere (Bernstein et al., 1995; Allamandola et al., 1988; Allamandola and Sandford, 1988).

Figure 13 shows a collage of comparisons between the spectrum of W33A, a protostar deeply embedded within a molecular cloud (Willner, 1977; Capps et al., 1978; Tielens et al., 1984) with the laboratory spectra of interstellar ice analogs. With the exception of the strong absorption near 10 μm for which the silicate Si-O stretch overwhelms the overlapping ice features (Allamandola and Sandford, 1988; Bowey et al., 1998), all the absorptions in the spectrum of W33A are readily assigned to ice components. Excellent matches between the interstellar absorption features with laboratory spectra of the type shown in Figure 13 represent the basis of our knowledge of the composition of interstellar ice particles. Interestingly, until quite recently, much more was known of interstellar ice grain composition - particles hundreds of light years away - than of cometary ices in our own Solar System!

Over the past five years, deeper insight into the nature of interstellar ice and dust has been achieved through analysis of data from the Infrared Space Observatory (ISO). That mission has enabled measurement of the complete mid-IR spectrum with one instrument, eliminating the need to piece together bits from different telescopes taken under different conditions and with different spectral parameters. The IR spectra obtained with this telescope has enabled this field to advance beyond analysis of the strongest spectral features, revealing important subtleties in the spectra that probe details of interstellar chemistry and grain evolution (Ehrenfreund, 2002; Ehrenfreund et al., 1998; Schutte et al., 1999; Teixeira et al., 1999).

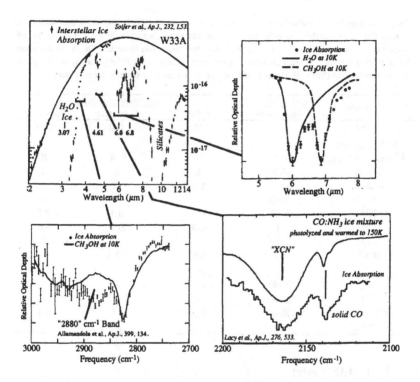

Figure 13. Comparisons of laboratory analog spectra with spectra from the object W33A, a protostar deeply embedded in a dense molecular cloud. Upper left: The dots trace out the interstellar spectrum and the solid line corresponds to the quasi-blackbody emission spectrum thought to be produced by the protostar. The strong absorption near 10 μm is due to the silicate grains and the excess absorption labeled "2880 cm^{-1} Band" visible in the lower left-hand panel is thought to arise from interstellar microdiamonds in the cloud. All the remaining absorptions are produced by interstellar ices. These features are highlighted and compared to laboratory spectra on expanded scales in the surrounding frames. Lower left: The solid line is due to methanol in a laboratory ice. Lower right: The upper smooth line corresponds to a laboratory analog comprised of CO (sharp band) and OCN$^-$ (XCN, broad band). Upper right: The solid and dashed lines correspond to spectra of H_2O and CH_3OH respectively (Allamandola et al., 1992; Willner, 1977; Tielens et al., 1984; Lacy et al., 1984).

Table 7 lists the dominant interstellar ice constituents and gives their average abundance with respect to H_2O. A detailed discussion of this topic can be found elsewhere (Ehrenfreund and Fraser, 2002). Here, only a very brief discussion of the most important components will be given in order to provide the basis for the subsequent discussion of more complex species. More information on each of these components can also be found in the literature (Allamandola et al., 1997).

Table 7. Composition and abundances of interstellar ice (relative to H_2O) compared to that deduced in Comets Halley, Hyakutake, and Hale-Bopp. The species listed above the dashed line in the second column have been definitely detected, those below have been tentatively identified. The evidence is good for all of these species. Cometary abundances taken from: Altwegg et al. (1999) (Halley) and Crovisier and Bocklee-Morvan (1999) (Hyakutake and Hale-Bopp). See text for interstellar ice references.

Molecule	*Interstellar Ice* Abundance	*Comet Parent Molecules* Halley	Hyakutake	Hale-Bopp
H_2O	100	100	100	100
CO(polar ice)	1-10			
CO(non-polar ice)	10-40	17	6-30	20
CH_3OH	<4-30	1.25	2	2
CO_2	1-10	3.5	<7	6
XCN	1-10	---	---	---
NH_3	5-10	1.5	0.5	0.7
CH_4	~1	<0.8	0.7	0.6
HCO	~1	---	---	---
H_2CO	1-4	3.8	0.2-1	1
H_2	~1	---	---	---
N_2	10-40	---	---	---
O_2	10-40	---	---	---
OCS or CO_3	Few	0.2	0.1	0.3
HCOOH	Low	---	---	---

2.1.1 H_2O (water)

H_2O is the dominant ice component in dense clouds (Tielens et al., 1984; Gillett and Forrest, 1973; Gillett et al., 1975; Cox, 1989; Omont et al., 1990). This identification is as secure as any based on infrared spectral matching since at least five interstellar features fit laboratory H_2O ice spectra very well. Since the strong 3280 cm^{-1} (3.07 µm) band typically dominates the interstellar spectra (see Figure 13), it was one of the first features to be detected and identified in dense clouds. There is an extensive astronomical and laboratory literature dealing with this feature (Hagen et al., 1983; Knacke et al., 1982; Leger et al., 1983; Smith et al., 1989; Smith et al., 1993; Sato et al., 1990).

2.1.2 CH₃OH (methanol)

It was suggested early on that part of the prominent interstellar absorption near 1460 cm⁻¹ (6.85 μm) shown in Figure 13 might be due to the CH deformation vibration of methanol (Tielens et al., 1984; Hagen et al., 1980; Tielens and Allamandola, 1987). However, a one band match in the IR is not sufficient for a firm identification in most cases and the unequivocal identification of methanol frozen on dust grains was not secure until the subsequent detection of other bands (Allamandola et al., 1992; Grim et al., 1991; Skinner et al., 1992; Sandford et al., 1993; Dartois et al., 1999b). However these bands were weaker than expected based on the depth of the 1460 cm⁻¹ band, implying that other species contribute to the latter interstellar absorption. Aliphatic organic compounds and carbonates are reasonable candidates (Allamandola and Sandford, 1988). When present, methanol is often the second or third most abundant component of the ice after H₂O, with relative concentrations lying between about 5 and 30%. The highest CH3OH ice abundances are seen toward these warmer regions.

2.1.3 CO (carbon monoxide)

After H₂O, carbon monoxide is the most studied interstellar ice component. CO has a characteristic absorption feature near 2140 cm⁻¹ (4.67 μm), as shown in Figure 13. Its position, width, and profile are a sensitive function of the ice matrix in which it is frozen (Sandford et al., 1988; Elsila et al., 1997; Palumbo, 1997; Palumbo and Strazzulla, 1993; Palumbo and Strazzulla, 2000; Ehrenfreund et al., 1996). Many, but not all, of the lines-of-sight that contain H₂O ice also contain CO ice and the relative strengths of the H₂O and CO bands indicate CO/H₂O ratios ranging from 0.0 to as much as 0.3 (Ehrenfreund and Fraser, 2002; Lacy et al., 1984; Geballe, 1986; Eiroa and Hodapp, 1989; Tielens et al., 1991; Whittet et al., 1989; Chiar et al., 1995). Although the CO band in a few objects has a position and profile consistent with CO frozen in H₂O-rich matrices, most lines-of-sight exhibit profiles indicative of CO frozen in *non-polar* matrices, i.e., ices thought to be dominated by molecules such as CO, CO₂, O₂, and N₂ rather than H₂O (Sandford et al., 1988; Elsila et al., 1997; Tielens et al., 1991; Chiar et al., 1995; Chiar et al., 1998b). These are precisely the two sorts of mantles predicted on the basis of the H/H₂ ratio discussed earlier and sketched in Figure 12.

2.1.4 OCN- (formerly XCN)

The spectra of a limited number of lines-of-sight through dense clouds contain a broad, often weak, feature near 2165 cm^{-1} (4.62 μm) which was originally attributed to an unspecified nitrile (-CN) designated as XCN (Figure 13) (Lacy et al., 1984; Tegler et al., 1993; Weintraub et al., 1994). Although the number of examples is small , there is an indication that this feature is present only in the spectra of protostellar sources embedded within clouds and not in the quiescent regions sampled by background stars (Tegler et al., 1995; Pendleton et al., 1999; Whittet et al., 2001). This suggests that the carrier of the feature is associated with the local environment of the embedded star. Laboratory work shows that the responsible species contains O, C, and N; and that energetic processing is required for it's production. Although many laboratory studies have been carried out and excellent fits to the astronomical data have been realized, the precise molecular identity of the carrier remains elusive and contradictory conclusions abound (Tegler et al., 1993; Grim and Greenberg, 1987; Schutte and Greenberg, 1997; Demyk et al., 1998; Hudson et al., 2001; Palumbo et al., 2001; Bernstein et al., 2001b; d'Hendecourt and Jordain de Muizon, 1989). Nevertheless, the case in favor of an assignment to OCN$^-$ perturbed by the local ice mantle material is quite compelling (Schutte and Greenberg, 1997; Demyk et al., 1998; Hudson et al., 2001; Palumbo et al., 2001).

2.1.5 CO$_2$ (carbon dioxide)

This species was long expected to be common in interstellar ices due to its ready production by energetic processing of interstellar ice analogs. However, due to screening by the its atmospheric absorptions, frozen interstellar carbon dioxide detection had to await a spaceborn spectrometer. The first detection of this species was made by d'hendecourt and Jourdain de Muizon using IRAS LRS data (Bernstein et al., 2001b; d'Hendecourt and Jordain de Muizon, 1989). The sensitivity and spectral resolution limitations of this instrument however, permitted a serious study of only a few lines of site. This situation was vastly improved with the ISO SWS instrument. It is now known that frozen interstellar CO$_2$ is similar in abundance to water ice, with the very strong CO stretch band near 2340 cm^{-1} often as deep as the 3400 cm^{-1} water ice band. The ISO spectra combined with extensive laboratory studies focusing on both the CO stretching and bending modes of this molecule have provided great insight into the various forms of mixed molecular ices present in dense clouds (Ehrenfreund and Fraser, 2002; DeGrauw et al., 1996; Ehrenfreund et al., 1998; Ehrenfreund et al., 1997;

Gerakines et al., 1999; Sandford and Allamandola, 1990; Sandford et al., 2001; Dartois et al., 1999a).

2.1.6 Other IS Ice Species

In addition to the species considered specifically above, frozen CH_4 (methane) (Lacy et al., 1991; Boogert et al., 1997) and NH_3 (ammonia) (Lacy et al., 1998) have also been clearly identified in interstellar ices. Furthermore, based on limited telescopic observations, laboratory studies of ice analogs, and theoretical chemistry models, a number of other molecular species are suspected of being present in interstellar ices in quantities on the order of a few percent relative to H_2O. These include HCO (formyl radical) (Tielens and Allamandola, 1987); H_2CO (formaldehyde) (Schutte et al., 1996); HCOOH, $HCOO^-$, and CH_3COH (formic acid, formate ion, and actetaldehyde, respectively) (Schutte et al., 1999); OCS (carbonyl sulfide) (Palumbo et al., 1995) or CO_3 (carbon trioxide) (Elsila et al., 1997); and possibly ketones and/or aldehydes (Tielens and Allamandola, 1987). N_2 and O_2 might also be present as inferred by the profile of the CO band in certain lines-of-sight (Elsila et al., 1997). However, direct detection of these two important species in the IR will be very difficult, if possible at all (Ehrenfreund and Fraser, 2002; Sandford et al., 2001).

2.2 PHOTOCHEMICAL EVOLUTION OF THE ORGANIC MATERIAL IN INTERSTELLAR/PRECOMETARY ICES AND THE PRODUCTION OF COMPLEX ORGANIC MOLECULES.

The picture of mixed molecular interstellar ice described up to this point is supported by direct spectroscopic evidence (e.g. Figure 13). The identities, relative amounts and absolute abundances of the ice species listed in Table 7 and also discussed in Ehrenfreund and Fraser (2002) are sound. However, this is not thought to be the entire story. Indeed, from the origin of life perspective, this is only the beginning of the story. Here's why: the ices in dense molecular clouds are irradiated by UV photons and cosmic rays, breaking and rearranging chemical bonds within the ice to form new species. Although the abundance of this new material is only a few percent, this is a critically important process since it can create remarkably complex chemical groups and molecular species that cannot be made via gas phase and gas-grain reactions at the low temperatures and pressures characteristic of dense clouds. This is because the protection and ready availability of reaction partners in the solid phase favors chemical complexity and diversity while the energetics, radiation fields, and low densities of the gas-phase favor simplicity.

The interstellar/precometary ice composition in protostellar regions is also of particular relevance from the origin of life perspective since this is where new planets are formed. These are the regions in which the OCN^- (formerly XCN) and CH_3OH bands are present, the tracers of *in-situ* energetic processing. The presence of methanol in these ices is of pivotal importance since it drives a rich interstellar ice photochemistry (Bernstein et al., 1995; Allamandola et al., 1988; Dworkin et al., 2001) and plays an important role in gas phase chemistry (Charnley et al., 1992). Furthermore, methanol has profound effects on the physical behavior of H_2O-rich ices (Blake et al., 1991). Since CH_3OH is often an abundant ice component in comets (Table 7, cf. Mumma et al., 1993; Mumma, 1997), this may impact their structural (Blake et al., 1991) and vaporization (Sandford and Allamandola, 1993) behavior as well.

Figure 14 shows the spectral evolution of an interstellar ice analog comprised of $H_2O:CH_3OH:CO:NH_3:C_3H_6$ (100:50:10:10:10) as a function of UV photolysis. Except for the C_3H_6, this analog mixture reflects the major interstellar ice components associated with protostellar environments. The exposure to UV results in the destruction of several species (particularly methanol) and the creation of others. The simplest and most abundant include HCO, H_2CO, CH_4, CO, and CO_2. As shown in Table 7, all of these new species have been identified in interstellar ices. The detection of these ice components in the dense molecular clouds does not necessarily imply radiation processing is responsible for their production since many of these molecules can also be formed by gas-phase or gas-grain chemistry. So, at present, all we can say is that observations are consistent with energetic processing. To reiterate, the strongest evidence that energetic processing is important, at least in some locations within dense clouds, is provided by the OCN^- (formerly XCN) feature (e.g. Figure 13) and perhaps by the ubiquity of CO_2. The OCN^- feature cannot be explained by any of the more abundant species predicted by gas and gas-grain chemical models, but is readily made by the radiative processing of laboratory ices containing C, N and O. Excellent, detailed descriptions of the UV induced chemical evolution of interstellar ice analogs can be found in the literature (Bernstein et al., 1995; Gerakines et al., 1995a; Allamandola et al., 1988; Hagen et al., 1979; Agarwal et al., 1985; d'Hendecourt et al., 1986; Briggs et al., 1992). The role of methanol in the ice photochemistry discussed here is very critical to the kinds of compounds produced.

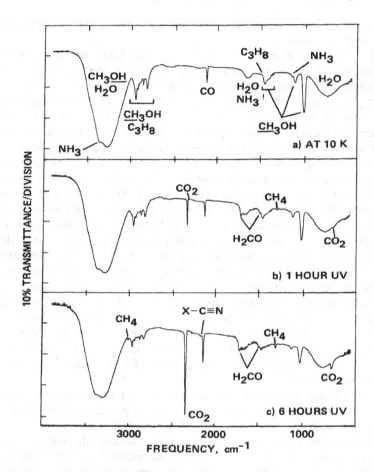

Figure 14. The photochemical evolution of an $H_2O:CH_3OH:NH_3:CO:C_3H_8$ (100:50:10:10:10) interstellar ice analog as traced by infrared spectra measured at 10 K. The spectra were taken (a) before and after (b) 1 hour and (c) 6 hours of UV irradiation. Note the ready formation of CO_2, H_2CO, CH_4, and XCN at the expense of CH_3OH. Figure adapted from Allamandola et al. (1988).

In contrast to photochemistry expected in the water -rich ices of dense star forming regions is that which characterizes the non-polar ices of quiescent regions. Here, since the polar species H_2O, CH_3OH, and NH_3 aren't abundant, simple, low polarity species such as CO_2, N_2O, O_3, CO_3, HCO, H_2CO, and possibly NO and NO_2 are important photoproducts in laboratory simulations (Elsila et al., 1997). To date, ice composition in quiescent regions is not as well characterized as that in denser, high mass star forming regions.

For the remainder of this paper we will focus our attention on photochemical processing of the water-rich, polar ices associated with protostellar environments. The ice chemistry so far considered involves the photoproduction of the most abundant, simple species in the solid state at 10 K. However, the full scope of the chemistry is much more complex and perhaps biologically relevant. Upon warm-up to about 200 K under vacuum one observes that the parent compounds and most volatile ice constituents sublime, leaving a residual mixture of less volatile species on the substrate. Of the staggering array of compounds produced from even the simplest starting ice containing H_2O, CH_3OH, NH_3, and CO, only a few have been identified. These are presented in Figure 15. In keeping with their expected low abundance, clear-cut spectroscopic evidence for these types of compounds in interstellar ices is presently lacking, although some of the weak spectral structure detected in the 2000-1250 cm^{-1} region is consistent with their presence. Additionally, spectral screening by the much more abundant, simpler ice species will likely represent an important, long-term obstacle. Higher quality astronomical spectra than those currently available will be needed to probe the species present at this level of concentration. Even then, identifications will likely be limited to chemical classes. Nonetheless, since many of these compound classes are of exobiological interest, a spectroscopic search for evidence of these materials in dense clouds would be of value.

The residue that remains on the window at room temperature is of particular interest from a prebiotic perspective. It is quite plausible that this type of material is closely related to that preserved in comets, meteorites and interplanetary dust particles (IDPs) and it is believed that these sources deliver between 12 and 30 *tons* of organic material to Earth monthly. During the period of great bombardment some 4 billion years ago, the amount of extraterrestrial organic material brought to the prebiotic early Earth was many orders of magnitude greater. Thus, this type of material could have been an important source of prebiotic organic matter on the early Earth and perhaps played an important role in the emergence of life.

Now, let's consider more closely the molecules of biogenic interest which are produced upon photolysis of *realistic* interstellar ice analogs. The word realistic is used to indicate that the laboratory ice composition reflects the interstellar polar ice composition shown in Table 7. The standard ice we study has a starting composition of $H_2O:CH_3OH:NH_3:CO$ (100:50:1:1) or $H_2O:CH_3OH:NH_3:CO$ (100:50:5:5). As evident from Table 7, these have higher methanol concentrations (H_2O/CH_3OH ~2/1) than is currently considered (H_2O/CH_3OH ~5/1) appropriate.

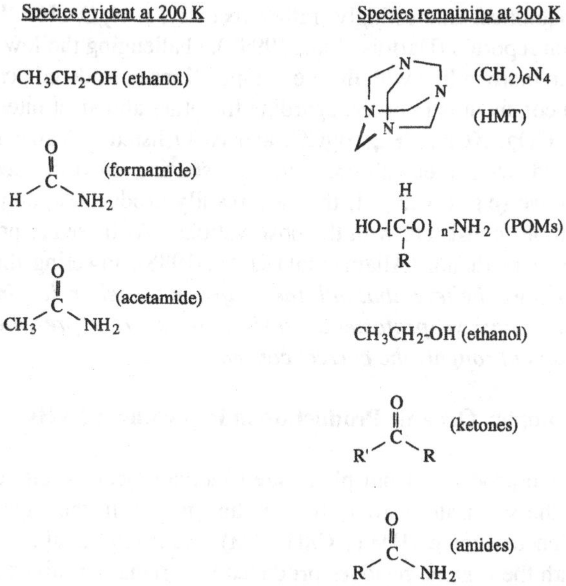

Species evident at 200 K   Species remaining at 300 K

CH_3CH_2-OH (ethanol)

$(CH_2)_6N_4$

(HMT)

formamide structure:

$$\underset{H}{\overset{O}{\underset{\|}{C}}}NH_2 \quad \text{(formamide)}$$

acetamide structure:

$$\underset{CH_3}{\overset{O}{\underset{\|}{C}}}NH_2 \quad \text{(acetamide)}$$

POMs structure:

$$HO\text{-}[\underset{R}{\overset{H}{\underset{|}{C}}}\text{-O}]_n\text{-}NH_2 \quad \text{(POMs)}$$

CH_3CH_2-OH (ethanol)

ketones structure:

$$\underset{R'}{\overset{O}{\underset{\|}{C}}}R \quad \text{(ketones)}$$

amides structure:

$$\underset{R}{\overset{O}{\underset{\|}{C}}}NH_2 \quad \text{(amides)}$$

Figure 15. Identified compounds produced by the 10 K UV photolysis and subsequent warm-up of the realistic interstellar ice analogs $H_2O{:}CH_3OH{:}NH_3{:}CO$ (100:50:10:10) and (100:50:1:1). Figure adapted from Bernstein et al. (1995).

There is, however, good reasons for this choice. It has been more than 15 years since we first undertook to study the photochemistry of interstellar mixed molecular ices. Since there was, at the time, no in-depth knowledge of the types of chemicals produced in this manner it was decided to focus our efforts on a consistent composition and at that time it was believed that the correct H_2O/CH_3OH ratio was 2/1. By the time the interstellar methanol concentration was called into question, a significant amount of work had already been undertaken and, a large body of knowledge and insight had already been gained into the chemistry of that ice, setting the stage for an attempt to analyze the minor photoproducts, a very challenging task. Moreover, experiments carried out at the lower methanol concentrations now thought to be more representative of interstellar ice composition have now been proven to yield essentially the same array of new compounds, albeit in significantly lower yields. As it is, even with the higher methanol concentrations, in order to produce sufficient material for one analytical run requires between two to three months of constant sample preparation. Thus, using a single apparatus for sample preparation it is typically possible to do only four to five experiments *per year*. Reducing the yield would shift this from a very difficult project to one that is impossible given our current analytic capabilities. Interestingly, rather recently, a H_2O/CH_3OH ratio of ~3/1 has been reported (Dartois et al., 1999b), challenging the low methanol paradigm and more in line with the ice compositions considered here.

Lastly, a comment is in order regarding the other abundant interstellar ice components CO_2, XCN, CH_4, H_2CO, and HCO listed in Table 7. We are not concerned with their absence in the starting mixture because, as explained above (e.g. Figure 14, they are readily produced upon photolysis at concentrations consistent with the observations. As these are produced at the expense of methanol (Allamandola et al., 1988), lowering that species concentration, *we believe that all the major interstellar ice ingredients known toward massive protostellar environments are represented in our experiment and at roughly the correct concentrations.*

2.2.1 Complex Organic Production in Ices without PAHs

The residue produced from photolyzed methanol-containing ices which remains on the substrate at room temperature is rich in the cage molecule Hexamethylenetetramine (HMT, $C_6H_{12}N_4$) (Bernstein et al., 1995). This contrasts with the organic residues produced by irradiating mixed molecular ices which do not contain methanol (Agarwal et al., 1985; d'Hendecourt et al., 1986; Gerakines et al., 1995b) and those produced in thermally promoted polymerization-type reactions in unirradiated realistic ice mixtures (Schutte et al., 1993a). In those experiments HMT is only a minor product in a residue dominated by a mixture of POM related species. (Interestingly, POMs already start to form in these ices at temperatures as low as 40 K. Further, POM-like species have been suggested as an important organic component detected in the coma of Comet Halley; Huebner et al., 1989). In addition to the possible prebiotic activity of HMT itself, its synthesis by photolysis implies the presence of several other biogenically interesting intermediates in these ices as well. For example, there are a variety of secondary carbon and nitrogen containing species (including amino acids) which are readily formed by HMT hydrolysis, thermalization, or photolysis, all processes which can occur during an interstellar ice grain, comet, or asteroid's lifetime. Since ultraviolet photolysis of HMT frozen in H_2O ice produces the OCN⁻ band (Bernstein et al., 1994), HMT may be related to the carrier of that band in dense clouds and, perhaps to the CN observed in cometary comae. These laboratory simulations of grain processing have also influenced our understanding of interstellar, gas phase chemistry. The HMT production pathway proposed in Bernstein et al. (1995) involves the intermediate methylimine (CH_2N). Armed with this information, radioastronomers searched for and found methylimine to be widespread throughout the gas in many molecular clouds (Dickens et al., 1997). Beyond

HMT and POMs, the involatile residue of photolyzed, methanol containing ices comprises lower concentrations of a bewildering array of organic compounds.

Due to the extreme complexity and analytical challenge posed by deeper analysis and encouraged by recent interest in the biological application of these results, effort has been redirected from solely establishing the chemical inventory of species produced in these interstellar/precometary ice analogs to searching for the presence of specific biogenically important species. This effort has required the use of two new techniques. These are High Performance Liquid Chromatography (HPLC) and laser desorption-laser ionization mass spectroscopy (L^2MS). Access to the latter technique has been made possible through a collaboration with Prof. R. Zare and his colleagues at Stanford University. While neither technique can provide an unequivocal identification directly, they are both particularly suited to microanalysis and give very valuable insight into the chemical properties of the compounds that make up the residue.

Figure 16 shows the HPLC chromatogram of one of the residues compared to a soluble extract from the primitive meteorite Murchison. There are two conclusions to be drawn from this figure. First, since each peak represents a different compound, or more likely a different family of compounds, both the laboratory residue and meteoritic extract are complex chemical mixtures. Second, the similarity in peak distributions between the two samples indicates that the kinds of chemicals present in each sample are similar. This similarity raises the interesting question, "Do the families of compounds in carbonaceous meteorites have an interstellar ice/cometary heritage?"

Figure 17 shows the L^2MS of the residue from a photolyzed $H_2O:CH_3OH:NH_3:CO$ (100:50:1:1) ice. Mass spectra such as this provide us with further critical insight into the nature of the residue, showing that there are hundreds of compounds produced. Further, this shows that they are far more complex than the starting materials. The new materials produced are responsible for the envelope spanning the mass range from about 100 M/Z to 350 M/Z. (M/Z is equivalent to an atomic mass unit). Given that none of the simple starting materials of $H_2O:CH_3OH:NH_3:CO$ has a mass greater than 32 nor do they contain a single CC bond, the complexity and extent of the photoproducts is staggering. Many of these compounds have molecular masses up to *ten times* larger than that of any of the starting materials.

When this mass spectrum is compared with those of two interplanetary dust particles (IDPs) there is an intriguing similarity between the mass envelope between roughly 150 to 300 M/Z and peaking near 220 M/Z characteristic of the interstellar ice residue shown in Figure 17 and the mass envelopes in the IDP spectra which span the range from about 200 M/Z to 300 M/Z and appear to peak near 250 M/Z (see e.g. Clemett et al., 1993).

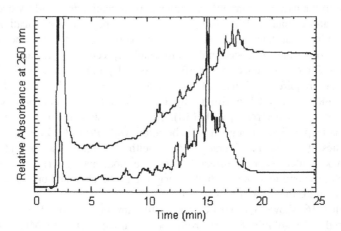

Figure 16. The high precision liquid chromatogram (HPLC) of (A) the room temperature residue produced by UV photolysis of an $H_2O:CH_3OH:NH_3:CO$ (100:50:1:1) ice (profile magnified 10x) and (B) mixed acid and base extracts of Murchison meteorite. HPLC conditions: 1mL/min 50% water, 50% methanol to 100% methanol in 15 minutes, C_{18} reverse phase analytical column (Dworkin et al., 2001).

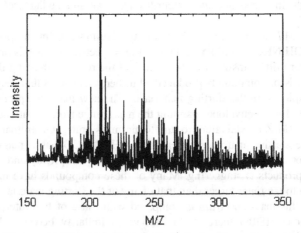

Figure 17. The laser desorption (L^2MS) mass spectrum of an interstellar ice analog residue produced by the photolysis of a $H_2O:CH_3OH:NH_3:CO$ (100:50:1:1) ice. The complexity of the interstellar ice analog residue is clearly evident. This work initially reported in Dworkin et al. (2000).

As with the HPLC results above, this resemblance between the laboratory ice residue and the extraterrestrial organics brought to Earth - this time in IDPs - raises the question, "Do the compounds in IDPs have an interstellar ice/cometary heritage?"

2.2.2 Complex Organic Production in Ices with PAHs

As described in §2, PAHs are one of the most abundant interstellar polyatomic molecules known in the gas and are widespread throughout the interstellar medium. As with all other polyatomic gas phase species in dense clouds, they should freeze out onto the grains and become part of the mixed molecular ice. During the past three years we also investigated the photochemical behavior of a few small PAHs in H_2O ices to lay the foundation for a study of PAH photochemistry in realistic, multicomponent IS ice analogs. While the UV photolysis of PAHs in interstellar ices modifies only a small fraction of the interstellar PAH population, this change is significant in terms of interstellar chemistry and interesting from the prebiotic chemistry perspective. Upon photolysis in pure H_2O ice at 10 K, simple PAHs are not destroyed. Rather, they are partially oxidized and/or reduced (hydrogenated) (Bernstein et al., 1999). These PAH structures are illustrated in Figures 18 and 19.

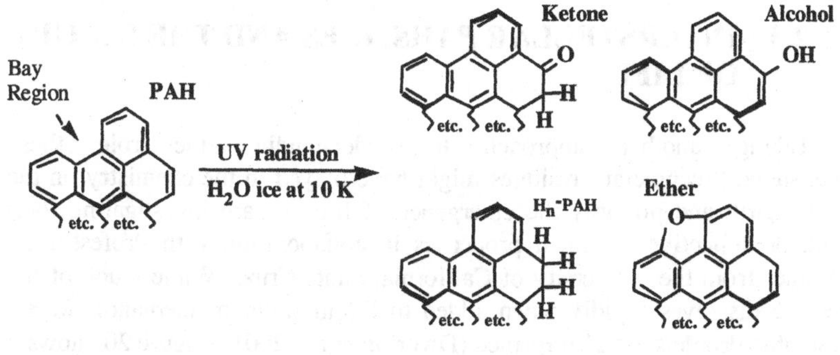

Figure 18. The types of PAH structures produced when PAHs are UV irradiated in water ice at 10 K (Bernstein et al., 1999).

If deuterium is present in the ice, deuterated aromatics are also produced (Bernstein et al., 1999; Sandford et al., 2000). These alterations have significantly different effects on the chemical nature of the parent. Hydrogen atom addition transforms some of the edge rings into cyclic aliphatic hydrocarbons, thereby creating molecules with both aromatic and aliphatic character and decreasing the overall degree of aromaticity. Oxygenation produces ketones or aldehydes, changes which open up an entire range of possible chemical reactions that were not available to the parent PAH. Aromatic ketones (quinones) are of particular prebiotic interest since they are widely used in current living systems for electron transport across cell membranes. In view of PAH photochemistry in pure H_2O ice, it seems plausible that aromatic structures decorated with alkyl, amino, hydroxyl, cyano, carboxyl, and other interesting functional groups may be produced when mixed-molecular ices containing PAHs are photolyzed.

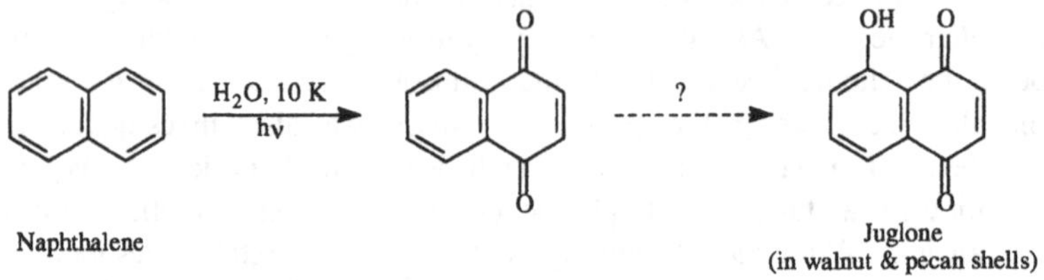

Naphthalene

Juglone
(in walnut & pecan shells)

Figure 19. Molecules very similar to juglone, a compound synthesized and used by walnut trees can be produced abiotically in water-rich ices containing the aromatic molecule naphthalene (Bernstein et al., 2000; Bernstein et al., 2001a). The starting PAH/H_2O concentration of a few percent is conservatively lower than their observed relative abundance in the interstellar medium (Sellgren et al., 1995; Brooke et al., 1999; Chiar et al., 2000).

2.3   INTERSTELLAR PAHS, ICES AND THE ORIGIN OF LIFE

Taking another approach to understanding the role these interstellar/precometary residues might have played in the chemistry on the early Earth and possibly the emergence of life, we are investigating their bulk or collective chemical properties in collaboration with Professor D. Deamer from the University of California, Santa Cruz. While much of the residue dissolves rapidly when added to liquid water or methanol, water-insoluble droplets are also formed (Dworkin et al., 2001). Figure 20 shows a micrograph of these non-soluble droplets in water. Many of the droplets also show intriguing internal structures. Droplet formation shows that some of the complex organic compounds produced in these interstellar ice analogs are amphipillic, i.e. they have both a polar and non-polar component, similar in structure to the molecules which comprise soap. These are also the types of molecules that make up cell membranes and membrane production is considered a critical step in the origin of life. These droplets can encapsulate hydrophyllic fluorescent dyes within their interior, demonstrating that they are true vesicles (hollow droplets) with their interiors separated from surroundings by their lipid multilayer (see Fig 5 in Dworkin et al., 2001).

Vesicle formation is thought critical to the origin of life since vesicles provide an environment in which life can evolve, isolating and protecting the process from the surrounding medium. For example, within the confines of a vesicle, pH can be moderated and held at a different value from that in the surrounding medium, and nutrients, catalysts, and other materials can be concentrated and held together. While it is uncertain where membrane formation falls in the sequence of events leading up to the origin of life, with some arguing that it must have been one of the first steps (Luisi, 1997), and others that it occurred at a later stage (DeDuve, 1997), it is considered a very crucial step.

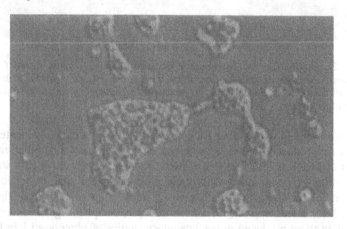

Figure 20. Phase micrograph of the water insoluble droplets formed when a few drops of water are placed on the organic residue produced from the UV photolysis of the realistic interstellar/precometary ice analog: $H_2O:CH_3OH:NH_3:CO$ (100:50:1:1) at 10 K. The addition of liquid water mimics what might happen within a comet or on a parent body when these materials are exposed to liquid water. The large structure is 10 μm across. These self-organizing droplets also trap luminescent materials in their structures as shown in Figure 21B (Dworkin et al., 2001).

Figure 21 shows that the membranes trap other, photoluminescent, molecules that are also produced within the ice by UV irradiation (Dworkin et al., 2001). Thus, not only are vesicle forming compounds produced from the simplest and most abundant IS starting materials, complex organics which absorb UV are also formed. This ability to form and trap energy receptors within these structures is considered another critical step in the origin of life as it provides the means to utilize energy available outside the system.

As shown in Figure 21, the residue droplet-forming behavior upon addition to liquid water is strikingly similar to that found for the organic components of the Murchison meteorite (Deamer, 1985), another indication of similarity between the laboratory ice residue and extraterrestrial organics in meteorites.

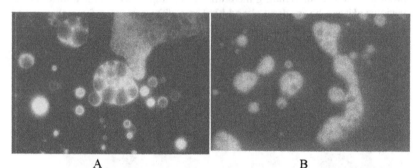

Figure 21. (A) Fluorescence micrograph of the water insoluble droplets formed from a Murchison meteorite extract (Deamer, 1985) compared to (B) the fluorescent droplets produced from the photolysis residue of the interstellar/precometary ice analog: $H_2O:CH-OH:NH:CO$ (100:50:1:1) at 10 K (Dworkin et al., 2001).

Before ending with a discussion on the possible role that these materials which are produced by the energetic processing of IS ices might play in the origin of life, it is useful to first consider the evidence in favor of a strong connection between interstellar organic materials and the carbonaceous fractions of meteorites and IDPs (and by implication comets) (Kerridge et al., 1987; Messenger et al., 1995). For example, complex organic molecules similar to those produced in our laboratory studies of photolyzed ices have been identified in meteorites (Hahn et al., 1998; Cronin and Chang, 1993) and many oxidized polycyclic aromatic hydrocarbons are present in organic extracts from the Murchison meteorite (Deamer, 1997). Furthermore, the PAHs in meteorites are deuterium enriched (Kerridge et al., 1987) as is the carbon in IDPs (Clemett et al., 1993; McKeegan et al., 1985). The formation histories of these extraterrestrial materials are not well understood, although the presence of deuterium enrichments in many of the classes of these compounds has been taken to implicate an interstellar origin (Kerridge et al., 1987; Cronin et al., 1993). Likewise, the similarity between the interstellar ice constituents and relative abundances with known cometary constituents listed in Table 7 is remarkable (see e.g. Mumma et al., 1993; Mumma, 1997). Recent work continues to support this picture.

These similarities between interstellar ices and PAHs with the extraterrestrial materials delivered to the Earth strengthens the case for taking interstellar PAHs and ices into account when pondering an exogenous origin of life. Three roles that these IS materials might have played in the exogenous origin of life on Earth considered are: (1) A supplier of the basic, prebiotic raw materials from which biotic compounds were eventually produced on Earth; (2) The source of complex prebiotic materials poised to play a direct role once in a favorable environment; and (3) The fountainhead of some organic species and mixtures capable of carrying out the basic processes of life.

(1) Supplier of basic, raw materials. First, comets were very likely responsible for raining a rich inventory of potentially important prebiotic molecules down onto the early Earth. HMT and the family of related species provides an example. Recall the variety of species that are formed by HMT hydrolysis, thermalization, or photolysis, all processes which can occur during a comet's lifetime (§3.2.1). For example, hydrolysis under acidic conditions yields ammonia and formaldehyde as well as amino acids. In addition to the possible prebiotic activity of HMT itself, its ice synthesis implicates the presence of several other biogenically interesting intermediates as well, including formaldehyde, ammonia, and cyano-containing species. The observed differential sublimation of these moderately volatile compounds in the laboratory may be similar to that which occurs when cometary ice fragments and dust grains are ejected into cometary comae (Allamandola et al., 1988). This process of differential sublimation in the coma followed by photolysis can regulate the distribution of some of the molecular photofragments seen in cometary comae. Perhaps since ultraviolet photolysis of HMT frozen in H_2O ice produces the OCN^- band observed in the spectra of protostellar objects and laboratory ices, HMT may contribute to some of the OCN^- in protostellar objects and, as such, contribute to the extended CN source in cometary comae.

(2) Source of complex prebiotic materials. Apart from supporting the hypothesis that meteorites, dust, and comets delivered organics to the early earth which served as a useful prebiotic chemical reservoir (Li and Greenberg, 2002; Oro and Cosmovici, 1997; Chyba et al., 1990), the laboratory work described here supports the idea that these extraterrestrial sources could have delivered species which were sufficiently complex to play a direct role in the chemistry of the origin of life and not only provided the raw materials. Section 3.2 described the intriguing behavior of some of the compounds produced by irradiation of realistic interstellar/precometary ice analogs. Some of these products have been found to self-organize to form membranes and ultimately vesicles when they are immersed in liquid water. Other species produced in the ice together with the vesicle forming compounds are trapped within these vesicles. These other organic molecules can harvest ultraviolet light, bringing energy from the surroundings into the vesicle and driving further chemistry. When PAHs, very abundant interstellar species, are included in the ice irradiation experiments, the level

of chemical complexity in both structure and behavior rises significantly. Substituted aromatic structures are readily produced which are common in today's biological systems. Quinones, for example, are used to transport electrons across membrane walls in living systems. Also, many alkaloids have aromatic base skeletons. In addition, other interstellar ice simulation experiments are known to produce amino acids (Briggs et al., 1992), as well as glycerol the precursors of simple sugars. Thus, comets might represent quite potent prebiotic cocktails. *All* of these complex materials are produced by the irradiation of simple ice mixtures *that are known* to be widespread throughout the galaxy in dense molecular clouds. Thus, it seems warranted to hypothesize that wherever a planet or satellite is formed in the habitable zone, it stands to accrete these chemical remnants from the nascent cloud from infalling IDPs, cometary (formerly interstellar) and meteoritic materials. If the emergence of life is indeed an increase in chemical complexity as is currently thought, one of the challenges before us is to understand this chemistry and to assess the relative importance of this input into the origin of life versus the build-up from the compounds already present on the primordial planet.

(3) Fountainhead of species undergoing the basic processes of life. Carrying this one step further, we end this article on a speculative note. These early results are quite startling when viewed from the perspective held not long ago that chemistry in space was very limited. In view of this and the fact that there are hundreds of residue compounds not yet analyzed, an intriguing possibility now supported by these results is the production, within the comet itself, of species poised to take part in the life process, or perhaps even at the earliest stages of what would be perceived of as a living system. This "jump-starting" of the life process by the introduction of such marginally biologically active species, into the "warm pond" on a habitable planet may not be as far fetched as it would at first seem. Although one could hardly imagine a better deep freeze than isolation within a comet, with temperatures measured in the ten's of degrees K, there are repeated episodes of warming for periodic comets (such as Halley) when they approach the Sun. Since these warming episodes, which can be repeated many times in a given comet, can last anywhere from periods of weeks to several months, there is ample time for a very rich mixture of complex organics to develop even though these warming episodes are sandwiched between long periods of extreme cold. Since the surface becomes the warmest, above 300 K for comets which pass closer to the Sun than the Earth, most of the volatiles within a few meters of the surface sublime during each passage. During these warming periods, regions within the comet nucleus will be warmed to different temperatures as well. As the surface heat wave propagates inward,

different processes can occur within different regions of the cometary nucleus depending on the maximum temperature and duration of the thermal pulse. In the deepest regions where the temperatures only rise to between 50 and 150 K, diffusion limited reactions will be favored. At higher temperatures, these as well as other reactions will be allowed. It is even conceivable that liquid water might be present for short periods within the larger comets. These low temperatures and the conditions that ensure slow processing and gradual periods of warming and cooling may actually serve to protect any larger species formed much in the way membranes are involved for aqueous systems. At this early stage in our understanding of cometary ices it already appears that comets contain many of the types of compounds which are considered important players in the origin of life, compounds which play central physical roles and compounds which play key chemical roles. Indeed there is evidence for nearly every class of organic chemical thought important in the "RNA world" in these irradiated ices and they are all held in close proximity to one another. Consider also the unexplored avenues afforded by the additional internal energy available to drive chemistry as the ice undergoes transitions from one phase to the next, the crystal field effects, the structural and templateing restrictions the ice places on reactions and the forced stereochemistry. We are just beginning to appreciate the complexity and possibilities of these ices. Thus, it is no longer inconceivable that comets played a far more active role in the origin of life than simply that of a spectator-delivery system of raw materials. Perhaps Darwin's 'warm little pond' is a warmed comet.

3. CONCLUSION

The last two and a half decades of infrared spectroscopic observations of cosmic ices and PAHs, the two classes of materials we have learned most about through these observations, has taught us that these species are an integral part of the rich and complex world of Interstellar Chemistry. This chemistry includes contributions from gas phase ion-molecule reactions, radical driven reactions, surface reactions on grains, and solid state reactions initiated by energetic processing of icy grain mantles, to name but a few of the processes at work. Further, cometary observations and meteoritic and interplanetary dust particle (IDP) isotopic anomalies now indicate that some interstellar material could be delivered to planetary surfaces essentially unmodified. Taken together, these results are driving a very large shift in our understanding of the kinds of materials that might fall on primordial planetary systems. As recently as the 1960's, it was thought that interstellar conditions were too harsh for any significant polyatomic chemistry to take place. It was thought that any compounds that could form under the extremely low densities of space would quickly be dissociated by the plentiful high-energy radiation. Factors such as the dust extinction which moderates UV in dense clouds, the protection and chemical productivity afforded by ices, and the ubiquity of the UV-hardy aromatic hydrocarbons were simply not considered. Furthermore, it was thought that even if there was some chemical complexity prior to the Solar System's formation, the extreme temperatures and kinetic violence of that process would have erased any significant chemical memory of interstellar input. Indeed, the aromatic hydrocarbons in meteorites were interpreted to be the end products of simpler hydrocarbons heated above 300 to 400 C on the various parent bodies during star and planet formation. Today we are beginning to see that PAHs and related materials are abundant throughout the ISM; that the simplest interstellar ices can foster the production of a stunning array of complex organic species; and that copious quantities of these materials can not only survive the rigors of the star formation process, but also deposition onto the resultant planetary bodies. Thus, we are in the midst of a major shift in our understanding of what contributed to the 'atmospheres' of the terrestrial planets. Instead of simply assuming that a planet's prebiotic chemistry had to 'start from scratch', evidence is building that a far more chemically complex interstellar broth was present right from the start.

This more recent perspective requires much work before its role can be fully assessed. For example, "How important is it?" is a key question. In view of the size of the terrestrial planets and the amount of materials initially present, how significant a role could this extraterrestrial input have played anyway? Conversely, the size of the terrestrial planets might be viewed as a disadvantage to generating chemical complexity just as the low densities of space do not favor the production of complex polyatomic molecules. In both cases, the low concentrations of organic materials hampers such reactions. From this perspective, the chemical intimacy of an interstellar and cometary ice might have great advantage.

Interstellar ice composition is most relevant for those comets which preserve their IS ice heritage and IDPs and meteorites which preserve interstellar organic refractory materials. IS ice composition depends on local conditions. In areas associated with star, planet, and comet formation, ices comprised of simple, polar species and entrapped volatiles such as H_2O, CH_3OH, CO, CO_2, NH_3, and H_2CO are most important. Photolysis of these ices produces the simple species H_2, H_2CO, CO_2, CO, CH_4, HCO as well as an extremely rich mixture of complex organic molecules. We now know that this prebiotic cocktail includes unsaturated fluorescent compounds; self-assembling, amphiphillic molecules; sugar precursors; and amino acids and we have just scratched the surface. These same lines of sight also show evidence for polycyclic aromatic hydrocarbons (PAHs), the organic molecules known from their characteristic emission to be widespread and abundant throughout the interstellar medium. Including PAHs in the irradiated ices produces aromatic alcohols, quinones and ethers, species similar to those found in meteorites and having similar chemical properties to functionalized aromatic species used in living systems today. All of the above compounds are readily formed and thus likely cometary constituents at the 0.1 to 1 percent level. Evidence is growing that this interstellar heritage did not become erased during the Solar Nebula phase, implying that this chemical inventory contributed to the chemistry on the primitive Earth. The ready formation of these organic species from simple starting mixtures under general interstellar conditions, the ice chemistry that ensues when these ices are mildly warmed, and the observation that the more complex refractory photoproducts form fluorescent vesicles upon exposure to liquid water underscore the possibility that IS and cometary ices could have played an important role in the origin of life. Three possibilities are considered, ranging from these interstellar materials simply providing the raw materials used for a completely endoginous origin to the opposite extreme that they delivered species poised to take part in the life process, or perhaps even at the earliest stages of what would be perceived of as a living system. While far-fetched to many of us only a few years ago, the evidence suggests that this is conceivable.

REFERENCES

A&A Special Edition: First Results from ISO (1996) *A&A*, **315**(2).

Agarwal, V.K. et al (1985) *Origins of Life*, **16**, 21.

Aitken, D.K. (1981), in Wynn-Williams, C.G. and Cruikshank, D.P. (eds.), *Infrared Astronomy*, D. Reidel Pub. Co., Dordrecht, p. 207.

Allamandola, L.J. (1996), in Greenberg, J.M. (ed.), *The Cosmic Dust Connection*, Kluwer Academic Pub., Dordrecht, p.81.

Allamandola, L.J., Bernstein, M.P. and Sandford, S.A. (1997), in Cosmovici, C.B., Bowyer, S. and Werthimer, D. (eds.), *Astronomical and Biochemical Origins and the Search for Life in the Universe*, Editrice Compositori of the Italian Physical Society, Bologna, p.23.

Allamandola, L.J., Hudgins, D.M. and Sandford, S.A. (1999), *ApJ*, **511**, L115.

Allamandola, L.J. and Sandford, S.A. (1988), in Bailey, M.E. and Williams, D.A. (eds.), *Dust in the Universe*, Cambridge Univ. Press, Cambridge, p.229.

Allamandola, L.J., Sandford, S.A., Tielens, A.G.G.M. and Herbst, T.M. (1992), *ApJ*, **399**, 134.

Allamandola, L.J., Sandford, S.A., Tielens, A.G.G.M. and Herbst, T.M. (1993), *Science*, **260**, 64.

Allamandola, L.J., Sandford, S.A. and Valero, G. (1988), *Icarus*, **76**, 225.

Allamandola, L.J., Sandford, S.A. and Wopenka, B. (1987a), *Science*, **237**, 56.

Allamandola, L.J. and Tielens, A.G.G.M. (eds.) (1989), *Interstellar Dust, Proceedings of IAU Symposium 135*, Kluwer Academic Pub., Dordrecht.

Allamandola, L.J., Tielens, A.G.G.M. and Barker, J.R. (1985), *ApJ*, **290**, L25.

Allamandola, L.J., Tielens, A.G.G.M. and Barker, J.R. (1987b), in Leger, A., d'Hendecourt, L.B. and Boccara, N. (eds.), *Polycyclic Aromatic Hydrocarbons and Astrophysics*, D. Reidel Pub. Co., Dordrecht, p.255.

Allamandola, L.J., Tielens, A.G.G.M. and Barker, J.R. (1989), *ApJSS*, **71**, 733.

Altwegg, K., Balsiger, H. and Geiss, J. (1999), in Altwegg, K., Ehrenfreund, P., Geiss, J. and Huebner, W. (eds.), *The Composition and Origin of Cometary Materials*, Kluwer Academic Pub., Dordrecht, p.3.

Barker, J.R. and Cherchneff, I. (1989), in Allamandola, L.J. and Tielens, A.G.G.M., (eds.), *Interstellar Dust: Proceedings of IAU Symposium 135*, Kluwer Academic Pub., Dordrecht, p.197.

Bauschlicher, C.W. (1998a), *Chem. Phys.*, **233**, 29.

Bauschlicher, C.W. (1998b), *Chem. Phys.*, **234**, 87.

Bauschlicher, C.W. Jr., private communication.

Bauschlicher, C.W. and Bakes, E.L.O. (2000), *Chem. Phys.*, **262**, 285.

Bauschlicher, C.W. and Langhoff, S.R. (1997), *Spectrochim. Acta A*, **53**, 1225.

Bauschlicher, C.W. and Langhoff, S.R. (1998), *Chem. Phys.*, **234**, 79.

Bauschlicher, C.W. Jr., Langhoff, S.R., Sandford, S.A. and Hudgins, D.M. (1997), *J. Phys. Chem. A*, **101**, 2414.

Bellamy, L.J. (1958), *The Infrared Spectra of Complex Organic Molecules*, John Wiley & Sons, New York.

Bernstein, M.P., Dworkin, J.P., Sandford, S.A. and Allamandola, L.J. (2001a), *Met. & Planet. Sci.*, **36**, 351.

Bernstein, M.P., Sandford, S.A. and Allamandola, L.J. (1996), *ApJ*, **472**, L127.

Bernstein, M.P., Sandford, S.A. and Allamandola, L.J. (2000), *ApJ*, **542**, 894.

Bernstein, M.P., Sandford, S.A. and Allamandola, L.J. (2001b), *ApJ*, **542**, 894.

Bernstein, M.P., Sandford, S.A., Allamandola, L.J. and Chang, S. (1994), *J. Phys. Chem.*, **98**, 12206.

Bernstein, M.P., Sandford S.A., Allamandola L.J., Chang, S. and Scharberg, M.A. (1995), *ApJ*, **454**, 327.

Bernstein, M.P., Sandford, S.A., Allamandola, L.J., Gillette, J.S., Clemett, S.J. and Zare, R.N. (1999), *Science*, **283**, 1135.

Blake, D., Allamandola, L.J., Sandford, S.A., Hudgins, D.M. and Freund, F. (1991), *Science*, **254**, 548.

Blake, G.A., Sutton, E.C., Masson, C.R. and Phillips, T.G. (1987), *ApJ*, **315**, 621.

Blanco, A., Bussoletti, E. and Colangeli, L. (1988), *ApJ*, **334**, 875.

Boogert, A.C.A. et al. (2000), *A&A*, **353**, 349.

Boogert, A.C.A., Schutte, W.A., Helmich, F.P., Tielens, A.G.G.M. and Wooden, D.H. (1997), *A&A.*, **317**, 929.

Bowey, J.E., Adamson, A.J. and Whittet, D.C.B. (1998), *MNRAS*, **298**, 131.

Bregman, J.D. (1989), in Allamandola, L.J. and Tielens, A.G.G.M. (eds.), *Interstellar Dust*, Kluwer Acabemic Pub., Dordrecht, p.109.

Bregman, J.D., Allamandola, L.J., Tielens, A.G.G.M., Geballe, T.R. and Witteborn, F.C. (1989), *ApJ*, **344**, 791.

Bregman, J., Rank, D., Sandford, S. and Temi, P. (1993a), *ApJ*, **410**, 668.

Bregman, J., Rank, D., Temi, P., Hudgins, D. and Kay, L. (1993b), *ApJ*, **411**, 794.

Brenner, J.D. and Barker, J.R. (1992), *ApJ*, **388**, L39.

Briggs, R., Ertem, G., Ferris, J.P., Greenberg, J.M., McCain, P.J., Mendoza-Gomez, C.X. and Schutte, W. (1992), *Orig. Life Evol. Biosphere*, **22**, 287.

Brooke, T.Y., Sellgren, K. and Geballe, T.R. (1999), *ApJ*, **517**, 883.

Buss, R.H. Jr. et al. (1993), *ApJ*, **415**, 250.

Cadwell, B.J., Wang, H., Feigelson, E.D. and Frenklach, M. (1994), *ApJ*, **429**, 285.

Capps, R.W., Gillett, F.C. and Knacke, R.F. (1978), *ApJ*, **226**, 863.

Charnley, S.B. (1997), in Cosmovici, C.B., Bowyer, S. and Werthimer, D. (eds.), *Astronomical and Biochemical Origins and the Search for Life in the Universe*, Editrice Compositori of the Italian Physical Society, Bologna, p.89.

Charnley, S. (2001), *ApJ*, **562**, 99.

Charnley, S.B., Rodgers, S.D. and Ehrenfreund, P. (2001), *A&A*, **378**, 1024.

Charnley, S.B., Tielens, A.G.G.M. and Millar, T.J. (1992), *ApJ*, **399**, L71.

Cherchneff, I., Barker, J.R. and Tielens, A.G.G.M. (1992), *ApJ*, **401**, 269.

Chiar, J.E., Adamson, A.J., Kerr, T.H. and Whittet, D.C.B. (1994), *ApJ*, **426**, 240.

Chiar, J.E., Adamson, A.J., Kerr, T.H. and Whittet, D.C.B. (1995), *ApJ*, **455**, 234.

Chiar, J.E. et al. (2000), *ApJ*, **537**, 749.

Chiar J.E., Gerakines P.A., Whittet D.C.B., Pendleton Y.J., Tielens A.G.G.M. and Boogert, A.C.A. (1998a), *ApJ*, **498**, 716.

Chiar, J.E., Pendleton, Y.J., Geballe, T.R. and Tielens, A.G.G.M. (1998b), *ApJ*, **507**, 281.

Chyba, C.F., Thomas, P.J., Brookshaw, L. and Sagan, C. (1990), *Science*, **249**, 366.

Clemett, S., Maechling, C., Zare, R., Swan, P. and Walker, R. (1993), *Science*, **262**, 721.

Cohen, M. et al. (1986), *ApJ*, **302**, 737.

Cohen, M. et al. (1989), *ApJ*, **341**, 246.

Cohen, M., Tielens, A.G.G.M. and Allamandola, L.J. (1985), *ApJ*, **299**, L93.

Cook, D.J. and Saykally, R.J. (1998), *ApJ*, **493**, 793.

Cox, P. (1989), *A&A*, **225**, L1.

Crawford, M.K., Tielens, A.G.G.M. and Allamandola, L.J. (1985), *ApJ*, **293**, L45.

Cronin, J.R. and Chang, S. (1993), in Greenberg, J. M., Mendoza-Gómez, C. X. and Pirronello, V. (eds.), *The Chemistry of Life's Origins*, Kluwer Academic Pub., Dordrecht, p. 209.

Cronin, J.R., Pizzarello, S., Epstein, S. and Krishnamurthy, R.V. (1993), *Geochim. Cosmochim. Acta*, **57**, 4745.

Crovisier, J. and Bocklee-Morvan, D. (1999), in Altwegg, K, Ehrenfreund, P., Geiss, J. and Huebner, W. (eds.), *The Composition and Origin of Cometary Materials*, Kluwer Academic Pub., Dordrecht, p.19.

Dartois, E., Demyk, K., d'Hendecourt, L.B. and Ehrenfreund, P. (1999a), *A&A*, **351**, 1066.

Dartois, E., Schutte, W., Geballe, T.R., Demyk, K., Ehrenfreund, P. and d'Hendecourt, L.B. (1999b), *A&A*, **342**, L32.

de Frees, D.J., Miller, M.D., Talbi, D., Pauzat, F. and Ellinger, Y. (1993), *ApJ*, **408**, 530.

Deamer, D.W. (1985), *Nature*, **317**, 792.

Deamer, D.W. (1997), *Microbio. & Molec. Bio. Rev.*, **61**, 239.

DeDuve, C. (1997), in Cosmovici, C.B., Bowyer, S. and Werthimer, D. (eds.), *Astronomical and Biochemical Origins and the Search for Life in the Universe*, Editrice Compositori of the Italian Physical Society, Bologna, p.391.

DeGrauw , T. et al. (1996), *A&A*, **315**, L345.

Demyk, K., Dartois, E., d'Hendecourt, L.B., Jourdain de Muizon, M., Heras, A., Breitfellner, M. (1998), *A&A*, **339**, 553.

DeVries, M.S. et al. (1993), *Geochim. Cosmochim. Acta*, **57**, 933.

d'Hendecourt, L.B., Allamandola, L.J. and Greenberg, J.M. (1985), *A&A*, **152**, 130.

d'Hendecourt, L.B., Allamandola, L.J., Grim, R.J.A. and Greenberg, J.M. (1986), *A&A*, **158**, 119.

d'Hendecourt, L.B. and Jordain de Muizon, M. (1989), *A&A*, **223**, L5.

d'Hendecourt, L.B. and Leger, A. (1987), *A&A*, **180**, L9.

Dickens, J.E., Irvine, W.M., DeVries, C.H. and Ohishi, M. (1997), *ApJ*, **479**, 307.

Dorschner, J. and Henning, Th. (1995), *A&A Rev.*, **6**, 271.

Dworkin, J.P., Deamer, D.W., Sandford, S.A. and Allamandola, L.J. (2001), *Pub. Nat. Acad.*, **98**, 815.

Dworkin, J.P., Sandford, S.A., Allamandola, L.J., Deamer, D., Gilette, J.S. and Zare, R.N. (2000), *Orig. Life Evol. Biosphere*, **30**, 228.

Ehrenfreund, P., Boogert, A.C.A., Gerakines, P.A., Jansen, D.J., Schutte, W.A., Tielens, A.G.G.M. and van Dishoeck, E.F. (1996), *A&A*, **315**, L341.

Ehrenfreund, P., Boogert, A.C.A., Gerakines, P.A., Tielens, A.G.G.M. and van Dishoeck, E.F. (1997), *A&A*, **328**, 649.

Ehrenfreund, P., Dartois, E., Demyk, K. and d'Hendecourt, L.B. (1998), *A&A*, **239**, L17.

Ehrenfreund, P. and Fraser, H. (2002), in this volume.

Eiroa, C. and Hodapp, K.-W. (1989), *A&A*, **210**, 345.

Elsila, J., Allamandola, L.J. and Sandford, S.A. (1997), *ApJ*, **479**, 818.

Flickinger, G.C. and Wdowiak, T.J. (1990), *ApJ*, **362**, L71.

Flickinger, G.C., Wdowiak, T.J. and Gomez, P.L. (1991), *ApJ*, **380**, L43.

Gauger, A., Gail, H.P. and Sedlmayr, E. (1990), *A&A*, **235**, 345.

Geballe, T.R. (1986), *A&A*, **162**, 248.

Genzel, R. et al. (1998), *ApJ*, **498**, 579.

Gerakines, P.A. et al. (1999), *ApJ*, **522**, 357.

Gerakines, P.A., Schutte, W.A. and Ehrenfreundt, P. (1995a), *A&A*, **312**, 289.

Gerakines, P.A., Schutte, W.A., Greenberg, J.M. and van Dishoeck, E.F. (1995b), *A&A*, **296**, 810.

Gillett, F.C. and Forrest, W.J. (1973), *ApJ*, **179**, 483.

Gillett, F.C., Forrest, W.J. and Merrill, K.M. (1973), *ApJ*, **183**, 87.

Gillett, F.C., Jones, T.W., Merrill, K.M. and Stein, W.A. (1975), *A&A*, **45**, 77.

Greenberg, J.M. (1971), *A&A*, **12**, 240.

Greenberg, J.M. (1978), in McDonnell, J.A.M. (ed.), *Cosmic Dust*, John Wiley and Sons, New York, p.187.

Greenberg, J.M. (1989), in Allamandola, L.J. and Tielens, A.G.G.M. (eds.), *Interstellar Dust*, Kluwer Academic Pub., Dordrecht, p. 345.

Grim, R.J.A., Baas, F., Geballe, T.R., Greenberg, J.M. and Schutte, W.A. (1991), *A&A*, **243**, 473.

Grim, R.J.A. and Greenberg, J.M. (1987), *ApJ*, **321**, L91.

Hagen, W., Allamandola, L.J. and Greenberg, J.M. (1979), *Astrophys. Spa. Sci.*, **65**, 215.

Hagen, W., Allamandola, L.J. and Greenberg, J.M. (1980), *A&A*, **86**, L3.

Hagen, W., Tielens, A.G.G.M. and Greenberg, J.M. (1983), *A&A Supp. Ser.*, **51**, 389.

Hahn, J.H., Zenobi, R., Bada, J.L. and Zare, R.N. (1998), *Science*, **239**, 1523.

Halasinski, T., Hudgins, D.M., Salama, F., Allamandola, L.J. and Bally, T. (2000), *J. Phys. Chem.*, **104**, 7484.

Herbst, E. (2002), in this volume.

Hiraoka, K., Miyagoshi, T., Takayama, T., Yamamoto, K. and Kihara, Y. (1998), *ApJ*, **498**, 710.

Hony, S., Van Kerckhoven, C., Peeters, E., Tielens, A.G.G.M., Hudgins, D.M. and Allamandola, L.J. (2001), *A&A.*, **370**, 1030.

Hudgins, D.M. and Allamandola, L.J. (1995a), *J. Phys. Chem.*, **99**, 3033.

Hudgins, D.M. and Allamandola, L.J. (1995b), *J. Phys. Chem.*, **99**, 8978.

Hudgins, D.M. and Allamandola, L.J. (1997), *J. Phys. Chem.*, **101**, 3472.

Hudgins, D.M. and Allamandola, L.J. (1999), *ApJ*, **516**, L41.

Hudgins, D.M., Bauschlicher, C.W. and Allamandola, L.J. (2001), *Spectrochim. Acta A*, **57**, 907.

Hudgins, D.M., Bauschlicher, C.W. Jr., Allamandola, L.J. and Fetzer, J.C. (2000), *J. Phys. Chem. A*, **104**, 3655.

Hudgins, D.M. and Sandford, S.A. (1998a), *J. Phys. Chem. A*, **102**, 329.

Hudgins, D.M. and Sandford, S.A. (1998b), *J. Phys. Chem. A*, **102**, 344.

Hudgins, D.M. and Sandford, S.A. (1998c), *J. Phy .Chem. A*, **102**, 353.

Hudgins, D., Sandford, S.A. and Allamandola, L.J. (1994), *J. Phys. Chem.*, **98**, 4243.

Hudson, R.L., Moore, M.H. and Gerakines, P.A. (2001), *ApJ*, **550**, 1140.

Huebner, W.F., Boice, D.C. and Korth, A. (1989), *Adv. Space Res.*, **9**, 29.

Irvine, W.M. (1998), *Orig. Life Evol. Biosphere*, **28**, 365.

Joblin, C., Boissel, P., Leger, A., d'Hendecourt, L.B. and Defourneau, D. (1995), *A&A*, **299**, 835.

Kerridge, J.F., Chang, S. and Shipp, R. (1987), *Geochim. Cosmochim. Acta*, **51**, 2527.

Kim, H.-S., Wagner, D.R. and Saykally, R.J. (2001), *ApJ*, in press.

Knacke, R.F., McCorkle, S., Puetter, R.C., Erickson, E.F. and Kratschmer, W. (1982), *ApJ*, **260**, 141.

Kurtz, J. (1992), *A&A*, **255**, L1.

Kwok, S., Hrivnak, B.J. and Geballe, T.R. (1995), *ApJ*, **454**, 394.

Lacy, J., Carr, J., Evans, N., Baas, F., Achtermann, J. and Arens, J. (1991), *ApJ*, **376**, 556.

Lacy, J.H. et al. (1984), *ApJ*, **276**, 533.

Lacy, J., Faraji, H., Sandford, S.A. and Allamandola, L.J. (1998), *ApJ*, **501**, L105.

Langhoff, S.R. (1996), *J. Phys. Chem.*, **100**, 2819.

Langhoff, S.R. Bauschlicher, C.W. Jr., Hudgins, D.M., Sandford, S.A. and Allamandola, L.J. (1998), *J. Phys. Chem. A*, **102**, 1632.

Leger, A. and d'Hendecourt, L.B. (1985), *A&A*, **146**, 81.

Leger, A., d'Hendecourt, L.B. and Defourneau, D. (1989), *A&A*, **216**, 148.

Leger, A., Gauthier, S., Defourneau, D. and Rouan, D. (1983), *A&A*, **117**, 164.

Leger, A. and Puget, J.L. (1984), *A&A*, **137**, L5.

Lepp, S. and Dalgarno, A. (1988), *ApJ*, **324**, 553.

Li, A. and Greenberg, J.M. (2002), in this volume.

Luisi, P.L. (1997), in Cosmovici, C.B., Bowyer, S. and Werthimer, D. (eds.), *Astronomical and Biochemical Origins and the Search for Life in the Universe*, Editrice Compositori of the Italian Physical Society, Bologna, p.461.

Mattioda, A., Hudgins, D.M., Bauschlicher, C.W. Jr. and Allamandola, L.J. (manuscript in preparation).

McKeegan, K.D., Walker, R.M. and Zinner, E. (1985), *Geochim. Cosmochim. Acta*, **49**, 1971.

Messenger, S., Clemett, S.J., Keller, L.P., Thomas, K.L., Chillier, X.D.F. and Zare, R.N. (1995), *Meteoritics*, **30**, 546.

Moore, M.H., Donn, B., Khanna, R. and A'Hearn, M.F. (1983), *Icarus*, **54**, 388.

Moore, M.H., Ferrante, R.F. and Nuth, J.A. (1996), *Planet. Spa. Sci.*, **44**, 927.

Moutou, C., Verstraete, L., Leger, A., Sellgren, K. and Schmidt, W. (2000), *A&A*, **354**, L17.

Mumma, M.J. (1997), in Cosmovici, C.B., Bowyer, S. and Werthimer, D. (eds.), *Astronomical and Biochemical Origins and the Search for Life in the Universe*, Editrice Compositori of the Italian Physical Society, Bologna, p.121.

Mumma, M.J., Stern, S.A. and Weissman, P.R. (1993), in Levy, E.H., Lunine, J.I. and
    Matthews, M.S. (eds.), *Protostars and Planets III*, University of Arizona Press, Tucson,
    p.1177.
Ohishi, M., Irvine, W.M. and Kaifu, N. (1992), in Singh, P.D. (ed.), *Astrochemistry of Cosmic
    Phenomena*, Kluwer Academic Pub., Dordrecht, p.171.
Omont, A. (1986), *A&A*, **164**, 159.
Omont, A. et al. (1990), *ApJ*, **355**, L27.
Oro, J. and Cosmovici, C.B. (1997), in Cosmovici, C.B., Bowyer, S. and Werthimer, D.
    (eds.), *Astronomical and Biochemical Origins and the Search for Life in the Universe*,
    Editrice Compositori of the Italian Physical Society, Bologna, p.97.
Palumbo, M.E. (1997), *J. Phys. Chem. A*, **101**, 4298.
Palumbo, M.E., Pendleton, Y.J. and Strazzulla, G. (2000), *ApJ*, **542**, 890.
Palumbo, M.E., Pendleton, Y.J. and Strazzulla, G. (2001), *ApJ*, **542**, 890
Palumbo, M.E. and Strazzulla, G. (1993), *A&A*, **269**, 568.
Palumbo, M.E. and Strazzulla, G. (2000), *ApJ*, **542**, 894.
Palumbo, M.E., Tielens, A.G.G.M. and Tokunaga, A.T. (1995), *ApJ*, **449**, 674.
Pauzat, F. and Ellinger, Y. (2001), *MNRAS*, **324**, 355.
Pauzat, F., Talbi, D. and Ellinger, Y. (1995), *A&A*, **293**, 263.
Pauzat, F., Talbi, D. and Ellinger, Y. (1997), *A&A*, **319**, 318.
Pauzat, F., Talbi, D. and Ellinger, Y. (1999), *MNRAS*, **304**, 241.
Pendleton, Y.J., Tielens, A.G.G.M., Tokunaga, A.T. and Bernstein, M.P. (1999), *ApJ*, **513**,
    294.
Phillips, M.M., Aitken, D.K. and Roche, P.F. (1984), *MNRAS*, **207**, 25.
Piest, J.A., Oomens, J., Bakker, J., von Helden, G. and Meijer, G. (2001), *Spectrochim. Acta
    A*, **57A**, 717.
Piest, J.A., von Helden, G. and Meijer, G. (1999), *ApJ*, **520**, L75.
Pirronello, V. (1997), in Cosmovici, C.B., Bowyer, S. and Werthimer, D. (eds.), *Astronomical
    and Biochemical Origins and the Search for Life in the Universe*, Editrice Compositori of
    the Italian Physical Society, Bologna, p.77.
Pirronello, V. and Averna, D. (1988), *A&A*, **196**, 201.
Pirronello, V., Krelowski, J. and Manicò, G. (eds.) (2002), *Solid State Astrochemistry*, Kluwer
    Academic Pub., Dordrecht.
Pirronello, V., Liu, C., Shen, L. and Vidalli, G. (1997), *ApJ*, **475**, L69.
Puget, J.L. (1989), in Allamandola, L.J. and Tielens, A.G.G.M. (eds.), *Interstellar Dust*,
    Kluwer Academic Pub., Dordrecht, p.119.
Puget, J.L. and Leger, A. (1989), *Ann. Rev. Astron. Astrophys.*, **27**, 161.
Purcell, E.M., (1976), *ApJ*, **206**, 685.
Roche, P.F., Aitken, D.K. and Smith, C.H. (1989), *MNRAS*, **236**, 485.
Roelfsma, P.R. et al. (1996), *A&A*, **315**, L289.
Russell, R.W., Soifer, B.T. and Willner, S.P. (1977), *ApJ*, **217**, L149.
Salama, F., Bakes, E., Allamandola, L.J. and Tielens, A.G.G.M. (1996), *ApJ*, **458**, 621.
Salama, F., Galazutdinov, G.A., Krelowski, J., Allamandola, L.J. and Musaev, F.A. (1999),
    ApJ, **526**, 265.
Sandford, S.A. (1996), *Met. & Planet. Sci.*, **31**, 449.
Sandford, S.A. and Allamandola, L.J. (1990), *ApJ*, **355**, 357.
Sandford, S.A. and Allamandola, L.J. (1993), *ApJ*, **417**, 815.
Sandford, S.A., Allamandola, L.J. and Geballe, T.R. (1993), *Science*, **262**, 400.
Sandford, S.A., Allamandola, L.J., Tielens, A.G.G.M. and Valero, G. (1988), *ApJ*, **329**, 498.
Sandford, S.A., Bernstein, M.P., Allamandola, L.J., Gilette, J.S. and Zare, R.N. (2000), *ApJ*,
    538, 691.

Sandford, S.A., Bernstein, M.P., Allamandola, L.J., Goorvitch, D. and Teixeira, T.C.V.S. (2001), *ApJ*, **548**, 836.

Sato, S., Nagata, T., Tanaka, M. and Yamamoto, T. (1990), *ApJ*, **359**, 192.

Schlemmer, S., Cook, D.J., Harrison, J.A., Wurfel, B., Chapman, W. and Saykally, R.J. (1994), *Science*, **265**, 1686.

Schutte, W. A. (1999), in Ehrenfreund, P. et al. (ed.), *Laboratory Astrophysics and Space Research* , Kluwer Academic Pub., Dordrecht, p.69.

Schutte, W.A., Allamandola, L.J. and Sandford, S.A. (1993a), *Icarus*, **104**, 118.

Schutte, W. A. et al. (1999), *A&A*, **343**, 966.

Schutte, W.A., Gerakines, P.A., Geballe, T.R., van Dishoeck, E.F. and Greenberg, J.M. (1996), *A&A*, **309**, 633.

Schutte, W.A. and Greenberg, J.M. (1997), *A&A.*, **317**, L43.

Schutte, W., Tielens, A.G.G.M. and Allamandola, L.J. (1993b), *ApJ*, **415**, 397.

Sellgren, K. (1984), *ApJ*, **235**, 138.

Sellgren, K., Brooke, T.Y., Smith, R.G. and Geballe, T.R. (1995), *ApJ*, **449**, L69.

Skinner, C.L., Tielens, A.G.G.M., Barlow, M.J. and Justtanont, K. (1992), *ApJ*, **399**, L79.

Sloan, G.C., Bregman, J.D., Geballe, T.R., Allamandola, L.J. and Woodward, C.E. (1996), *ApJ*, **474**, 735.

Sloan, G., Grasdalen, G. and LeVan, P. (1993), *ApJ*, **409**, 412.

Sloan, G.C., Hayward, T.L., Allamandola, L.J., Bregman, J.D., DeVito, B. and Hudgins, D.M. (1999), *ApJ*, **513**, L65.

Smith, R.G., Sellgren, K. and Brooke, T.Y. (1993), *MNRAS*, **263**, 749.

Smith, R.G., Sellgren, K. and Tokunaga , A.T. (1989), *ApJ*, **344**, 413.

Szczepanski, J., Chapo, C. and Vala, M. (1993a), *Chem. Phys.*, **205**, 434.

Szczepanski, J., Drawdy, J., Wehlburg, C. and Vala, M. (1995a), *Chem. Phys. Lett.*, **245**, 539.

Szczepanski, J., Roser, D., Personette, W., Eyring, M., Pellow, R. and Vala, M. (1992), *J. Phys. Chem.*, **96**, 7876.

Szczepanski, J. and Vala, M. (1993a), *ApJ*, **414**, 179.

Szczepanski, J. and Vala, M. (1993b), *Nature* **363**, 699.

Szczepanski, J., Vala, M., Talbi, D., Parisel, O. and Ellinger, Y.J. (1993b), *J. Chem. Phys.*, **98**, 4494.

Szczepanski, J., Wehlburg, C. and Vala, M. (1995b), *Chem. Phys. Lett.*, **232**, 221.

Talbi, D., Pauzat, F. and Ellinger, Y. (1993), *A&A*, **268**, 805.

Tegler, S.C., Weintraub, D.A., Allamandola, L.J., Sandford, S.A., Rettig, T.W. and Campins, H. (1993), *ApJ*, **411**, 260.

Tegler, S.C., Weintraub, D.A., Rettig, T.W., Pendleton, Y.J., Whittet, D.C.B. and Kulesa, C.A. (1995), *ApJ*, **439**, 279.

Teixeira, T.C., Devlin, J.P., Buch, V. and Emerson, J.P. (1999), *A&A*, **347**, L19.

Tielens, A.G.G.M. and Allamandola, L.J. (1987), in Morfill, G.E. and Scholer, M. (eds.), *Physical Processes in Interstellar Clouds*, D. Reidel Pub. Co., Dordrecht, p. 333.

Tielens, A.G.G.M., Allamandola, L.J., Bregman, J., Goebel, J., d'Hendecourt, L.B. and Witteborn, F.C. (1984), *ApJ*, **287**, 697.

Tielens, A.G.G.M. et al. (1993), *Science*, **262**, 86.

Tielens, A.G.G.M. and Hagen, W. (1982), *A&A*, **114**, 245.

Tielens, A.G.G.M., Tokunaga, A.T., Geballe, T.R. and Baas, F. (1991), *ApJ*, **381**, 181.

Vala, M., Szczepanski, J., Pauzat, F., Parisel, O., Talbi, D and Ellinger, Y.J. (1994), *J. Phys. Chem.*, **98**, 9187.

van der Zwet, G.P. and Allamandola, L.J. (1985), *A&A*, **146**, 76.

van Dishoeck, E.F., Blake, G.A., Draine, B.T. and Lunine, J.I. (1993), in Levy, E.H. and Lunine, J.I. (eds.), *Protostars and Planets III*, Univ. of Arizona Press, Tucson, Arizona, p. 163.

van Dishoeck, E.F. et al. (1996), *A&A*, **315**, L349.

Van Kerckhoven, C. et al. (2000), *A&A*, **357**, 1013.

Wagner, D.R., Kim, H.-S. and Saykally, R.J. (2000), *ApJ*, **545**, 854.

Weintraub, D.A., Tegler, S.C., Kastner, J.H. and Rettig, T. (1994), *ApJ*, **423**, 674.

Whittet, D.C.B., Adamson, A.J., Duley, W.W., Geballe, T.R. and McFadzean, A.D. (1989), *MNRAS*, **241**, 707.

Whittet, D.C.B., Pendleton, Y.J., Gibb, E.L., Boogert, A.C.A., Chiar, J.E. and Nummelin, A. (2001), *ApJ*, **550**, 793.

Williams, D.A. (2002), in this volume.

Willner, S.P. (1977), *ApJ*, **214**, 706.

Willner, S.P. (1984), in Kessler, M.F. and Phillips, J.P., (eds.), *Galactic and Extragalactic Infrared Spectroscopy*, D. Reidel Publishing Co., Dordrecht, , p.37.

Winnewisser, G. (1997), in Cosmovici, C.B., Bowyer, S. and Werthimer, D. (eds.), *Astronomical and Biochemical Origins and the Search for Life in the Universe*, Editrice Compositori of the Italian Physical Society, Bologna, p.5.

Witteborn, F.C., Sandford, S.A., Bregman, J.D., Allamandola, L.J., Cohen, M., Wooden, D. and Graps, A.L. (1989), *ApJ*, **341**, 270.

ICE CHEMISTRY IN SPACE

Pascale Ehrenfreund and Helen Fraser

Leiden Observatory, P O Box 9513, 2300 RA Leiden, The Netherlands

Keywords: Interstellar ices – Solar System Ices – Laboratory simulations – photo-chemistry

Abstract   In this chapter we review the current knowledge on ice chemistry in space. We describe the basic processes of ice formation and destruction on interstellar grains and discuss the role of icy particles as catalysts in dense molecular clouds. In the following sections we describe laboratory simulation techniques such as low temperature infrared spectroscopy and surface reaction techniques which are used to obtain information on low temperature ices. We briefly review "observations" of ices in the interstellar medium and in solar system objects obtained from ground and space-based infrared data. Finally, we emphasize future studies, including infrared space missions, large telescope facilities and studies of ice properties in microgravity.

Introduction

The voids between the stars of our and external galaxies are filled with tenuous material. These voids are also called 'interstellar medium (ISM)' and are composed of several types of interstellar clouds (Spaans & Ehrenfreund 1999). Molecular clouds which are dense ($\sim 10^6$ H atoms per cm^{-3}) and very cold (3-90 K) harbor icy particles which play a crucial role in the structure and evolution of the ISM. Stars form by contraction and fragmentation of such molecular clouds and thus the icy material gets incorporated into planetesimals which grow into planets or small remnant objects, such as comets and asteroids. The temperature gradient in the solar system determines that ices cover the surfaces of many outer solar system bodies, including planets, moons and comets as well as a large number of Kuiper belt objects located beyond Neptune. Ices are also present on the poles of planet's Earth and Mars, outer main belt asteroids and even on permanently shaded areas of the innermost

V. Pirronello et al. (eds.), Solid State Astrochemistry, 317–356.
© 2003 *Kluwer Academic Publishers. Printed in the Netherlands.*

planet Mercury. The only planet which definitely does not show any evidence for ice is our sister planet Venus.

The most dominant ice species throughout the universe is water ice, which is also the most intensively studied species in the laboratory. However, open questions remain whether ices produced in Earth laboratories are indeed good analogs for ices detected in space environments. Whereas temperature and pressure conditions can be relatively well controlled in the laboratory it is obviously impossible to simulate the time-scales in space environments. The bulk structure of ice, which also determines the catalytic properties of the surface, could be rather different when formed on timescales of millions of years in zero gravity in space.

1.   BASIC PROCESSES OF ICE FORMATION AND DESTRUCTION IN INTERSTELLAR CLOUDS

There are four basic processes which influence the ice chemistry in the ISM. Figure 1 shows a schematic overview of an icy grain and its catalytic surface. Accretion is a very efficient process in cold environments because most of the gaseous species (with the exception of H_2 and He) stick onto the grains with almost 100% efficiency. This accretion process occurs on a timescale of $\tau \sim 2 \times 10^9/n(H_2)$ yr, assuming a sticking efficiency of unity. The accretion process occurs to specific grain sites, known as binding sites.

Whenever gas phase species 'freeze-out' onto grains, the process they actually undergo is a direct phase change from the gas to the solid phase. Therefore during the freeze-out process energy is released in the form of sublimation energy. Although this physics is well understood, it is still not clear where this energy might be dissipated in interstellar grains, nor the effect it may have on the chemistry in the surrounding ice mantle. It is also possible that some of the excess energy could be reabsorbed by the "freeze-out" molecule. As a rule of thumb, surface diffusion energies are expected to be (and in some cases have been measured as) around 1/3 of the sublimation energy. Consequently, some of the accreted species are sufficiently mobile on the icy surface to diffuse across the ice, to new binding sites, or even to diffuse into pores and voids deep inside the bulk of the solid structure. The grain surface acts as a catalyst for the accreting atoms, molecules and radicals, brings them into close contact with each other, and assists chemical reactions that are unable to occur in the gas phase. This leads to the formation of new molecules, which may later desorb and return to the gas phase. Indeed, with roughly

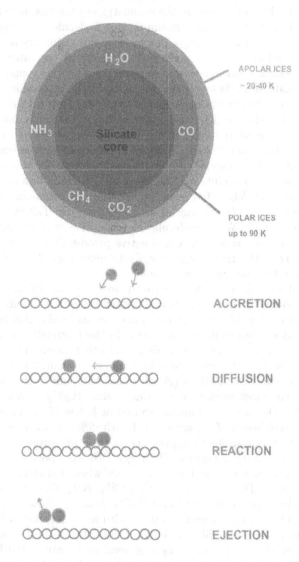

Figure 1 A schematic view of an interstellar grain covered with polar (water-dominated) and/or apolar (highly volatile, H-deficient) ice layers. Four major processes act to form and equilibrate interstellar icy grain mantles. A more detailed explanation is given in the text.

10^6 surface sites (on a grain with a diameter of ~ 0.1 μm) and with an abundance of 10^{-3} condensable species (O,C,N) respective to H, surfaces must play an important role in the chemistry of dark clouds, see Figure 2. An efficient desorption mechanism returns molecules to the gas phase (see Schmitt 1994 for a review). The basis of this chemistry is that new molecules are formed on the grain surface which are thereafter released in the interstellar gas (Tielens & Whittet 1997). The evaporation of icy grain mantles leads to further reactions in the interstellar gas and may drive the complexity of molecule formation even further (Charnley 1997). The chemistries of gases and grains are strongly interlinked, and observations of molecules in the interstellar gas thus also provide constraints on the solid phase. Additional processes such as cosmic ray bombardment and temperature variations determine the rates of grain mantle growth and evolution. The effect of UV irradiation on icy particles is currently debated, due to the low flux of photons in such cold, dense environments (Gerakines et al. 1999, van der Tak et al. 2000, Roser et al. 2001). When molecular clouds evolve from an initial cold quiescent phase to warm, dense and active proto-stellar regions, shocks which can raise the temperature locally to more than 2,000 K lead to grain processing and grain destruction.

Astronomical observations indicate the existence of different types of ices in proto-stellar environments and toward field stars. Figure 3 shows a scheme of the line of sight conditions in dense molecular clouds. Water ice is the most abundant ice component in the interstellar medium as well as in the solar system. Abundances of other species are therefore scaled relative to water ice. Water ice sublimates at around $\sim$ 100-120 K under interstellar conditions (Fraser et al. 2001) and is therefore considered as the most refractory ice component. Highly volatile species such as CO, O_2 and N_2 sublimate around or below 20 K. This implies that successive layers of ice are formed with different ice compositions according to the prevailing temperature and gas pressures (or absolute density) in protostellar regions. Hydrogen-rich ices (polar ices), dominated by H_2O ice, are formed in dense clouds when H is abundant in the interstellar gas. They also contain CO, CH_4, NH_3, CO_2 and CH_3OH, and possibly traces of HCOOH and H_2CO. Some of those species (e.g. CH_3OH, CO_2) are not intimately mixed with water ice, but rather exist in separate phases (e.g. Skinner et al. 1992, Ehrenfreund et al. 1998, Gerakines et al. 1999). Polar ices in general evaporate $\sim$ 120 K under astrophysical conditions (Fraser et al. 2001) and can therefore survive in higher temperature regions close to the star. Apolar or hydrogen-poor ices are formed far away from the proto-star and are composed of molecules with high volatility (evaporation temperatures of < 20 K)

ICY DUST PARTICLES AS CATALYSTS

Accretion of atoms on interstellar grains slow: ~ 1 atom/day

At low temperatures the sticking coefficients are close to unity!

Accreted atoms such as H, C, N, O are very mobile:

- migration time scales range from 10^{-12} to 10^{-3} s
- quantum tunnelling or thermal hopping

Simple molecules H_2O, NH_3, CH_4 can be formed by exothermic H addition reactions without activation energy barrier !

Many neutral molecules, such as CO, possess activation barriers!

$$CO \xrightarrow{H} HCO \xrightarrow{H} H_2CO \xrightarrow{H} CH_3O \xrightarrow{H} CH_3OH$$

Addtional factors for reactions are thermal processing and cosmic rays!

Figure 2   Basic parameters for grain surface reactions.

such as CO, O_2 and N_2 (Ehrenfreund et al. 1997a). However recent work by Collings and co-workers in Nottingham has shown that the evaporation kinetics of even a binary layered ice (e.g. CO + H_2O) is much more complex than the current grain models suggest. The evaporation kinetics between polar and apolar layers may crucially depend on the pre-processing that has occurred to the polar layer prior to the CO adsorption (Collings et al. 2002). As a consequence of the accretion process and the lifecycle of icy grains through turbulent interstellar clouds, a multilayer ice structure dominates. However, spectroscopy can not distinguish distinct grain populations from such 'onion-ring' layers on grains. Recent results from ISO (Infrared Space Observatory) show a strong indication for ice segregation in the vicinity of proto-stars (see section 6 and 7).

2.   WATER ICE

Water ice is detected toward a large number of interstellar targets and a few circumstellar dust shells (Smith et al. 1989, Omont et al. 1990,

Line of sight conditions in dense molecular clouds

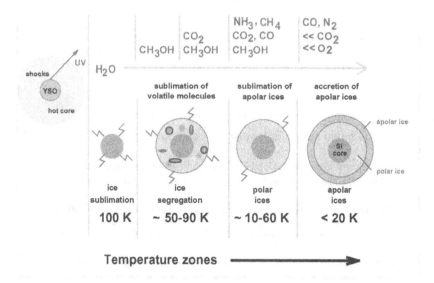

Figure 3 The line of sight conditions in molecular clouds. Different ice types are present in proto-stellar environments. Apolar ices, containing volatile species, reside in cold environments far from the proto-stellar source. Thermal processing in temperature zones above 50 K shows only polar and segregated ice layers. By ∼ 100-120 K all ices are sublimating under interstellar conditions (Ehrenfreund et al. 1998).

Whittet 1993, Tanaka et al. 1994, Dartois et al. 1998a). The most prominent water ice band is the 3.07 μm O-H stretching mode of water ice which is typically one of the strongest bands in interstellar infrared (IR) spectra (e.g. Whittet 1993). With the Infrared Space Observatory ISO the 6 μm O-H water bending mode could be observed in more detail (Keane et al. 2001a) compared with previous airborne missions. The libration mode of water ice at 12 μm is blended with the strong silicate absorption band at 10 μm. Therefore much less information can be obtained from this weaker water ice feature. The abundance and morphology of H_2O ice in the ISM is generally inferred from the intensity and line profile of the 3.07 μm band. In astronomical environments water ice is formed at low temperatures and believed to be in amorphous form. Laboratory experiments have shown that water accreted under the conditions prevalent in dark clouds - temperatures of roughly 10

K, low adsorption rates and random trajectories of adsorbate molecules - will form with a highly porous, amorphous ice structure (Stevenson et al. 1999, Kimmel et al. 2001). In the laboratory, water vapor deposition at temperature < 65 K results in a high-density amorphous ice structure (I_{had}, density 1.1g cm^{-3}), which transforms irreversibly to low-density amorphous ice (I_{lad}, density 0.94 g cm^{-3}) at ~ 35-65 K (Jenniskens et al. 1995). Although this phase change is not evident from the infrared spectrum, it has been detected by several other techniques (Narten, Venkatesh & Rice 1976, Jenniskens & Blake 1994, Lu et al. 2001). Before water is crystallized into cubic ice (I_c) at temperatures ~ 120-140 K, it goes through the glass transition where it transforms from an amorphous solid to a viscous liquid (McMillan & Los 1965). Shortly thereafter the low-density amorphous ice I_{lad} transforms into cubic crystalline ice (Smith et al. 1997). As with any phase change, higher impurity concentrations raise the temperature range to which a single solid phase may persist. Monitoring the evolution of ice upon heating using x-ray diffraction studies has revealed that the crystallization occurs locally by nucleation of small domains, leaving the bulk of the ice still in amorphous state which is defined as 'restrained amorphous' ice (I_{ar}) (Jenniskens et al. 1996). Though this form of water ice is characterized as a 'liquid' it is thermodynamically distinct from liquid water. A short-range hexagonal stacking order of I_{ar} could be responsible for the ice resisting the crystallization process (Jenniskens & Blake 1994). Upon further heating, water ice restructures completely into the thermodynamically stable hexagonal polymorph, which is dominant in terrestrial environments. The hexagonal form of water ice, I_h, is the thermodynamically most stable form of ice and will not exist under interstellar conditions.

The restructuring of water ice has significant relevance for the trapping and release of volatiles within the matrix (Collings et al. 2002). At low pressures and temperatures, such as those in the ISM, the H_2O ice can exist in both amorphous and crystalline forms. It is generally assumed that the dominant morphology of icy mantles on interstellar grains resembles that of high density amorphous ice, with other 'impurity' molecules trapped in the H_2O ice matrix (Jenniskens et al. 1995). This is evident in comparisons of ISO-SWS (Short Wavelength Spectrometer) and ground-based observations with laboratory spectra of mixed ice analogues (Ehrenfreund & Schutte 2000 and references therein). However, laboratory spectra of pure H_2O ices show clear differences in the line profile of the 3.07 μm feature associated with different ice morphologies and temperatures (Maldoni et al. 1998). It is therefore likely that some H_2O ices have undergone 'mild processing' and re-cooling, (such as gentle heating below 80-100 K, or rapid localised heating due to shocks), and will therefore also exist in low-density amorphous and crystalline forms (Jenniskens & Blake 1994, Jenniskens et al. 1996). The physical properties of the H_2O ice, such as density, conductivity, vapour-pressure, and sublimation rate, are dictated by its structure. Significant differences are expected between the surface chemistry and bulk behaviour of the ice phases. In addition, the physical and chemical properties of the ice may also be affected by the way the ice film is deposited, its lifetime, and processing prior to thermal desorption (Sack & Baragiola 1993).

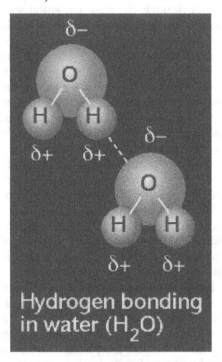

Figure 4 Water ice is the most abundant ice species in the universe. H_2O ice is a very dynamically system, with three amorphous and two crystalline forms, whose properties strongly influence processes on icy surfaces in the ISM and the solar system.

2.1.   MOLECULAR HYDROGEN

Molecular hydrogen is formed catalytically on the surfaces of dust grains (e.g. Black & Dalgarno 1977). Hydrogen atoms chemisorb onto dust grains and migrate across its surface. As two H atoms meet, a chemical bond is formed and the released energy is sufficient to desorb the H_2 molecule from the grain's surface. Although the magnitude of the formation rate and its general temperature dependence are believed to be $\sim 3 \times 10^{-18} T^{1/2}$ cm^3 s^{-1}, many uncertainties remain. The influence of the substrate, i.e. silicate or carbonaceous grains, is not accurately known nor is the influence of the grain size on the resulting formation efficiency. The sticking efficiency of H atoms (usually assumed to be unity) depends on temperature (Schmitt 1994) but also on the availability of adsorption sites on the grain. Important experiments are now underway to determine the exact chemistry of these reactions (Manico et al. 2001). Further details can be seen in this volume, in the chapter by Biham, Pirronello and Vidali. Other experiments ongoing in this area include a very exciting experiment at UCL in the UK, which predicts that the excess energy of the reaction will be channeled into the excited vibrational and rotational states of the H_2 molecule as it leaves the surface. The UCL experiment will attempt to observe these excited states. A number of examples are known in surface dynamics, where molecules desorb in 'hot' vibrational or rotational modes. This has a number of implications for the interstellar medium, since somehow these molecules have to lose their 'excess' energy, either by radiation, collisional de-excitation or by reacting with species they would not normally react with. This is an exciting new avenue of research in gas-grain chemistry.

3.   THERMAL AND ENERGETIC PROCESSING IN DENSE CLOUDS

The chemistry in interstellar clouds may be enriched by energetic processing such as cosmic rays and probably some ultraviolet (UV) irradiation. Cosmic rays can break bonds, ionize species and generate high-energy secondary electrons. The penetration of cosmic ray particles is limited by the energy of the particle and the stopping power of the ice. Penetration of cosmic rays particles can proceed deep into the ice or pass right through the grains. UV photons are limited by the optical depth (~ 0.2 μm). In the interstellar regions, where ices are efficiently accreted, the UV radiation field of a young protostar is strongly attenuated. However, cosmic rays can penetrate throughout the cloud and ionize H_2 molecules; the energetic electrons from these molecules can, in turn, excite H_2 to higher electronic states. These excited H_2 molecules subsequently decay by emitting UV photons. The cosmic ray induced UV radiation field is $\sim 1.4 \times 10^3$ photons cm^{-2} s^{-1} with energies above 6.2 eV (Prasad & Tarafdar 1983). The role of UV energetic processing of interstellar ices, however, is difficult to evaluate observationally in astronomical environments; in general, determining the UV flux incident on icy grain mantles is a nontrivial problem.

The cosmic ray ionization rate ζ_{CR} has been recently revisited by analyzing the H_3^+ absorption and $H^{13}CO^+$ submillimeter emission lines toward massive young stars (van der Tak & van Dishoeck 2000). No relation of ζ_{CR} with luminosity of the objects was found, indicating that local X-ray ionization and shielding against cosmic rays are unimportant for these sources. For low mass-star forming regions ζ_{CR} has still large uncertainties.

Thermal processing in dense clouds leads to processes such as crystallization, polymerization, clathrate formation and probably acid-base chemistry in grain mantles. It triggers selective desorption of mantle species as a function of grain temperature and hence of distance from the protostar. This temperature-selective outgassing gives rise to spatial variations in the mantle composition, as well as the gas, along the line-of-sight to the protostar. It leads to the removal of the outer, most volatile layers of the ices (see Figure 3). Second, heating of the mantles can cause structural and chemical changes (see for example Collings et al. 2002). In ices, both small, light molecules such as CO and reactive heavy species (such as CH_3OH or NH_3) will be efficiently trapped at 10 K, and the timescales for diffusion and subsequent reaction will be very long. Species which can integrate into the hydrogen bonded network of the H_2O-ice will be trapped with particularly high efficiency. Warming of the ices promotes the migration of reactive species since the energy barrier for diffusion is typically $\sim (0.3\text{-}0.5)$ sublimation energy (d'Hendecourt et al. 1985) and so, as long as the species in question can remain bound to the grain, it can diffuse and react.

Laboratory simulations of interstellar ice analogs show increasing complexity on the molecular scale when energetic and thermal processing is applied (Allamandola et al. 1997, Moore & Hudson 1998, Cottin et al. 2001). Insufficient knowledge about the precise conditions in dense interstellar clouds, and in protostellar environments, raises the issue of whether or not the evolution of icy grain mantles in these regions can indeed be simulated accurately in the laboratory. Most of the experiments in the search for complex organics have been conducted in such a way that ice mixtures have been irradiated and subsequently warmed up to room temperature (without continuous irradiation). Using starting compounds such as H_2O, CO, NH_3, CH_3OH and others, the formation of more complex organic species is definitely not surprising. However, such experiments are not representative of interstellar cloud conditions; no high temperature regions exist in the ISM which are devoid from radiation. In the contrary, grains which travel into regions of higher temperature are exposed to increasing irradiation doses!

Recent experiments show that both CO_2 and CH_3OH can be formed by O and H atom addition to CO at around 10 K *without* any form of additional energetic processing (Roser et al. 2001, Hiraoka et al. 1994). Recent theoretical work also supports this view (Charnley 2001, Charnley et al. 2001). Previous experiments required UV photolysis or heavy particle bombardment to form CO_2 ice (e.g. Allamandola et al. 1997) and it now appears that the importance of UV photolysis in the solid state chemistry of dark molecular clouds may have been greatly overestimated (see also Gerakines et al. 1999).

Electronic and ionic bombardment of grains is a major consideration. Many discussions persist as to whether or not grains carry charge, (either slightly positive or more probably slightly negative). In laboratory experiments electronic bombardment change the structure of the ices, can induce heating or UV fluorescence, and naturally sputter the ice surface on impact, generating impact craters at the surface which are regions of high chemical reactivity. MeV ions as well as keV projectiles are expected to modify ices in space (Kaiser & Roessler 1998). This research topic is further reviewed in detail in this volume by Johnson and Bringa.

4.   LABORATORY SIMULATIONS

Various laboratory studies are used to characterize the physical and chemical properties of ices in space environments. Electron microscopy and X-ray diffraction studies are crucial methods to study the ice structure, though there is debate in the scientific community whether those techniques introduce changes in the ice structure. In order to determine the composition of ice in space, infrared spectroscopy of ice analogs has proven to be very effective.

4.1.   LOW TEMPERATURE INFRARED SPECTROSCOPY

Most of the abundant ice species do have specific molecular vibration between 2.5 and 25 μm and can therefore be studied in detail. Laboratory simulations of low temperature ices are achieved by condensing ices as pure gas or gas mixtures in a high-vacuum chamber on the surface of a caesiumiodide (CsI) window, cooled by a closed cycle He refrigerator to 10 K. Infrared transmission spectra are usually obtained with a Fourier transform spectrometer at a resolution of 1 cm^{-1}. Stepwise annealing in temperature intervals of a few kelvin avoids explosive sublimation of the ices and enables us to monitor changes in the infrared spectra due to thermal processing. UV irradiation is generally performed using microwave-excited hydrogen flow lamps. These lamps have a sharp emission peak at 1216 Å (Lyman α) and additional bands between 1300 - 2800 Å, that produce a total UV flux of $\sim 10^{15}$ photons cm^{-2} s^{-1}.

Astronomical data which are compared to laboratory spectra give us crucial information on the ice composition and structure. Interactions with surrounding molecules are of chemical and physical nature and depend on parameters, such as size and shape of sites, dielectric constant, dipole moment, polarizibility etc. Those interactions lead to shifts in the peak position and line broadening, see Figure 5. Databases on ice analogs can be efficiently used to investigate those effects and to subsequently reconstruct the ice composition. The change in ice structure due to thermal or energetic processing can also be traced by infrared spectroscopy (see Figure 6). Those methods also allow us to determine the abundance of many ice species in space. The integrated absorption cross sections can be measured in the laboratory and have actually been refined over the last two decades for the most important species (e.g. Gerakines et al. 1995, Kerkhof et al. 1999). By measuring the band intensity of an absorption species in space and dividing it by the integrated absorption cross section measured in the laboratory, the column density of the given ice species in space can be measured to a very high accuracy.

4.2.   SURFACE REACTIONS

A number of astrochemistry groups around the world are now working on laboratory simulations in order to study surface effects in ices. In addition there is a wide and highly active solid state physics, and surface science community, which is working on astrochemical and related (usually atmospheric chemistry) issues; work mostly found in the chemistry and physics literature. Here, we describe two particular experiments in some detail, the Nottingham Surface Astrophysics Experiment (NOSAE) in Nottingham University, UK and the Surface Reaction Simulation Device (SURFRESIDE) at Leiden University, NL.

4.3.   NOSAE

The Nottingham Surface Astrophysics Experiment (NoSAE) has been designed and constructed to obtain qualitative spectroscopic data, and quantitative measurements of binding energies and sticking probabilities in astrophysically relevant surface-adsorbate systems. A review of the experiment has recently been published (Fraser et al. 2002). Operating

ICE MATRIX CONDITIONS

- chemical and physical interactions
- polarizability
- size and shape of sites
- dielectric constant
- dipole moment

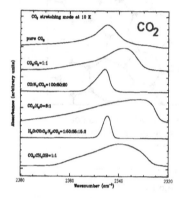

PEAK POSITION --> SHIFTS

- **attractive and repulsive interactions**

LINE WIDTH --> BROADENING

- **aggregates - complexes - sites**

--> COMBINED EFFECTS

Figure 5 An ice matrix is a dynamically system. Interactions with neighbouring molecules can be traced like a 'fingerprint' by infrared spectroscopy. The CO_2 stretching mode is shown here as an example. For this feature the band width ranges from 5-30 cm^{-1} and the band position shifts up to 20 cm^{-1} according to the pressure, temperature, structure and impurities included in the ice matrix.

at UHV pressures and cryogenic temperatures, the experiment mimics the harsh conditions found in ISM regions in a controlled environment. The system reaches temperatures from 10-500 K, similar to temperatures in ISM, protostellar and circumstellar disk environments. The base pressure is better than 2×10^{-10} mbar, only a few orders of magnitude higher than the typical pressure in the dense interstellar medium, and a few orders of magnitude lower than the base pressure in dusty protostellar and circumstellar disks. Furthermore, the experimental vacuum is dominated by H_2 - just like the ISM. The chamber is equipped with a number of traditional surface science instruments, including a Fourier Transform Reflection Absorption Infrared Spectrometer (FT- RAIRS), Quartz Crystal Microbalance (QCM) and a Quadrupole Mass Spectrometer (QMS). A schematic of the Nottingham Surface Astrophysics Experiment (NoSAE) is shown in Figure 7. The apparatus comprises a 30 cm diameter, cylindrical, stainless-steel chamber, pumped by a liquid N_2 trapped, 9" oil diffusion pump, and a Titanium sublimation pump, backed by a mechan-

ICE STRUCTURES

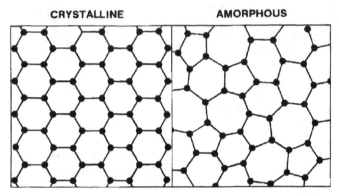

ATOMIC ARRANGEMENT IN A CRYSTALLINE AND AN AMORPHOUS SOLID.
DOTS REPRESENT THE EQUILIBRIUM POSITIONS ABOUT WHICH THE ATOMS
VIBRATE AND LINES INDICATE THE CHEMICAL BONDS.

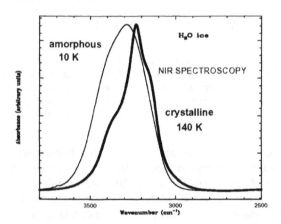

Figure 6   Low temperature ices in the interstellar medium are amorphous in nature but can be crystallized in higher temperature environments, in particular in circumstellar dust shells and solar system environment. The transition from a random network into an ordered structure changes the ice properties and can be monitored by infrared spectroscopy. The O-H stretching mode of crystalline H_2O ice at 3 μm is shifted in wavelength position and appears as a sharper peak compared to the signature of amorphous ice (e.g. Maldoni et al. 1998).

ical rotary pump. Access ports (directed at the center of the chamber) are provided at four levels on the main chamber, as well as at the top and base, for sample mounting, effusive gas dosing of substrates and adsorbates, surface analysis, and system monitoring. A base pressure of better than 2×10^{-10} mbar is routinely achieved after baking at 120 C for 48-72 hours. The chamber is equipped with a differentially-pumped, precision $XYZ\theta$ manipulator supporting a UHV compatible closed-cycle Helium cryostat that is capable of reaching temperatures below 7 K. The substrate is a gold-coated quartz crystal, which is one half of a heterogeneous quartz crystal microbalance (QCM), designed specifically for this experiment and capable of working at cryogenic temperatures. Molecular ices are grown in situ by direct vapor deposition from the gas phase. Gases are mixed on a dedicated gas handling line, and then introduced into the chamber via a pair of fine control leak valves. The mass of the deposited film can be measured directly using the QCM. The chamber is also equipped with two differentially-pumped IR windows for surface analysis by Fourier Transform Reflection-Absorption Infrared Spectroscopy (FT-RAIRS). A dedicated temperature control system and a line-of-sight configured quadrupole mass-spectrometer are fitted for temperature programmed desorption (TPD) measurements. A hot-cathode ion gauge and viewing port are also fitted to the chamber for monitoring purposes. This experiment has already been used to successfully study the IR-spectroscopy and desorption kinetics of the H_2O ice system, the H_2O -CO binary ice mixture and the NH_3-H_2O system. Further work is ongoing with a number of astronomically relevant molecules such as SO_2, CH_4, CH_3OH and CO_2.

4.4. SURFRESIDE

The Leiden experiment was designed to study atom-molecule, molecule-molecule and radical-molecule chemical reactions occurring at the surface or in the bulk of interstellar icy grains. The experiment combines UHV surface science techniques with an effusive atomic source, RAIRS and a Multiple-Ion Detection Quadropole Mass Spectrometer, see Figure 8. Its main aim is to identify the key reactions that generate simple molecular species in the gas phase. In addition to information concerning the kinetics, activation energies, and branching ratios of such reactions, it should also be possible, for the first time, to determine important astrophysical quantities relating to the chemical systems, e.g. sticking probabilities of atomic and radical species on ices, desorption enthalpy between mono and binary ice mixtures and atomic/radical species. An understanding of such reactions and empirical measurements of the physical data will provide important constraints in our understanding and

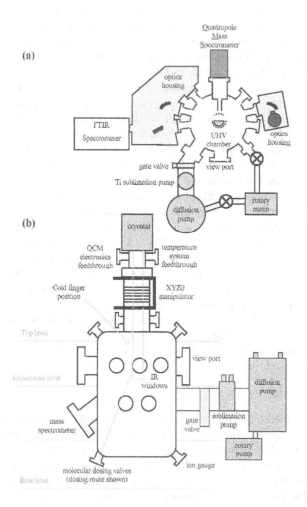

Figure 7 Experimental view of NOSAE, from (a) top and (b) side.

evaluation of the molecules found, and the chemistry occurring in interstellar, protoplanetary and circumstellar media.

Clean ultra high-vacuum conditions ($\sim$ few times 10^{-10} torr) are
achieved with a magnetically levitated turbomolecular pumping sys-

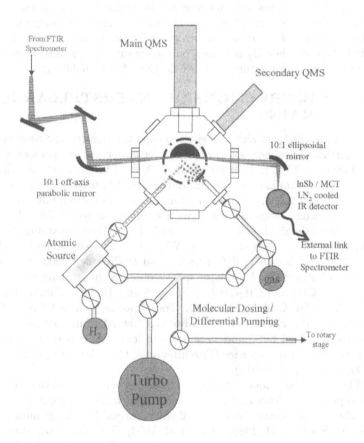

Figure 8 Schematic view of SURFRESIDE.

tem. The ice sample is prepared by slow deposition of molecular gas on an gold-coated OFHC (oxygen free high conductivity) copper substrate cooled to ∼ 10 K. The atom source is a differentially pumped microwave discharge, which is then cooled prior to being introduced to the chamber.

Grazing angle reflection absorption infrared spectroscopy and temperature programmed desorption, are combined on the system. Together with isotopic substitution and sophisticated analysis, these two techniques allow us to identify reaction products, and to calculate branching ratios, reaction rates and diffusion limits within the ice manatle analogues. Over the past 18 months the project has been designed, constructed and tested. It is now being used to determine the desorption rate of CO ice when H_2 is formed on the ice surface, and also to observe chemical reaction products, e.g. HCO, CO_2, H_2O, HCOOH.

5.   OBSERVATIONS OF INTERSTELLAR ICY GRAINS

Ground-based and airborne spectroscopy through telluric windows of the Earth's atmosphere allowed to monitor important ice species such as H_2O ice and CO (e.g. Whittet 1993). ISO, the European Infrared Space Observatory, enabled observations of the complete wavelength range between 2.5 and 200 μm, free of any telluric contamination, but with limited resolution. Below we list a small summary of recent ISO observations. More detailed reviews on ices, including ISO observations, can be found in d'Hendecourt et al. 1996, Whittet et al. 1996, Ehrenfreund et al. 1997b, Gibb et al. 2000, Ehrenfreund & Charnley 2000, Ehrenfreund & Schutte 2000. The major species observed in interstellar ices are H_2O, CO, CO_2 and CH_3OH (see Figure 9 and Table 1; Ehrenfreund & Charnley 2000, Gibb et al. 2000). Trace species such as OCS, H_2CO, HCOOH, CH_4, and OCN^- are observed toward some protostars and are characterized by abundances between less than a percent to a few percent relative to water ice (Ehrenfreund & Charnley 2000, Gibb et al. 2000, Keane et al. 2001a).

The stretching modes of H_2O and CO ice can be observed through atmospheric windows and have been studied from the ground for two decades. Those transitions are well characterized in a large number of sources (Smith et al. 1989, Chiar et al. 1998). The ubiquitous presence of CO_2 was one of the major discoveries of ISO. This molecule was observed through its stretching and bending vibrations at 4.27 and 15.2 μm, respectively (de Graauw et al. 1996, d'Hendecourt et al. 1996, Gürtler et al. 1996, Whittet et al. 1998, Gerakines et al. 1999). Even the NIR combination modes of the CO_2 molecule could be observed (Keane et al. 2001b). Observations of $^{13}CO_2$ led to an estimate of the isotopic $^{13}C/^{12}C$ ratio in the galaxy (Boogert et al. 2000). It has been shown that CO_2 ice is present in annealed (hot) form, which indicates an ice segregation process on grain mantles toward high-mass protostars.

Table 1 Millennium abundances of ices in high-mass (W33A) and low-mass (Elias 29) star-forming regions and gas phase molecules observed in the interstellar medium (Orion hot core) and in cometary comae. Abundances for the Orion hot core are normalized to CO. Abundances for interstellar ices and volatiles in comets are scaled to H_2O. For further details see Ehrenfreund & Charnley (2000) and references therein.

Molecule	W33A high	Elias29 low	Orion hot core	Comet Hyakutake	Comet Hale-Bopp
H_2O	100	100	> 100	100	100
CO	9	5.6	1000	6-30	20
CO_2	14	22	2-10	2-4	6-20
CH_4	2	< 1.6	-	0.7	0.6
CH_3OH	22	< 4	2	2	2
H_2CO	1.7-7	-	0.1-1	0.2-1	1
OCS	0.3	< 0.08	0.5	0.1	0.5
NH_3	15	< 9.2	8	0.5	0.7-1.8
C_2H_6	-	-	-	0.4	0.3
HCOOH	0.4-2	-	0.008	-	0.06
OCN^-	3	< 0.24	-	-	-
HCN	< 3	-	4	0.1	0.25
HNC	-	-	0.02	0.01	0.04
HNCO	-	-	0.06	0.07	0.06-0.1
C_2H_2	-	-	3-10	0.5	0.1
CH_3CN	-	-	0.2	0.01	0.02
$HCOOCH_3$	-	-	0.1	-	0.06
HC_3N	-	-	0.04	-	0.02
NH_2CHO	-	-	0.002	-	0.01
H_2S	-	-	1	0.8	1.5
H_2CS	-	-	0.01	-	0.02
SO	-	-	0.5	-	0.2-0.8
SO_2	-	-	0.6	-	0.1

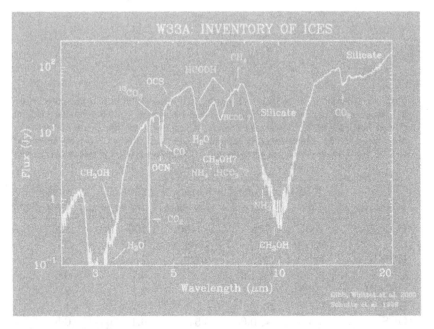

Figure 9 The ISO-SWS spectrum of the high-mass protostar W33A (Gibb et al. 2000). The two dominant spectral features are H_2O ice at 3 μm and silicates at 10 μm. A major ISO success was the detection of CO_2 and its isotope $^{13}CO_2$ (de Graauw et al. 1996, Boogert et al. 2000). Small peaks and shoulders give indications for the presence of acids and ions on interstellar icy grain mantles in the line of sight (Schutte et al. 2002).

Detailed spectroscopic studies also revealed intermolecular complexes of CO_2 with CH_3OH ice (Ehrenfreund et al. 1998, Dartois et al. 1999a), see Figure 10. These observations of extensive ice segregation in the vicinity of protostars and evidence for acid-base reactions (Schutte et al. 2002) occurring on interstellar grain surfaces indicate that thermal processing is a dominant factor in interstellar ice chemistry (Ehrenfreund et al. 2001).

The formation of CO_2 in dense clouds in such large quantities ($\sim$20% relative to water ice; de Graauw et al. 1996, Gerakines et al. 1999, new results even report up to 37% CO_2, Nummelin et al. 2001) remains a mystery. Recent grain surface reaction experiments involving the concurrent deposition of carbon monoxide and oxygen atoms on a copper substrate at 5 K showed that it is possible to form CO_2 in ice mantles on grains without the intervention of energizing (UV or energetic particles) agents (Roser et al. 2001). Abundances of gaseous CO_2 are surprisingly

Carbondioxide-Ice - ubiquitous in interstellar space

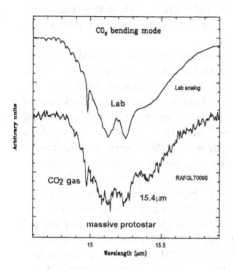

Abundance: ~ 20 % rel. to H2O

ISO spectra show:

CO_2-CH_3OH -H2O ice-mixtures

with a temperature of ~ 70-80 K

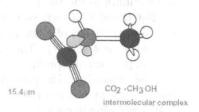

15.4µm

CO_2 -CH_3OH
intermolecular complex

Ehrenfreund et al. (1998, 1999), Dartois et al. (1999)

Figure 10 The 15.2 µm bending mode of CO_2 ice has been observed ubiquitously in interstellar space. The origin of its pronounced triple peak, as seen here through the line of sight toward the protostar RAFGL 7009S, was long unknown. Extended laboratory work revealed that this spectroscopic signature can be best reproduced with ice mixtures containing CO_2 and CH_3OH in equal proportions (and some water) at elevated temperatures. Whereas the two peaks with roughly equal intensity witness the segregation of CO_2 into boundary layers (splitting of the 15.2 µm CO_2 bending mode is only observed in pure CO_2 ice), CO_2/CH_3OH complexes are responsible for the shoulder at 15.4 µm (Dartois et al. 1999a). The sharp feature at 14.98 µm is due to gas phase CO_2.

low (van Dishoeck et al. 1996, Dartois et al. 1998b), probably a result of destruction of CO_2 in shocks (Charnley & Kaufman 2000).

The variable abundance of CH_3OH remains also another open question in interstellar ice chemistry. Recent ground-based observations of the 3.54 µm band indicate large amounts of CH_3OH in the line of sight toward some high massive protostars, but rather low CH_3OH ice abundances are observed toward low-mass protostars (Dartois et al. 1999b). Most of the spectroscopic analysis of CH_3OH indicate that this molecule is not mixed in with the bulk water ice. CH_3OH may be formed either by grain surface reactions (hydrogenation of CO, Hiraoka et al. 1994) or by energetic processing via cosmic rays (UV irradiation has very low efficiency to form CH_3OH). No local enhancement of X-ray ionization has been observed toward high-mass protostars (see section 5, van der Tak & van Dishoeck 2000), which in turn show a highly variable CH_3OH abundance in the line of sight. Therefore grain surface reactions seem to date the most plausible pathway for the formation of CH_3OH ice on interstellar grains.

Observations of the strong 6.0 μm absorption band assigned to the deformation mode of H_2O (amorphous and at a temperature < 50 K) show excess absorption on the blue as well as the red wing of this feature for several astronomical targets. This indicates the presence of additional molecules with a carbonyl and/or a carboxyl group. The most promising molecules which have been identified with the help of laboratory data to contribute to the short wavelength wing of the water bending mode, are H_2CO and $HCOOH$. Such species may be present in interstellar ices with an abundance of a few percent relative to water ice (see Keane et al. 2001a for a review). The 6.85 μm band displays systematic variations in position and profile along different lines of sight which indicates thermal processing. The 6.85 μm feature might be composed of several components, among those are CH_3OH and NH_4^+ (Keane et al. 2001a).

ISO was able to confirm the presence of CH_4 at 7.68 μm toward several objects, and showed an abundance relative to water ice of $\sim$2–4% (Boogert et al 1996, Dartois et al. 1998b). OCS is currently the only sulphur-containing species identified in interstellar ices, and shows an abundance of at maximum 0.2% relative to water ice (Palumbo et al 1995, 1997, d'Hendecourt et al 1996). OCS is only detected in a handful sources and its formation might be a localized phenomenon (Palumbo et al. 1997). Observations and theoretical models show that sulphur is heavily depleted in the interstellar gas. Therefore the identification of other sulphur-bearing species than OCS in the solid state are important future perspectives. Among the most important molecules, whose presence and abundance are difficult to determine, are the infrared inactive molecules H_2, O_2 and N_2. It has been shown that the infrared transitions of those molecules become weakly infrared active in the solid state due to the break in symmetry (Ehrenfreund et al. 1992, Sandford et al. 1993). None of those molecules has up to now been detected in interstellar ices. The search for O_2 ice at 6.45 μm led to upper limits which are compatible with SWAS data (Vandenbussche et al. 1999, Melnick et al. 2000). The fact that O_2 is strongly underabundant in the gas phase argues against its depletion on grains. Upper limits for N_2 (transition at 4.28 μm, very close to the strong CO_2 band) have been recently reported by Sandford et al. (2001). Finally, a broad band at 4.62 μm adjacent to the CO band, called the 'XCN' band, has been observed toward several targets (Tegler et al. 1995, Pendleton et al. 1999). The identification with OCN⁻ remains a debated subject in the literature (Schutte & Greenberg 1997, Demyk et al. 1998, Pendleton et al. 1999, Bernstein et al. 2000, Palumbo et al. 2000, Whittet et al. 2001).

The largest unambiguously identified molecule in interstellar ices remains CH_3OH. Weak features due to more complex molecules might be masked by other strong transitions. We will have to await higher sensitivity measurements of the infrared spectrum toward star-forming regions to identify more complex molecules which might be present in negligible abundances.

With the help of recent ISO observations and ground-based data it is clear that the observed spectra sample the spectral signature of many different grain populations in the line of sight through a molecular cloud. Additionally, those grain populations are mixed within regions of different temperature and density conditions. In brief, what we are observing is a dynamical system which is changing from one region to another. The spectral analysis of multi-component mixtures which are additionally processed in the laboratory by energetic sources have been used in the previous decade to explain astronomical data. Such an approach has proved very insightful as a first step, but has obviously no relevance for the interpretation of the complex spectral information we obtain today from satellite data. By far a more efficient way to enhance our understanding in the ice chemistry in star-forming regions is to study simultaneously the composition of the gas and solid phase, the geometry of the source and to obtain information on the temperature and density conditions in the line of sight. This combined information has to be used as a basis for future laboratory experiments.

6.   OBSERVATIONS OF SOLAR SYSTEM ICES

6.1.   PLANETS AND MOONS

Most of the outer solar system bodies are covered with ice (Schmitt, de Bergh & Festou 1998), predominantly water in crystalline phase, though at such low temperatures one would expect the presence of amorphous ice. Ices more volatile than water are found on Uranus and beyond, see Table 2 (Roush et al. 1995). Photolysis and radiolysis of icy surfaces should produce a variety of chemical products (Roush 2001 and references therein). Also the poles of the terrestrial planet's Earth and Mars are covered with ice. The neutron spectrometer on-board of Lunar Prospector measured enhanced hydrogen in some deep craters near the lunar polar regions, which are permanently shaded. This might be an indication for the presence of water ice on the Moon. Also polarized radar observations indicate possible deposits of ice in the polar craters of Mercury and the Moon. Venus is the only terrestrial planet where ice is not expected. Mars has permanent ice caps at both poles composed mostly of solid CO_2. The ice caps exhibit a layered structure with alternating layers of ice with varying concentrations of dust. On the north pole water ice exists below the CO_2 ice layers. Mars Global Surveyor brought back 3D images of the north pole which indicates an ice cap 1200 km wide and 3 km thick. The volume of the northern ice cap is $\sim$ 4 % of the Earth antarctic sheet. Mars, which will be visited by numerous spacecraft in this decade may hide permafrost or liquid water in its subsurface.

Table 2   Ices observed in the outer solar system (taken from Roush et al. 1995, Roush 2001).

Planet	Moons	Ice species
Jupiter	Io	SO_2, SO_3, H_2S?, H_2O?
Jupiter	Europa	HO, SO_2, SH, CO_2, CH, XCN, hydrous sulfate & carbonate salts, H_2O_2, H_2SO_4
Jupiter	Ganymede	H_2O, SO_2, SH, CO_2, CH, XCN, Hydrated & hydroxylated mineral, O_2, O_3
Jupiter	Callisto	H_2O, SO_2, SH, CO_2, CH, XCN, Hydrated & hydroxylated mineral
Saturn	Mimas	H_2O
Saturn	Enceladus	H_2O
Saturn	Tethys	H_2O
Saturn	Dione	H_2O, C, HC, O_3
Saturn	Rhea	H_2O, HC?, O_3
Saturn	Hyperion	H_2O
Saturn	Iapetus	H_2O, C, HC, H_2S?
Saturn	Phoebe	H_2O
Saturn	Rings	H_2O
Uranus	Miranda	H_2O, NH_3 hydrate, hydroxylated silicates
Uranus	Ariel	H_2O, OH ?
Uranus	Umbriel	H_2O
Uranus	Titania	H_2O, C, HC, OH?
Uranus	Oberon	H_2O, C, HC, OH?
Neptun	Triton	N_2, CH_4, CO, CO_2, H_2O
Pluto		N_2, CH_4, CO, H_2O
Pluto	Charon	H_2O, NH_3, NH_3 hydrate
KBO's		H_2O, HC-ices (CH_4, CH_3OH), HC, silicates

Among the better studied objects in the outer solar system are the Galilean satellites of Jupiter which have been visited by several spacecraft. Europa, which is covered by a thick water ice layer, may harbor an ocean below its crust and also Callisto may have a 200 km thick ice band just beneath the surface. The Saturn system with its 18 moons will be visited shortly by the CASSINI-HUYGENS mission. Saturn's complicate ring system contains innumerable small particles with independent orbit which are composed of water ice and rocky particles with ice coating. The European Huygens probe will enter the atmosphere of Saturn's moon Titan and will measure its properties during descent. High resolution images through the nearly opaque haze on Titan showed bright regions, which could be ice or rock continents (Meier et al. 2000). The Huygens probe may be able to confirm those observations. There are a large number of moons in the outer solar system which have interesting icy surfaces (see Schmitt et al. 1998). An intimate mixture of water ice, amorphous carbon, and a nitrogen-rich organic compound can be used to fit the entire composite dark side spectrum of Iapetus (Owen et al. 2001). Neptune's largest moon Triton has a southern hemisphere covered with an ice cap of frozen N_2 and CH_4. N_2, CH_4 and CO cover the surface of Pluto, the coolest and farthest planet in our solar system. Here we focus our discussion on two prominent icy moons, Europa and Triton.

Jupiter's moon Europa is the second of the large Galilean satellites. Previous ground-based observations showed that the surface is composed dominantly of water ice. Europa was visited recently by the Galileo spacecraft. The high resolution images from the spacecraft revealed structures with morphologies reminiscent of those seen on terrestrial sea ice (Greeley et al. 1998a). Besides H_2O ice, hydrated materials are also present, as well as traces of SO_2, CO_2 and H_2O_2. Europa is subjected to intense bombardment by Jovian magnetosphere particles and the presence of H_2O_2 proves that the surface chemistry is dominated by radiolysis (Carlson et al. 1999a). NIMS (the spectrometer on-board Galileo) observations indicate the presence of hydrated H_2SO_4, a product of a radiolytic sulphur cycle (Carlson et al. 1999b). Europa probably hosts a subsurface water ocean beneath its outer ice crust. What geological processes create the ice rafts and other ice-tectonic processes that are at the origin of prominent surface features on Europa? Europa's surface might be renewed by solid state convection (Pappalardo et al. 1998) or extrusion of solid or liquid material from below (Greeley et al. 31998b). A future mission currently in the planning stage is to visit Jupiter's moon Europa in order to study the properties of the ice crust with radar measurements. Europa seems to have an internal energy source provided by tidal friction through its interaction with Jupiter and could keep water in the liquid state below the crust (see Figure 11). Europa provides therefore key ingredients for life (water, energy and possibly organic molecules) and will certainly be a future target for several space missions.

The infrared spectrum of Neptun's moon Triton has been studied in great detail. Triton exhibits numerous bands in the 1.4-2.5 μm region (Cruikshank et al. 1993). A surface temperature of 34-41 K implies that the observed molecules exist in an icy phase. Water ice is clearly detected, but it can't be unambiguously be determined whether Triton's water ice is crystalline or amorphous (Cruikshank et al. 2000). The presence of CH_4, N_2, CO and CO_2 ice molecules has been revisited by UKIRT observations and recent laboratory experiments, see Figure 12 (Quirico et al. 1999). In this way, a two region surface of Triton was defined composed of a N_2:CH_4:CO terrain, where N_2 is the dominant molecule and of a H_2O+CO_2 terrain, composed of a mixture of pure crystalline H_2O and CO_2 grains. Also a three region surface composed of a N_2:CH_4:CO terrain and two geographically separated H_2O and CO_2 terrains have been invoked (Quirico et al. 1999). Some new bands, observed in the spectrum of Triton, are not yet identified, which may indicate the presence of new compounds. Cryo-volcanic activity has been observed on Triton. The eruptive material is probably liquid nitrogen, dust or methane from beneath the surface. These eruptions are driven by seasonal heating from the Sun.

6.2.   COMETS

Comets are the most primitive bodies in the solar system and their composition encodes information on their origin. They are composed of frozen gases, ices and rocky debris. The solar radiation heats the icy material and forms a gaseous cloud, the coma. Among the sublimation products one has to distinguish 'parent' and 'daughter' molecules, the latter being produced by photolysis in the coma. Since comets are formed in the outer solar system, a large part of their nucleus might be composed of relative pristine interstellar matter. A way to test this hypothesis is to compare the distribution and abundances of species observed on interstellar icy grains to cometary observations. Thanks to the chance appearances of two exceptionally bright comets in the last few years (Hyakutake in 1996 and Hale-Bopp in 1997), the chemical composition of cometary comae has been more effectively determined (Crovisier & Bockelée-Morvan 1999). Those two long-period comets are similar in composition to star-forming clouds (Ehrenfreund et al. 1997c,

THE FROSTY PLAINS OF EUROPA

The large Jovian moon Europa may be hiding
liquid water beneath the frozen crust

The dark linear ridges extending across the scene
are probably frozen remnants of cryo-volcanic
activity

Tidal friction from Jupiter and neighboring moons
could keep large parts of Europa's ocean liquid

Figure 11 The moon Europa was discovered in 1610 and is the second of the Galilean moons. Europa has a diameter $\sim$ 3120 km and it has attracted large interest due to the possible presence of an ocean below its ice crust. Credit: Galileo Project, JPL, NASA.

Bockelée-Morvan et al. 2000, Irvine et al. 2000, Ehrenfreund & Charnley 2000), see Table 1. However, the detection of crystalline silicate in

TRITON: Neptune's largest moon

Southern hemisphere is covered with an "ice cap" of frozen nitrogen and methane

ICE VOLCANOES: liquid N, dust, CH_4

New laboratory data revealed a:

- N_2: CH_4: CO terrain (100:11:0.05)
- H_2O and CO_2 are also present

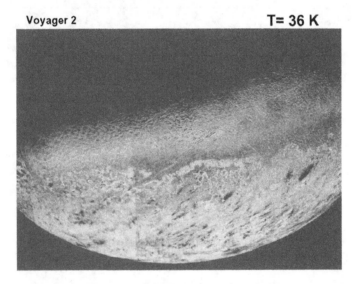

Voyager 2   T= 36 K

Figure 12 Triton, Neptune's largest moon has a ice cap on the southern hemisphere which is covered with N_2 and CH_4, credit: Voyager 2, NASA. Recent laboratory studies have revisited Triton's surface composition in detail (Quirico et al. 1999). Triton is also one of the few bodies which are volcanically active at present time. Triton's eruptions are of volatile compounds and are seasonal.

comets (e.g. Crovisier et al. 1996) indicates an effective mixing with material from the solar nebula.

It is now becoming more apparent that specific chemical differentiation exists amongst the comet population. This was first demonstrated for C_2 and C_3 by A'Hearn et al. (1995) who showed that short-period comets are much more likely to be depleted in the small carbon chains. More recently, ethane has also been found to be depleted in the short period comet Giacobini-Zinner (Mumma et al. 2000). Comet S4 LINEAR shows depletions in CH_3OH and CO (Bockelée-Morvan et al. 2001, Mumma et al. 2001a), whereas Comet Lee has a normal CH_3OH abundance but CO is strongly depleted (Biver et al. 2000, Mumma et al. 2001b). Understanding this differentiation can put constraints on the place of cometary origin and the chemical history of the organic material which formed them (Mumma et al. 2001b).

Water ice, which makes up $\sim$ 30 % of a comet (Greenberg 1998), was detected by the ISO satellite in the coma of Hale-Bopp at 2.9 AU from the Sun through its emission features at 44 and 65 μm (Lellouch et al. 1998). Far from the nucleus water ice may sublimate from icy dust particles rather than from the nucleus itself. Comets are bodies of low bulk density that contains a lot of voids (Rickman 1994). As discussed above, water ice is formed at low temperatures in the interstellar medium and believed to be in the high-density amorphous form, an ice structure with a lower H-bonding density. If this is the predominant material comets are made of, the transformation into the low-density amorphous ice upon heating will increase the H-bonding density. This may lead to cracks in the cometary nucleus, which subsequently could result in active vents or in the worst case to fractionation of the nucleus (Jenniskens & Blake 1996). The structure of the nuclear ice component which may have coexisting amorphous/crystalline phases and clathrates will also strongly influence the cometary outgassing properties (Blake et al. 1991, Jenniskens & Blake 1996).

Future spacecraft, such as ROSETTA, which will land on a comet and study directly the nuclear composition, will improve significantly our knowledge on the origin and structure of comets.

6.3.   KUIPER BELT OBJECTS

The outer regions of the solar system are densely populated by a large number of bodies orbiting the Sun beyond Neptune (Jewitt & Luu 2000). Up to 10^5 of those objects are estimated to orbit at a distance of 30-50 AU from the Sun. Currently $\sim$ 400 Kuiper belt objects (KBO's) have been detected and most of them have a diameter between 100-1200 km.

A few KBO's are bright enough to permit observations in the NIR. In the region between 1.4-2.5 μm many absorption bands of abundant ices can be found. Among them are H_2O, CO_2, CO, CH_4, N_2 and NH_3. Some KBO's show weak absorption bands of H_2O ice, which is expected to constitute at least 35 % of the bulk of KBO's. Since the IR spectrum only samples the very upper surface, the non-detection of water ice features does not imply the absence of H_2O ice on those bodies. Impurities of minerals and organic material as well as ice processing may additionally mask the signature of ice.

Perturbation of KBO's due to gravitational influence of the outer planets can inject some of these bodies into giant planet crossing orbits, which are characteristic of the Centaurs. The Centaurs are characterized by strong spectral diversity. Water ice has been reported in several Centaurs (Luu et al. 2000). Examples are 2060 Chiron, which shows cometary activity, neutral colours and water ice. 5145 Pholus, currently the reddest object of our solar system, shows the presence of CH_3OH ice or a similar light hydrocarbon (Cruikshank et al. 1998). KBO objects are found in the outer protoplanetary disks and may be little altered, they are likely the source of short-period comets (Duncan et al. 1988).

7.   FUTURE PERSPECTIVES

7.1.   SPACE MISSIONS WITH INFRARED FACILITIES

Figure 13 shows a summary of space missions which will be able to investigate ices in different space environments, in particular by providing infrared (IR) data (Foing 2001). Data from the ISO satellite, in operation between 1995-1998, made an important contribution to our understanding of the ice chemistry in the interstellar medium and in the solar system, see section 7. The next infrared satellite, the Space Infrared Telescope Facility (SIRTF) is already on the start-ramp. SIRTF is a cryogenically-cooled infrared observatory, with a 85 cm telecope, capable of studying objects ranging from our solar system to the distant reaches of the universe. Its estimated lifetime is $\sim$ 5 years and it will cover the wavelength range 3-180 μm. The airborne observatory SOFIA will be launched in 2004 and will be able to observe the infrared range 5-300 μm. SOFIA is a Boeing 747SP aircraft which is currently rebuilt by NASA and DLR to accomodate a 2.5 m reflecting telescope. SOFIA is the successor of the previous KAO (Kuiper Airborne Observatory) and will be the largest airborne observatory in the world. HERSCHEL is one of the cornerstone missions of ESA's Horizons 2000 programme and will be launched in 2007. The HERSCHEL Space Observatory will be the only space facility ever developed covering the far infrared to submillimetre range of the spectrum (from 80 to 670 μm). The HER-

SCHEL satellite is approximately 7 metres high and 4.3 metres wide, with a launch mass of around 3.25 tons. It will carry the infrared telescope and three scientific instruments and will be located 1.5 million km away from Earth. HERSCHEL has an operational lifetime of three years minimum. It is a multiuser observatory accessible to astronomers from all over the world. HERSCHEL will (among other topics) study the processes by which stars, their surrounding protoplanetary disks and planets themselves are made.

ASTRONOMY SPACE MISSIONS SUPPORTING ASTROBIOLOGY

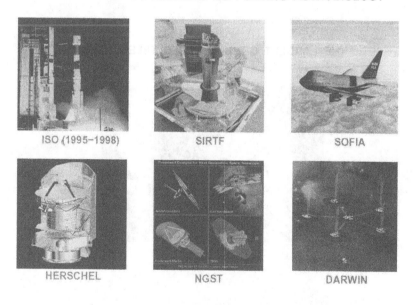

Figure 13 Past and future infrared space missions.

The next generation Space Telescope NGST will be able to penetrate the dusty envelopes around new-born stars and take a closer look at the stars themselves. The largest gain in sensitivity and spatial resolution at mid-infrared wavelengths will be provided by a camera and spectrometer on-board NGST. How planetary systems form and evolve, and whether habitable or life-bearing planets exist around nearby stars, are major objectives to be studied with DARWIN after 2010. With the help of nulling interferometers in the thermal infrared, to remove the parent star light, IRSI-DARWIN will search for the spectral signature of gases such as CH_4 and O_3 in the atmosphere of extra-solar planets in order to identify Earth-like planets capable of sustaining life.

7.2.   LARGE TELESCOPE FACILITIES

The last decade was marked by the construction of large 8-10 m class telescopes, such as the VLT (Very Large Telescope) of the European Southern Observatory (Paranal, Chile) and the twin KECK telescopes (Mauna Kea, Hawaii), see Figure 14. Both facilities are situated at high altitude, best suited for infrared observations.

THE LARGEST GROUND-BASED TELESCOPES

VLT, Paranal, Chile

Keck, Hawaii

Figure 14 The largest ground-based telescopes.

The ESO Very Large Telescope consists of an array of four 8-meter telescopes which can work independently or in combined mode. In this latter mode the VLT provides the total light collecting power of a 16 m single telescope. The telescopes may also be used in interferometric mode providing high resolution imaging. The useful wavelength range extends from the near UV up to 25 μm in the infrared.

The twin KECK Telescopes are the world's largest optical and infrared telescopes. At the heart of each KECK telescope is a revolutionary primary mirror. Ten meters in diameter, the mirror is composed of 36 hexagonal segments that work in concert as a single piece of reflective glass. Together, the twin KECK telescopes have the resolving power of a single telescope 90m in diameter, able to discern sources just milliarcseconds apart. KECK I is operational since 1992. KECK II (completed in 1996) and its twin are located on the dormant volcano Mauna Kea, Hawaii, USA.

Both of these large telescopes have infrared facilities, which allow observations through atmospheric windows. ISAAC was the first Infrared instrument to be installed on the VLT and has been in operation since its 'first light' in 1999. It operates in two specific wavelength ranges, i.e. 1-2.5 μm and 2.5-5 μm. There are two cameras in the instrument, situated at the top and bottom of the optical path and optimized for use over only one of the two wavelength ranges. They are used to either re-image the telescope focal plane or the intermediate spectrum produced by the grating spectrometer. At short wavelengths the instrument is employed for imaging, polarimetry, (both over a 2.5 x 2.5 arcmin field), and low to medium resolution spectroscopy, (R= 500-3000). At longer wavelengths the imaging field is reduced to 1.25 x 1.25 arcmin. One of the key science objectives for ISAAC within our own galaxy is to survey and map embedded objects within dense molecular clouds: just the type of region where ices are abundant! The advent of 8 m class telescopes and large infrared arrays allows sensitive searches for minor species. Thi et al. (2002) reported the first observations with ISAAC of H_2O, CO and CH_3OH toward intermediate-mass stars in the Vela cloud. Together with high resolution infrared space instruments which become operational in this decade, facilities such as VLT/ISAAC will provide important information on the ice chemistry in dense clouds.

7.3. ICES IN MICROGRAVITY

As these previous sections show, despite the wealth of evidence we have amassed, it is still not possible for us to describe the exact formation mechanisms or behaviour of ices in space or in planetary atmospheres.

Consequently, it is difficult to assess the chemical and physical role of ices associated with their environment. It is clear that in the different regions where ices are found, pressure, temperature, and gravitational conditions differ significantly from those on Earth. The ice morphologies in these environments reflect the prevailing formation conditions. Many of the ice's physical properties are directly dependent on this morphology, e.g. density, porosity, vapour pressure, free energy, conductivity, sublimation rates and stability. In addition, the structure, number of dislocations, and defect sites at the ice surface will affect its chemical reactivity. It is also challenging to study secondary effects, such as gas diffusion into bulk ice, outgassing, aggregation, ionisation potential and optical properties of icy particles. Our current understanding of extraterrestrial ices relies entirely on data accrued from laboratory-based studies (on Earth and therefore at 1g). Experiments that evaluate the physical and chemical properties of the ice surface or bulk are generally performed with thin film ice layers, supported by a chemically inert substrate. Experiments on aggregation effects, or optical properties require small particles or clouds. In both instances ice studies under microgravity conditions offer significant advantages: Previous studies of molecular solids in microgravity (e.g. proteins and semiconductors) demonstrated that molecular solids whose crystal structure is determined by weak inter-layer or inter-molecular interactions, exhibit different optical properties, different physical properties, and larger void volumes (porous pockets) than those grown on Earth (Ahari et al. 1997). These differences have been attributed to the profound influence of gravity on the long range forces associated with Van-der Waals and hydrogen-bonding interactions. Since these types of bonding are also dominant in solid-phases of molecular ices, one would expect similar effects to be observed. Secondly, under long-duration microgravity conditions icy particles are isolated and free-floating for a sufficient time that their spectroscopic and optical properties can be measured (Levasseur-Regourd et al. 1998, 2001). During aggregation, size-discrimination effects between particles, and convection heat transfer processes are suppressed. In microgravity surface tension effects are particularly prevalent during melting, and may also play a role in determining the reactivity and ionization potential of icy particles.

By studying the surface and bulk morphology of molecular ices in microgravity over a range of pressure and temperature conditions, we will be able to emulate ice morphologies in other regions of our universe. There are currently two ESA funded projects focused directly on the physical and chemical properties of molecular ices formed under microgravity conditions. A thorough search of the NASA and MICREX microgravity research experiment database reveals only one previous study of water-ice under microgravity conditions, on Skylab in 1973! (Otto & Lacy 1973). The Japanese Society for the Promotion of Science, NASDA, and the Japanese Space Forum are funding research on the effects of gravity on the growth of ice crystals in thin cells from the solution phase (Nagashima & Furukawa 2000). This research provides key insights into nonlinear and nonequilibria phenomena in solution, but is not focused and does not address questions concerning condensation of ices from the gas phase.

THE INTERNATIONAL SPACE STATION

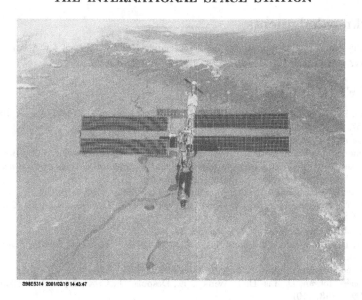

S98E5314 2001/02/16 14:43:47

Figure 15 The growing International Space Station ISS as of February 2001.

The planned Interactions in Cosmic and Atmospheric Particle Systems (ICAPS) facility on board the International Space Station (ISS) (see Figure 15) is one of the ESA funded projects (http://www.icaps.org). Its objectives are to characterise the interaction physics of small solid and liquid particles with an ambient gaseous atmosphere, with electromagnetic radiation, and with other particles in their vicinity. The central research topics highlighted on this facility include physics of protoplanetary disks, aerosol and haze physics, regoliths, and light scattering in particulate media. The experimental programme is expected to comprise four types of experiments: aggregation, regolith, aggregate collision, and cloud experiments. It is envisaged that the basic experiment will be modified slightly when changing from one experimental domain to another. In its current configuration ICAPS is not suitable for ice studies, which require cryogenic and UHV facilities. The basis of the second ESA project, a topical team on the 'physico-chemistry of ices in space', is to investigate the scientific gains and technological requirements for adapting an ICAPS-type experiment to ice morphology studies. This is an exciting and novel area of research, which will no doubt grow over the coming years and shed some interesting light on the true nature of interstellar ices.

Acknowledgments

This work was supported by the Netherlands Research School for Astronomy (NOVA). We thank B.H. Foing and F. van der Tak for discussion.

REFERENCES

Ahari, H., Bedard, R.L., Bowes, C.L., Coombs, N., Dag, O. et al. (1997), *Nature*, **388**, 857.

A'Hearn, M.F. et al. (1995), *Icarus*, **118**, 223.

Allamandola, J.M., Bernstein, M.P., Sandford, S.A. (1997), in Cosmovici, C.B., Bowyer, S. and Werthimer, D. (eds.), *Astronomical and Biochemical Origins and the search for Life in the Universe*, Editrice Compositori, p.23.

Bernstein, M.P., Sandford, S.A., Allamandola, L.J. (2000), *ApJ*, **542**, 894.

Biver, N., Bockelée-Morvan, D., Crovisier, J., Henry, F., Davies, J.K. et al. (2000), *AJ*, **120**, 1554.

Black, J.H., Dalgarno, A. (1977), *ApJSS*, **34**, 405.

Blake, D., Allamandola, L., Sandford, S., Hudginds, D., Freund, F. (1991), *Science*, **254**, 548.

Bockelée-Morvan, D. et al. (2001), *Science*, **292**, 1339.

Bockelée-Morvan, D., Lis, D.C., Wink, J.E., Despois, D., Crovisier, J. et al. (2000), *A&A*, **353**, 1101.

Boogert, A.C.A., Ehrenfreund, P., Gerakines, P.A., Tielens, A.G.G.M., Whittet, D.C.B. et al. (2000), *A&A*, **353**, 349.

Boogert, A.C.A. et al. (1996), *A&A*, **315**, L377.

Carlson, R.W. et al. (1999a), *Science*, **283**, 2062.

Carlson, R.W., Johnson, R.E. and Anderson, M.S. (1999b), *Science*, **286**, 97.

Charnley, S.B. (1997), in Cosmovici, C.B., Bowyer, S. and Werthimer, D. (eds.), *Astronomical and Biochemical Origins and the search for Life in the Universe*, Editrice Compositori, p.89.

Charnley, S.B. (2001), *ApJ*, **562**, L99.

Charnley, S.B., Kaufman, M.J. (2000), *ApJ*, **529**, L111.

Charnley, S.B., Rodgers, S.D., Ehrenfreund, P. (2001), *A&A*, **378**, 1024.

Chiar, J.E., Gerakines, P.A., Whittet, D.C.B., Pendleton, Y.J., Tielens, A.G.G.M. et al. (1998), *ApJ*, **498**, 716.

Collings, M. P., Dever, J.W., Fraser, H. J., McCoustra, M. R. S., Williams, D. A. (2002), *ApJ*, in press.

Cottin, H., Gazeau, M.C., Raulin, F. (1999), *Planet. Space Sci.*, **47**, 1141.

Cottin, H., Szopa, C., Moore, M.H. (2001), *ApJ*, **561**, 139.

Crovisier, J., Bockelée-Morvan, D. (1999), *Space Science Reviews*, **90**, 19.

Crovisier, J. et al. (1996), *A&A*, **315**, L385.

Cruikshank, D. et al. (2000), *Icarus*, **147**, 309.

Cruikshank, D.P., Roush, T.L., Bartholomew, M. et al. (1998), *Icarus*, **135**, 389.

Cruikshank, D.P., Roush, T.L., Owen, T.C., Geballe, T.R., de Bergh et al. (1993), *Science*, **261**, 742.

Dartois, E., Cox, P., Roelfsema, P.R., Jones, A.P., Tielens, A.G.G.M. et al. (1998a), *A&A*, **338**, 21.

Dartois, E., Demyk, K., d'Hendecourt, L., Ehrenfreund, P. (1999a), *A&A*, **351**, 1066.

Dartois, E., d'Hendecourt, L., Boulanger, F., Jourdain de Muizon, M., Breitfellner, M., Puget, J.-L., Habing, H.J. (1998b), *A&A*, **331**, 651.

Dartois, E., Schutte, W.A., Geballe, T.R., Demyk, K., Ehrenfreund, P. et al. (1999b), *A&A*, **342**, L32.

de Graauw, Th. et al. (1996), *A&A*, **315**, L345.

Demyk, K., Dartois, E., d'Hendecourt, L., Jourdain de Muizon, M., Heras, M.A., Breitfellner, M. (1998), *A&A*, **339**, 553.

d'Hendecourt, L.B., Allamandola, L.J., Greenberg, J.M. (1985), *A&A*, **152**, 130.

d'Hendecourt, L.B., Jourdain de Muizon, M., Dartois, E., Breitfellner, M., Ehrenfreund, P. et al. (1996), *A&A*, **315**, L365.

Duncan, M., Quinn, T., and Tremaine, S. (1988), *ApJ*, **328**, L69.

Ehrenfreund, P., Boogert, A., Gerakines, P., Tielens, A.G.G.M., van Dishoeck, E. (1997a), *A&A*, **328**, 649.

Ehrenfreund, P., Breukers, R., D'Hendecourt, L., Greenberg, J. M. (1992), *A&A*, **260**, 431.

Ehrenfreund, P. and Charnley, S.B. (2000), *Ann. Rev. A&A*, **38**, 427.

Ehrenfreund, P., Dartois, E., Demyk, K., d'Hendecourt, L. (1998), *A&A*, **339**, L17.

Ehrenfreund, P., d'Hendecourt, L., Charnley, S.B., Ruiterkamp, R. (2001), *J. Geophys. Research*, **106**, 33291.

Ehrenfreund, P., d'Hendecourt, L., Dartois, E., Jourdain de Muizon, M., Breitfellner, M. et al. (1997c), *Icarus*, **130**, 1.

Ehrenfreund, P. et al. (1997b), in *First ISO Workshop on Analytical Spectroscopy*, ESA Publications, p. 3, October 1997, Madrid.

Ehrenfreund, P., Kerkhof, O., Schutte, W. et al. (1999), *A&A*, **350**, 240.

Ehrenfreund, P., Schutte, W.A. (2000), in Minh, Y.C. and van Dishoeck, E.F. (eds.), *Astrochemistry: From Molecular Clouds to Planetary Systems*, IAU Symp. 197, ASP, San Francisco, p.135.

Foing, B.H. (2001), in Horneck, G. and Baumstark-Khan, C. (eds.), *Astrobiology: The quest for the conditions of life*, Springer Verlag, p.389.

Fraser, H.J., Collings, M.P., McCoustra, M.R.S. (2002), *Rev. Sci. Instr.*, in press.

Fraser, H.J., Collings, M.P., McCoustra, M.R.S., Williams, D.A. (2001), *MNRAS*, **327**, 1165.

Gerakines, P. A., Schutte, W. A., Greenberg, J. M., van Dishoeck, E. F. *A&A*, **296**, 810.

Gerakines, P.A., Whittet, D.C.B., Ehrenfreund, P., Boogert, A.C.A., Tielens, A.G.G.M. et al. (1999), *ApJ*, **522**, 357.

Gibb, E., Whittet, D.C.B., Schutte, W.A., Chiar, J., Ehrenfreund, P. et al. (2000), *ApJ*, **536**, 347.

Greeley, R. et al. (1998a), *Icarus*, **135**, 25.

Greeley, R. et al. (1998b), *Icarus*, **135**, 4.

Greenberg, J.M. (1998), *A&A*, **330**, 375.

Gürtler, J., Henning, Th., Koempe, C., Pfau, W., Kraetschmer, W. et al. (1996), *A&A*, **315**, L189.

Hiraoka, K., Ohashi, N., Kihara, Y., Yamamoto, K., Sato, T. et al. (1994), *Chem. Phys. Lett.*, **229**, 408.

Irvine, W.M., Schloerb, F.P., Crovisier, J., Fegley, B., Mumma, M.J. (2000), in Mannings, V., Boss, A. and Russell, S. (eds.), *Protostars and Planets IV*, Univ. Ariz. Press, Tucson, p. 1159.

Jenniskens, P., Banham, S.F., Blake, D.F., McCoustra, M.R.S. (1996), *J. Chem. Phys.*, **107**(4), 1232.

Jenniskens, P., Blake, D.F. (1994), *Science*, **265**, 753.

Jenniskens, P., Blake, D.F. (1996), *Meteoritics and Planetary Science*, **31**, 177.

Jenniskens, P., Blake, D.F., Wilson, M. A., Pohorille, A. (1995), *ApJ*, **455**, 389.

Jewitt, D. and Luu, J. (2000), in Mannings, V., Boss, A. and Russell, S. (eds.), *Protostars and Planets IV*, Univ. Ariz. Press, Tucson, p. 1201.

Kaiser, R.I., Roessler, K. (1998), *ApJ*, **503**, 959.

Keane, J.V., Boogert, A.C.A., Tielens, A.G.G.M., Ehrenfreund, P., Schutte, W.A. (2001b), *A&A*, **375**, L43.

Keane, J.V., Tielens, A.G.G.M., Boogert, A.C.A., Schutte, W.A., Whittet, D.C.B. (2001a), *A&A*, **376**, 254.

Kerkhof, O., Schutte, W. A., Ehrenfreund, P. (1999), *A&A*, **346**, 990.

Kimmel, G. A., Stevenson, K. P., Dohnalek, Z., Smith, R. S., Kay, B. D. (2001), *J. Chem. Phys.*, **114**, 5284.

Lellouch, E. et al. (1998), *A&A*, **339**, L9.

Levasseur-Regourd, A.C. , Cabane, M., Haudebourg, V., Worms, J.C. (1998), *Earth, Moon, Planets*, **80**, 343.

Levasseur-Regourd, A.C. , Haudebourg, V., Cabane, M., Worms, J.C. (2001), *ESA SP-454*, 797.

Lu, Q.-B., Madey, T. U., Parenteau, L., Weik, F., Sanche, L. (2001), *Chem. Phys. Lett.*, **342**, 1.

Luu, J., Jewitt, D.C., Trujillo, C. (2000), *ApJ*, **531**, L151.

Maldoni, M.M., Smith, R.G., Robinson, G. and Rookyard, V.L. (1998), *MNRAS*, **298**, 251.

Manicò, G., Raguní, G., Pirronello, V., Roser, J.E., Vidali, G. (2001), *ApJ*, **548**, L253.

McMillan, J.A. and Los, S.C. (1965), *Nature*, **206**, 806.

Meier, R., Smith, B.A., Owen, T., Terrile, R. (2000), *Icarus*, **145**, 462.

Melnick, G., Stauffer, J.R., Ashyby, M.L.N. et al. (2000), *ApJ*, **539**, L77.

Moore, M.H., Hudson, R.L. (1998), *Icarus*, **135**, 518.

Mumma, M.J., DiSanti, M.A., Dello Russo, N., Magee-Sauer, K., Rettig, T.W. (2000), *ApJ*, **531**, 155.

Mumma, M.J. et al. (2001a), *ApJ*, **546**, 1183.

Mumma, M.J. et al. (2001b), *Science*, **292**, 1334.

Nagashima, K., Furukawa, Y. (2000), *Physica D*, **147**, 177.

Narten, A.H., Venkatesh, C.G., Rice, S.A. (1976), *J. Chem. Phys.*, **64**, 1106.

Nummelin, A., Whittet, D.C.B., Gibb, E.L., Gerakines, P.A., Chiar, J.E. (2001), *ApJ*, **558**, 185.

Omont, A. et al. (1990), *ApJ*, **355**, L27.

Otto, G.H., Lacy, L.L. (1973), Ice Melting (SD16-TV111), on The Microgravity Research Experiments (MICREX) Database.

Owen, T., et al. (2001), *Icarus*, **149**, 160.

Palumbo, M.E., Geballe, T.R., Tielens, A.G.G.M. et al. (1997), *ApJ*, **479**, 839.

Palumbo, M.E., Pendleton, Y. and Strazzulla, G. (2000), *ApJ*, **542**, 890.

Palumbo, M.E., Tielens, A.G.G.M., Tokunaga, A.T. (1995), *ApJ*, **449**, 674.

Pappalardo, R.T. et al. (1998), *Nature*, **391**, 365.

Pendleton, Y.J., Tielens, A.G.G.M., Tokunaga, A.T., Bernstein, M.P. (1999), *ApJ*, **513**, 294.

Prasad, S., Tarafdar, S., Villere, K.R., Huntress, W.T. (1987), in Hollenbach, D. and Thronson, H. (eds.), *Interstellar Processes*, Reidel, Dordrecht, p.631.

Quirico, E. et al. (1999), *Icarus*, **139**, 159.

Rickman, H. (1994), in Milani, A., Di Martino, M., Cellino, A. (eds.), *Asteroids, comets, meteors 1993*, Kluwer Academic Publishers, Dordrecht, p.297.

Roser, J.E., Vidali, G., Manico, G., Pirronello, V. (2001), *ApJ*, **555**, L61.

Roush, T., (2001), *J. Geophys. Research*, **106**, 33315.

Roush, T., Cruikshank, D., Owen, T. (1995), *Am. Inst. Phys. Conf. Proc.*, **341**, 143. AIP Press New York.

Sack, N.J., Baragiola, R.A. (1993), *Phys. Rev. Lett.*, **48**, 9973.

Sandford, S.A., Allamandola, L.J. (1993), *ApJ*, **409**, 65.

Sandford, S.A., Bernstein, M.P., Allamandola, L.J., Goorvitch, D. Teixeira, T.C.V.S. (2001), *ApJ*, **548**, 836.

Schmitt, B. (1994), in Nenner, I. (ed.), *Molecules and Grains in Space*, AIP Press, New York, p.735.

Schmitt, B., de Bergh, C., Festou, X. (1998), *Solar System Ices*, ASSL 227, Dordrecht Kluwer Academic Publisher.

Schutte, W.A. et al. (2002), *A&A*, submitted.

Schutte, W.A., Greenberg, J.M. (1997), *A&A*, **317**, L43.

Skinner, C.J., Tielens, A.G.G.M., Barlow, M.J., Justtanont, K. (1992), *ApJ*, **399**, L79.

Smith, R.G., Sellgren, K., Tokunaga, A.T. (1989), *ApJ*, **344**, 209.

Smith, R.S., Huang, C., Kay, B.D. (1997), *J. Phys. Chem. B*, **101**, 6123.

Spaans, M., Ehrenfreund, P. (1999), in Ehrenfreund, P., Krafft, K., Kochan, H., Pirronello, V. (eds.), *Laboratory Astrophysics and Space Research*, Kluwer, Dordrecht, p.1.

Stevenson, K.P., Kimmel, G.A., Dohnalek, Z., Smith, R.S., Kay, B.D. (1999), *Science*, **283**, 1505.

Tanaka, M., Nagata, T., Sato, S., Yamamoto, T. (1994), *ApJ*, **430**, 779.

Tegler, S.C., Weintraub, D.A., Rettig, T.W., Pendleton, Y.J., Whittet, D.C.B., Kulesa, C.A. (1995. *ApJ*), **439**, 279.

Thi, W.F. et al. (2002), *A&A*, submitted.

Tielens, A.G.G.M., Whittet, D.C.B. (1997), in in van Dishoeck, E.F. (ed.), *Molecules in Astrophysics: Probes and Processes*, IAU Symp. 178, Kluwer, Dordrecht, p.45.

van der Tak, F.F.S., van Dishoeck, E.F. (2000), *A&A*, **358**, L79.

van der Tak, F.F.S., van Dishoeck, E.F., Caselli, P. (2000), *A&A*, **361**, 327.

van Dishoeck, E.F. et al. (1996), *A&A*, **315**, L349.

Vandenbussche, B., Ehrenfreund, P., Boogert, A.C.A., van Dishoeck, E.F., Schutte, W.A. et al. (1999), *A&A*, **346**, L57.

Whittet, D.C.B. (1993), in Millar, T.J. and Williams, D.A. (eds.), *Dust and Chemistry in Astronomy*, IOP Publ. Ltd., Bristol, p.9.

Whittet, D.C.B., Gerakines, P.A., Tielens, A.G.G.M., Adamson, A.J., Boogert, A.C.A et al. (1998), *ApJ*, **498**, L159.

Whittet, D.C.B., Pendleton, Y., Gibb, E.L., Boogert, A.C.A., Chiar, J.E., Nummelin, A. (2001), *ApJ*, **550**, 793.

Whittet, D.C.B., Schutte, W.A., Tielens, A.G.G.M., Boogert, A.C.A., de Graauw T. et al. (1996), *A&A*, **315**, L357.

ION INTERACTIONS WITH SOLIDS: ASTROPHYSICAL APPLICATIONS

E. M. Bringa

Lawrence Livermore National Laboratory, Chemistry and Material Sciences Dir., P.O. Box 808 L-353 Livermore CA 94550 USA

R. E. Johnson

University of Virginia, Engineering Physics Department Charlottesville VA 22903 USA

Keywords: ion bombardment – desorption – cosmic rays

Abstract   Energetic plasma ions and electrons can significantly alter surfaces in the solar system and the interstellar medium (ISM) and produce observable gas-phase atoms and molecules. In order to understand recent observations, the physics and chemistry of sputtering and radiolysis are reviewed. Emphasis is on recent molecular dynamics simulations of sputtering and their relevance to the desorption of molecules from grains in the ISM by heavy cosmic-ray ions and the effect of energetic ions and electrons on the icy surfaces in the magnetospheres of Jupiter and Saturn.

Introduction

A description of ion-solid interactions is necessary to understand the effects produced by energetic plasma ions and electrons incident on surfaces in the solar system and the interstellar medium (ISM). This interaction can lead to surface charging, sputtering, and to chemical and physical alterations of the exposed surfaces. The materials of interest are refractory, organic and low-temperature condensed-gas solids (ices). In addition to the presence of ices as mantles on grains in the ISM, Table 1 show where "ices" exist in the solar system. It has been shown that these ices are often exposed to, and modified by, photons and charged particles from the solar wind, planetary magnetospheres and background cosmic rays. Recent Galileo and Hubble Space Telescope data for the

V. Pironello et al. (eds.), Solid State Astrochemistry, 357–393.
© 2003 *Kluwer Academic Publishers. Printed in the Netherlands.*

Saturnian and Jovian satellites have shown dramatic effects produced by the energetic particles trapped in these planetary magnetospheres.

Table 1   Condensed Volatiles in the Solar System. T_f is the freezing temperature at normal pressure (1 atm) and the value in parentheses indicates the temperature in degrees Celsius. The coldest recorded temperature on Earth was -89 Celsius in Antarctica in 1983. Adapted from W. M. Calvin (Calvin, 1999).

"Ice"	T_f (K)	Where is it found
water (H_2O)	273 (0)	Earth, Mars, comets, outer planet satellites
carbon dioxide (CO_2)	215 (-58)	Mars, Triton, comets, Callisto
sulfur dioxide (SO_2)	200 (-73)	Io, Europa
ammonia (NH_3)	195 (-78)	Comets, maybe some outer planet satellites
methane (CH_4)	91 (-182)	Triton, Pluto, comets, Kuiper belt objects
ozone (O_3)	80 (-193)	Ganymede, Rhea, Dione
carbon monoxide (CO)	68 (-205)	Triton, Pluto, comets
nitrogen (N_2)	63 (-210)	Triton, Pluto, comets
oxygen (O_2)	55 (-218)	Ganymede, Europa

A fast ion transfers energy to atoms in a solid by direct momentum transfer to the nuclei or by energy transfer to the electrons. When a swift ion, with velocity greater than the mean speed of outer shell electrons, bombards a solid, a track of electronic excitations is produced along the path of the ion. This is typically referred to as an ion track. By excitations we mean promotion of electrons to excited estates, including ionizations (electron-hole pair formation). Both the momentum transfer collisions and the excitations can lead to ejection of atoms/molecules from the solid, a phenomenon that is called sputtering (Johnson and Schou, 1993). Figure 1 shows schematically a sequence of events that lead to "electronic" sputtering. The sputtering yield, Y, is the number of atoms/molecules that are ejected per individual bombarding ion. When the energy deposited by the ion near the surface is large the surface can be strongly modified with changes in crystallinity and the appearance of craters, rims and hillocks (Bringa and Johnson, 2002a). Table 2 shows some of the ions that can be found in the interstellar medium and the solar system.

In the following a brief overview of ion-solid interactions is presented, with emphasis on sputtering processes. First the energy loss and range of ions is discussed, then the formation of the ion track. Several sputtering yield models are discussed, and simulation of ion-solid interactions is shown to be necessary for numerous cases. The disagreements between analytical models and MD simulations are described. Finally, observations and applications to materials in the ISM and the solar sys-

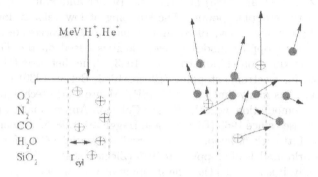

Figure 1 Left frame: The incident ion produces a track of excitations (+ circles); right frame, atoms can be ejected from the solid in the ground state (solid circles) or in excited states (encircled +) due to the energy release.

Table 2 Ions present in the ISM and in the solar system. Energies only give representative values from broad energy spectra.

Ions	Energy	Where is it found
H, He	0.5 eV	cometary atmospheres
H, He	1-10 eV	ISM shocks
H,He,...	keV	solar wind
H,O,S,...	keV-MeV	planetary magnetospheres
H,Fe,O, ...	GeV	Cosmic Rays

tem are described. Readers are strongly encouraged to browse through complementary information on the web, including several movies of the simulations (movies), at
http://dirac.ms.virginia.edu/~emb3t/research/research.html.

1.   STOPPING POWER AND RANGES

As a moving ion of velocity $\mathbf{v}$ and energy E traverses a material it experiences a force $\mathbf{F}$ that leads to energy dissipation, $dE/dt = \mathbf{v} \cdot \mathbf{F}$. If x is the direction of motion this can be written as $dE/dt = -(dx/dt)\,F$.

The energy loss per unit path length, also called the "instantaneous" stopping power of the ion, is given by $dE/dx = F$. There are entire books (Ziegler et al., 1985) and reviews (Berger and Paul, 1995) dedicated to the stopping power. The stopping of low velocity ions was studied by Firsov (Firsov, 1958), and Linhard and coworkers developed an electronic stopping model at low velocities based on the Thomas-Fermi approximation (Lindhard et al., 1963). The dielectric ("linear") formalism for electronic stopping (Bloch, 1933; Bethe, 1930) is widely used but it has many limitations (CASP). Monte Carlo techniques employed in combination with the Binary Collision Approximation (BCA) have been used since 1963 (Biersack and Haggmark, 1980). Ziegler, Biersack and Littmark developed a "universal" potential (ZBL) for use in Monte Carlo models of stopping in 1985 (Ziegler et al., 1985).

Typically, it is assumed that the stopping power can be separated into several components according to the physics of the energy loss processes:

$$(dE/dx) \approx (dE/dx)_n + (dE/dx)_e \tag{1}$$

$$(dE/dx)_e \approx (dE/dx)_{excit.} + (dE/dx)_{charge-exch.} +$$
$$(dE/dx)_{ioniz.} \tag{2}$$

These are the elastic collision (momentum transfer to nuclei), $(dE/dx)_n$, and electronic contributions, $(dE/dx)_e$. These can be further broken down as indicated above. Although used by everyone for convenience, this separation is only approximate since the momentum transfer and electronic processes are correlated. For scaling purposes, (dE/dx) is often given as a function of the dimensionless energy, $\epsilon = M_T E_P a / [(M_T + M_P) Z_P Z_T e^2]$, which depends on the projectile energy E_P, projectile and target mass $M_{P/T}$, charge $Z_{P/T}$, and screening length a. Nuclear stopping dominates for $\epsilon \leqslant 10$, and the electronic stopping dominates for $\epsilon \gtrsim 100$. At low velocities both nuclear and electronic stopping are roughly proportional to $\sqrt{E_P}$. They then reach a maximum [at $\epsilon \approx 0.5$ for $(dE/dx)_n$ and at $\epsilon \approx 100$ for $(dE/dx)_e$], and at high velocities, they decay as $(\ln E_P)/E_P$ (for non-relativistic ions). The stopping of ions in water is shown in Fig. 2.

Once the stopping power is known the range of the projectile within the material is calculated as:

$$R = \int_0^{E_o} |dE/dx|^{-1} dE \tag{3}$$

The energy deposition versus depth for a high velocity projectile is roughly constant except at the end of the particle's path, where it increases significantly (Bragg's peak). Since the range of 1 MeV He$^+$ in

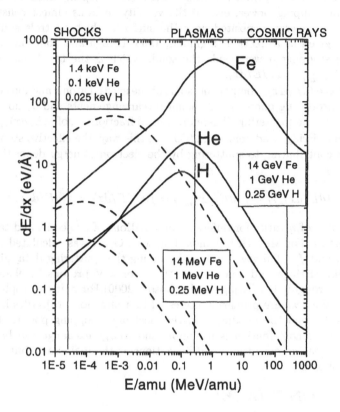

Figure 2 Stopping power of H, He and Fe in water, obtained from TRIM. $(dE/dx)_n$ (dashed lines) and $(dE/dx)_e$ (solid lines).

solid O_2 is 6 μm it is reasonable to consider the energy deposition to be constant along the cylindrical track. On the other hand, the range of a 10 keV C ion in carbon is only 240 Å and one has to take into account variations in the energy deposition with depth.

There are two main limitations in models for stopping power. First, all analytical and semi-analytical models of stopping deal with monoatomic materials. In order to calculate the stopping power of a multicomponent solid, corrections have to be made. The Bragg rule and bonding corrections (Ziegler et al., 1985; Biersack and Eckstein, 1984) can be applied but their success is limited. Second, the energy deposited near the surface is different from the value measured at depth, called the equilibrium stopping power, even if the velocity remains almost constant. Often the energy deposited near the surface, which controls sputtering, is given as a fraction (Sigmund, 1969; Bringa and Johnson, 2000) of the stopping power inside the solid, which is the tabulated value, $(dE/dx)_{surf} = \alpha \, (dE/dx)$.

For the $(dE/dx)_e$ component, some of the excited atoms and molecules decay producing photons and chemical rearrangements and do not contribute to the sputtering. Therefore, only a fraction, f, of $(dE/dx)_{surf}$ is relevant (Bringa and Johnson, 2000). In this way, the effective stopping power contributing to sputtering by the electronic energy deposition is given by:

$$(dE/dx)_{eff} = f \, (dE/dx)_{surf} = f\alpha \, (dE/dx) = \eta \, (dE/dx) \, . \qquad (4)$$

The value of η varies for different combinations of projectile and target properties. For MeV He^+ bombarding solid O_2, α was calculated to be ~ 0.4, and $f \sim 0.5$ was obtained by fitting the sputtering data, giving $\eta \sim 0.2$ (Bringa and Johnson, 2000) . For keV protons bombarding O_2, η is roughly 0.5 (Bringa and Johnson, 2000).For many astrophysics applications we are primarily interested in atomic ion projectiles in the regime beyond the maximum in the electronic stopping power. Here the nuclear contribution is negligible and charge exchange can be neglected, leaving excitations and ionizations as the main contribution to the stopping power.

2.   ION TRACKS

When energetic electrons, photons or ions bombard a solid they produce a number of excitations (including ionizations) along their path through the material. A fast ion typically creates a track of excitations along its path. In this track the excitation density depends on the incident ion energy and the target material. Fast electrons (velocity $> 2v_o$, with v_o the Bohr velocity) either directly incident or from an ionization produced by an energetic photon or an ion will, on the average, produce an additional ionization at some distance from the track. These electrons suffer significant angular scattering and produce a number of events near the end of their track, a region often called a 'spur' in which multiple ionizations are produced in a small volume.

We have been mainly interested in ions that penetrate a solid without large deflections creating a nearly cylindrical track of excitations. This is the case for swift light ions, such as H^+ or He^+, penetrating condensed

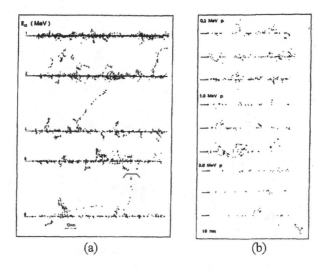

Figure 3 (a) α particle tracks (1, 2, 4, 6 and 8 MeV) in water. (b) Proton tracks (0.3, 1.0 and 3.0 MeV) in water. All tracks calculated with the Monte Carlo code MOCA-14, from Paretzke *et al.* (Paretzke et al., 1995). Notice that the He tracks are nearly continuous, and that there are significant statistical variations for the same energy for proton tracks.

gas solids. As can be seen for MeV α particle tracks in water in Fig. 3 (a), typically more than one excitation per monolayer is produced, corresponding to what we refer to as the high excitation density regime.

The maximum of the electronic stopping power for protons in water is at ~100 keV (Fig. 2). The stopping power decreases as energy increases in Fig. 3 (b) and it is seen that the probability of excitation also decreases leading to a sparse track. The transition from dilute excitations produced by medium energy protons to a high density of excitations for MeV He$^+$ has been studied for atomic targets (Bringa et al., 1999). Although we usually consider average quantities, like dE/dx, the tracks have large statistical fluctuations in excitation density. For a given energy, two tracks can have very different excitation distributions, as can be seen in Fig. 3 (b) for proton tracks in water in which excitations are sparse. This is important when describing the effect of cosmic-ray ions on grains.

A schematic view of the track of excitations produced by a fast ion can be seen in figure 4 (a). There are primary excitations produced directly by the ion and secondary excitations produced after the passage of the ion by energetic secondary electrons and by excited state diffusion. The

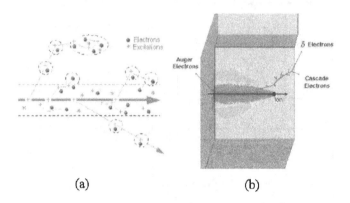

(a)   (b)

Figure 4 Tracks produced by a heavy ion. (a) The infratrack is marked with dashed lines. (b) Track structure showing secondary electrons. Colors indicate the strength of the track potential felt by the electrons (darker indicates higher value).

region in which the primary excitations are produced is often called the infratrack. The size of the infratrack is equal to a lattice spacing at low velocities, but at high velocities is given roughly by the Bohr adiabatic radius (Paretzke et al., 1995; Johnson, 1993), $r_B = \pi v/\omega$, where the mean excitation energy is $\hbar\omega$.

In the infratrack an energetic incident ion produces Auger electrons, low energy secondary electrons ($\lesssim 10$ eV), and more energetic electrons often called δ rays. The range of these δ electrons can be quite large ($\sim$100 Å) and determines an outer limit to the track in the absence of exciton diffusion. This region of excitation determined by the δ electrons is often called the ultratrack (Paretzke et al., 1995; Johnson, 1993). The size of the ultratrack is proportional to the energy of the incident ion and the spurs (or "blobs") are enclosed by dashed spheres or ellipses in Fig. 3(a). As the ion slows down the energy of the δ electrons decreases giving rise to the conical ultratrack seen in Fig. 4 (b). For surface modification and sputtering phenomena the energy deposited near the surface is important. In cases where $Y \sim 1$, the first 2-3 layers typically control sputtering, but for much larger yields energy deposition from deeper layers participates. Therefore, even when the ion penetration is not large, the conical shape of the ultra-track can often be neglected and considered roughly cylindrical.

Energy relaxation in the solid is a complex process involving several possible pathways in which the excitation energy is released to the lattice. This is shown in Table 3 for a solid with atoms "A". Repulsive states lead to extra kinetic energy (ΔE) deposited in the lattice which can cause sputtering and surface modification. The complexity of the relaxation process increases for molecular solids or multicomponent solids in which chemical reactions can occur.

Table 3 Track processes.

Coulomb Repulsion	$A^+ + A^+ \rightarrow A^+ + A^+ + \Delta E$
Charge Exchange	$A^{++} + A \rightarrow A^+ + A^+$
De-Excitation	$A^{**} + e \rightarrow A^* + e(\text{hot}) \rightarrow A^* + e + \Delta E$
Repulsive Decay	$(AA)^* \rightarrow A + A + \Delta E$
Lattice Relaxation	$A^{**} + \text{lattice} \rightarrow A^* + \text{lattice} + \Delta E$
Dielectronic Recombination	$A^+ + e + e \rightarrow A^* + e(\text{hot}) \rightarrow A^* + e + \Delta E$

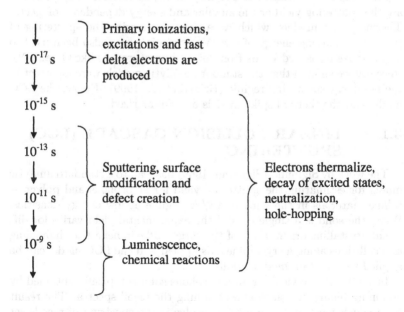

Figure 5 Timeline of the events produced by the bombarding ion.

A time line for the track processes in the physical stage is given in Fig. 5. Early track evolution is relatively well understood (Paretzke et al., 1995; Sanche et al., 1995) as is the other extreme, chemical effects in times of the order of μs or longer (Cobut et al., 1998). However, events occurring on the scale of ps, including the coupling of electronic excitations to the lattice, are not easily calculated (Ryazanov et al., 1995). In the next section models and data for sputtering and surface modification are described.

3.   SPUTTERING

Three different regimes can be found for the collisional sputtering process. The single knock-on regime occurs when a primary recoil receives enough energy to be ejected. In the linear collision cascade (LCC) regime a number of recoils can lead to sputtering and collisions occur mainly between fast atoms and atoms at rest. When many atoms within the "cascade" volume are moving, collisions among moving atoms can not be neglected and the LCC model breaks down (Sigmund and Claussen, 1981; Bringa and Johnson, 2000; Urbassek and Michl, 1997; Watson and Tombrello, 1985; Seiberling et al., 1980). This is the "non-linear" or "spike" regime, for which there is, at present, no standard way to evaluate the sputtering yield or the angular and energy dependence of ejecta. Thermal spike models, which were used to describe the sputtering of refractory (binding energy of few eVs) materials have also been used to parametrize measured yields from low-temperature ices like O_2 and N_2. Recently we showed that the standard analytic models were incorrect in the nonlinear sputtering regime (Bringa et al., 1999). Below, the LCC model and the thermal spike models are summarized.

3.1.   LINEAR COLLISION CASCADE (LCC) SPUTTERING

The work by Sigmund (Sigmund, 1969) has lead to a standard analytic model for estimating the sputtering yield for amorphous and polycrystalline materials in the nuclear dE/dx regime at low energy densities. When the solid is a single crystal the experimental yield varies for different crystalline orientations of the target. Phenomena like channeling and collisions along a crystalline axis appear and the LCC model can be applied with certain modifications.

In the LCC, the yield for an amorphous solid is typically obtained by assuming binary collisions and obtaining the recoil spectra. The result from which most of the properties evolve is the spectrum of recoils set in motion with energy between E' and $E' + dE'$. This varies roughly as $[dE' / E'^2]$ with only a weak dependence on the form of the interaction potential between the atoms. However, the binary collision assumption in the LCC model is not necessary as this energy spectrum also applies to low energy recoils and collisions involving molecules (Johnson and Liu, 1996). The form for the yield has been derived often

$$Y \approx C\,(l/U)\,(dE/dx)_n \tag{5}$$

$$(dY/d\theta)\,/Y \approx (\cos\theta)^{-1} \tag{6}$$

$$(dY/d\Theta)\,/Y \approx (\cos\Theta)^{-b} \tag{7}$$

$$(dY/dE)\,/Y \approx \frac{2UE}{(E+U)^3} \tag{8}$$

Here U is the surface binding energy, Θ is the angle to the surface normal, θ is the ejection angle measured from the surface normal, E is the energy of the an ejected atom, and C is a parameter that depends on the interaction potential and projectile-target parameters. The yield in the LCC regime is seen to be proportional to the energy deposited in the surface region and inversely proportional to the binding energy U. Sigmund (Sigmund, 1969) estimated the parameters using the first few moments of the transport equation for the recoils. He wrote $C\,(l/U) = \Lambda\alpha$ where Λ depends only on target properties such as U and α depends the mass ratio M_T/M_P and weakly on the electronic energy loss and the interaction potential. $\alpha\,(dE/dx)_n$ gives the energy deposited at the surface, as discussed, and is sometimes written as $F_D\,(0)$. LCC fits the energy dependence of the experimental data in the regime at low $(dE/dx)_n$ but the parameter α, adjusted to fit the magnitude of the data, agrees with the model values to about 50%.

The angular distribution of the ejecta in Eq. 6 was assumed to be $\cos\theta$ for a flat surface, although this form is not found experimentally or in detailed calculations. The change of the yield for different angles of incidence of the projectile with respect to the surface normal, Θ, was estimated for not-too-oblique incident angles using $b \sim 5/3$ for $M_T/M_P \lesssim 3$ and $b \sim 1$ for $M_T/M_P > 8$. At large angles the yield decreases abruptly due to projectile scattering and surface roughness.

The energy spectrum of the ejecta in Eq. 8, called the Thompson spectrum (Thompson, 1968), is obtained assuming a planar binding energy and the recoil spectrum $\propto E'^{-2}$. This can change slightly according to the interaction potential among the target atoms and due to emission from deep layers (Sigmund, 1981a). The Thompson distribution has a maximum located at $U/2$ and roughly applies up to the maximum possible energy of a primary recoil.

3.2. SPUTTERING MODELS IN THE NON-LINEAR REGIME

In the nuclear stopping (elastic collision) regime, the cascade of collisions directly generates a hot region, a "spike", that leads to ejection. In the electronic stopping regime, the mechanism for converting electronic energy into lattice motion can be either the transfer of energy to the lattice by the secondary electrons (Toulemonde et al., 1999; Ritchie and Claussen, 1982) or by a repulsion process because the electrons do not fully screen the nuclear charge. The repulsion may occur on electronic recombination ($O_2^+ + e \rightarrow$ repulsive state $\rightarrow O+O+\Delta E$), by a net repulsive force in the track, or by the short lived, partially screened Coulomb force (Bringa and Johnson, 2002). Once the energy is in the lattice motion, due either to electronic or collisional processes, then the spike models (thermal spikes, shock models, etc.) can be applied to treat the evolution of the energy and to calculate sputtering.

3.2.1   Analytical Spike Models for Sputtering.   A dense collision cascade is often described as a "spike" and analytical models have been applied to defect production and track formation (Toulemonde et al., 1999), ion-beam mixing (Urbassek, 1997), cratering (Bringa and Johnson, 2002a), and sputtering of metals and insulators (Johnson and Evatt, 1980). The sputtering yield is obtained by calculating the temperature profile at the surface and evaluating the yield as:

$$Y = \int_0^\infty dt \int dA \ \Phi\left(T_{surf}, U_s\right) \qquad (9)$$

where the integral is over time, t, and surface area, A. Here Φ is the evaporative flux which depends on the surface temperature T_{surf} and binding energy U_s. When the yields are large the cohesive energy (sublimation energy), U, is generally used for U_s and in most cases $T_{surf} \approx T_{bulk}$ is also assumed. Models dealing with spherically symmetric spikes (Johnson and Evatt, 1980) are often appropriate for very low energy ions in the nuclear stopping regime. Below we focus on cylindrical spikes produced by a fast ion.

The formula for the evaporation flux is often taken to be that for a hot gas,

$$\Phi\left(T\right) = n\sqrt{\frac{k_B T}{2\pi m}} \exp\left(-\frac{U_s}{k_B T}\right); \qquad (10)$$

Assuming that ejection occurs at some average effective temperature T_{eff} from a region of radius R_{sput}, for a time t_{sput}, then Eq. 9 can be approximated as:

$$Y \approx \pi R_{sput}^2 t_{sput} \Phi \left(T_{eff}, U \right)$$

$$= n\pi R_{sput}^2 t_{sput} \sqrt{\frac{k_B T_{eff}}{2\pi m}} \exp\left(-\frac{U_s}{k_B T_{eff}} \right) \tag{11}$$

Such a model also results in a Maxwellian energy distribution for the ejecta, which has been seen in some experiments (Seiberling et al., 1980) and applies roughly at relatively low dE/dx. In general, T will vary during sputtering and ejection will be largest at the center of the track.

For a narrow, high (dE/dx) spike it has been shown that Eq. 9 gives a yield of the form,.

$$Y \propto (dE/dx)_{eff}^2 \tag{12}$$

Eq. 12 is a very robust result verified by analytic and numerical (Jakas, 2000; Bringa et al., 1999a) models. The corresponding energy spectrum, $(dY/dE)/Y$, has less steep dependence at large E than in Eq. 8. Since ejection is assumed to be driven by evaporation, an isotropic $\cos\theta$ angular spectrum of ejecta applies. For a narrow track (Gibbs et al., 1988; Johnson, 1989) the incident angle dependence is $\sim (\cos\Theta)^{-1.68}$, with a steeper dependence at low dE/dx and large track radii (Claussen, 1982).

A large spike temperature at high excitation density can cause an increased pressure in the track, an effect neglected in most thermal spike models. There are also models in which thermal diffusion is assumed to be a smaller effect than the large pressure buildup. In such models the yield is estimated from the pressure or shock wave formed. This has been applied to sputtering of macromolecules, like amino acids, for which the yield respect to dE/dx appears to be roughly cubic instead of quadratic (Fenyö and Johnson, 1992) and has been suggested to be important in the ISM (Johnson et al., 1991). Details about pressure effects can be found elsewhere (Jakas et al., 2002).

3.2.2   Computer Simulation Models for Ion-Solid Interaction.
Computer simulation methods for ion-solid interactions are described in a number of texts (Eckstein, 1991; Smith, 1997) and in a number of reviews (Urbassek, 1987; Shapiro, 1997). There are three main groups of models: hydrodynamic models, MD models and the Monte Carlo binary collision (BCA) models mentioned earlier. The latter can only be used in the LCC regime and the first can only be used in the non-linear regime. MD can be used to cover the full range and has been applied to sputtering since the early work of Harrison and coworkers (Jakas and Harrison, 1985). BCA Monte Carlo codes, like TRIM

(Biersack and Eckstein, 1984) , can only be applied to amorphous materials, whereas MARLOWE (MARLOWE) is for an ordered target. MD and BCA codes agree reasonably well for amorphous targets and recoil energies much larger than the binding energy of the crystal. For crystalline targets and low primary energies ($E \leqslant 15U$) Monte Carlo codes fail (Sigmund et al., 1989). MD codes are especially suited to study nonlinear sputtering and sputtering of molecular solids, crystalline alloys, and effects related to low energy recoils.

4.   MOLECULAR DYNAMICS (MD)

Molecular Dynamics (MD) is a simulation technique widely used to deal with many-body problems and there are several good texts which contain Fortran or C codes. (Allen and Tildesley, 1987; Rapaport, 1995). A monograph can be downloaded from the web (Ercolesi) giving a good overview of simulation potentials. MD simulations give insights into material behavior at small dimensions and times which are inaccessible experimentally. The main advantage of MD over Monte Carlo is that it provides true dynamical information on the system, because one solves the equations of motion and does not just advance the system based on transition probabilities. A collection of useful MD-related bookmarks can be found in the web at
http://dirac.ms.virginia.edu/~emb3t/Public/MDbook.html.

In this section bold symbols represent vectors, while italics represent scalars. For instance, the distance between two points $\mathbf{r}_i$ and $\mathbf{r}_j$ is given by $r_{ij} = |\mathbf{r}_i - \mathbf{r}_j|$. Classical MD consists in solving Newton's Equations of motion for a collection of N particles with positions $\mathbf{r}_i$ (where $\mathbf{r}_i$ is a 3D vector) and with a potential energy function ϕ, giving $3N$ coupled second order differential equations:

$$m_i \, \ddot{\mathbf{r}}_i = \mathbf{F}_i = -\nabla_{\mathbf{r}_i} \phi \qquad i = 1, .., N \tag{13}$$

The above set of equations is equivalent to solving $6N$ first order differential equations for both the positions and the momenta $\mathbf{p}_i$:

$$\begin{aligned} \dot{\mathbf{r}}_i &= \mathbf{p}_i / m_i \\ \dot{\mathbf{p}}_i &= -\nabla_{\mathbf{r}_i} \phi \end{aligned} \tag{14}$$

Classical forces can be used when the wavelength of moving atoms, $\lambda = \sqrt{2\pi\hbar^2/Mk_BT}$, is shorter than the mean particle distance, $l = n^{-1/3}$, where n is the number density in the material. The zero point energy of crystals can be roughly incorporated into the classical simulation but the discreteness of energy levels poses a problem at low local temperatures in the simulation of vibrational and rotational excitation of molecules and clusters.

The collisional energy transfer regime can be directly simulated by bombarding a sample with the ion and choosing appropriate interaction potentials. In the electronic regime, an energetic, non-radiative repulsive decay may be represented by giving an atom or neighboring atoms in the lattice an extra kinetic energy, $\Delta E \approx E_{exc}$, such that the total extra energy per unit length is equal to $(dE/dx)_{eff}$. In reality, excitation relaxation processes are more complex but can be described accurately if necessary (Dutkiewicz et al., 1994).

MD has several limitations regarding the size of samples and simulation times. To calculate the evolution of a weakly bound L-J solid with a ~0.5 μm side ($4\,10^9$ atoms) for 20 ns (10^6 steps for a typical time step of 0.02 ps) even the most optimized MD codes have calculation times of 0.5 μs/atom/step/CPU. Therefore, it would take 63 years on a single CPU, assuming it has enough memory! If the code is running in parallel using 100% of CPU time in 1000 CPU's it would still take 23 days, neglecting overhead due to communications. Currently most MD simulations are used for relatively small volumes ($\sim 10^5$ atoms) and short times (~ 10 ps). They can then be combined with MC or continuum codes to expand the time and length scales (Insepov and Yamada, 1997; Zhigilei and Garrison, 1999; Abraham et al., 1999).

4.1.1   Analytic spike models versus MD simulations.   MD simulations of cylindrical spikes show that analytic models work reasonably well for energy densities $(dE/dx)_{eff} / \pi r_{cyl}^2$ that are much less than the cohesive energy density nU, but fail at high energy densities (Bringa et al., 1999). If these models are corrected by solving the full set of hydrodynamic (HD) equations using realistic transport properties and allowing for flow toward the surface as well as radial flow, the results agree reasonably well with MD, with some discrepancies because escape from the surface is not easily included (Jakas et al., 2002). Below we parametrized the MD results for the sputtering yield at high $(dE/dx)_{eff}$:

$$Y \approx C\,(r_{cyl}/l)^x \left[(l/U)\,(dE/dx)_{eff}\right]^y, \quad [(dE/dx)_{eff} / \pi\,r_{cyl}^2] > nU$$
(15)

For atomic solids $C \approx 0.14$, $x = 0.9 \pm 0.2$, and $y = 1.1 \pm 0.2$. For solid O_2 ((Bringa and Johnson, 2000)) $C \approx 0.092$, $x = 1.0 \pm 0.3$, and $y = 1.0 \pm 0.2$.

5.   EXPERIMENTAL DATA ON SPUTTERING AND SURFACE MODIFICATION

There are many experiments on sputtering by ion bombardment and results differ greatly depending on the nature of the material. However, two broad regimes can be identified. One in which the number of excitations or collisional energy transfer events is small but are sufficiently energetic to cause sputtering independently (dilute excitations, typically low dE/dx). This results in a yield that is linear in dE/dx. The other regime occurs when excitations can not be assumed to act independently (either low energy events and low sputtering yields or a high density of excitations at high dE/dx) giving a yield which depends non-linearly on dE/dx. These regimes, discussed below, apply when either collisional and electronic energy deposition dominate.

Electronic sputtering is significant in relatively few materials because the energy release in an event is generally of the order of the binding energy of many materials (few eVs). Although electronic energy deposition can lead to defects, amorphization, desorption of adsorbed species, and radiolytic decomposition, bulk atom ejection is problematic. Therefore, unless the electronic excitation density is extremely large sputtering of bulk refractory material is produced mainly by collisional energy deposition. This is often smaller than electronic energy deposition, but high energy recoils are produced by a single collision event. Electronic sputtering yields are large principally for the condensed gas solids, which have low binding energies (0.5 eV for water; 0.32 eV for SO_2; $\sim$0.08 eV for O_2, N_2 and Ar; and $\sim$0.01 eV for H_2 and D_2) (Johnson and Schou, 1993), but are also non-negligible for organic solids, relevant in many regions of space, and ejection of the most volatile radiolytic decomposition products such as oxygen and alkalis (Madey et al., 2002).

5.1.   DILUTE EXCITATIONS

5.1.1   Collisional Regime.   A large portion of available sputtering data for refractory solids falls into the dilute regime (Betz and Wien, 1994). Programs like TRIM (Biersack and Eckstein, 1984) and MARLOWE (MARLOWE) give reasonable agreement with experiments, especially when the surface binding energy in the code is adjusted to fit the data. For heavy ions the experimental results depart from the LCC model at high dE/dx.

5.1.2   Electronic Regime.   When excitations act independently, the energy release by non-radiative decay of individual excitations in a material with a very low U can lead directly to desorption or can initiate a mini-cascade below the surface giving a yield roughly linear in $(dE/dx)_e$. This happens for three different cases: 1. When the mean free path for producing excitations λ_{exc} is large compared with the mean lattice spacing $l = n^{-1/3}$ where $\lambda_{exc} = W/(dE/dx)$, and W is the average energy required to produce an excitation (Bringa et al., 1999); 2. The excitations are initially close together but exciton diffusion is rapid so the non-radiative decay events are again spatially dilute (Johnson and Schou, 1993); 3. Excitations decay over long times (up to μs) and are well separated temporally.

5.2.   HIGH EXCITATION DENSITY

In this regime the excitation density is important. For electronic sputtering of the molecular condensed gas solids, this occurs for excitation densities ranging from $\sim$1 to several per monolayer. The energy from the closely-spaced individual decay events can create a hot track where moving particles collide with other moving particles or the closely-spaced excitations overlap creating a more robust release of energy. Also, when the density of electron-hole pairs is large enough and the electronic screening and recombination are small enough "Coulomb explosion" can heat the lattice (Bringa and Johnson, 2002). In these cases a region in which the local temperature is much higher than the temperature of the rest of the solid is produced. It can be higher than the melting temperature and even higher than the binding energy of the solid. Coulomb explosion was first proposed for track formation in insulators (Fleischer et al., 1967). Tracks have been seen to be produced by solar flare helium ions in lunar grains and collected interplanetary grains. With the advent of new accelerators tracks can be produced even by electronic excitation of metals at enormous $(dE/dx)_{elec}$ (tens of keV/Å) (Lesueur and Dunlop, 1993).

At high excitation density the dependence of the yield on dE/dx in both the nuclear stopping regime and electronic stopping regime is similar because the high temperature spike evolves similarly. A "typical" energy spectra of the ejected particles for high energy densities (Kelly, 1990) can be seen in Fig. 6. The spectrum labeled as thermal is the one from Urbassek and Michl (Urbassek and Michl, 1997). The maxima in the spectrum is shifted towards lower ejecta energies than is the case for LCC. Measured energy distributions are often fit using a Thompson distribution (Thompson, 1968) with an artificial U (Reimann et al., 1984; Pedrys et al., 2000). That is, the process can still exhibit the cascade energy spectrum ($\sim 1/E^2$) but with a lower effective U because of

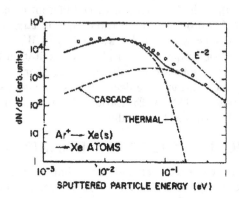

Figure 6 3 keV Ar bombardment of polycristalline Xe, at incident angle $\Theta = 60°$, from Kelly (Kelly, 1990). Cascade + thermal contributions. Cascade spectrum uses $U = 0.16\,\mathrm{eV}$ and thermal spectrum uses $T = 200\,\mathrm{K}$.

surface damage (Jakas and Harrison, 1985). However, the shift of the maximum can also be due to late evaporative contributions. In Fig.6, a Thomson spectrum does not fit the data well. Therefore, a realistic model must include the weakened binding in the track region (Watson and Tombrello, 1985) and late thermal effects (Dutkiewicz et al., 1995). The energy distribution is usually measured at a fixed angle and can differ from the angular-integrated distribution, leading to apparent shifts in the binding energy. The fact that the distribution often goes as E^{-2} at large E is general, as it is determined by the energy partition among the recoils (Johnson and Liu, 1996) so that E^{-2}-like tails are also observed in the "non-linear" regime (Dutkiewicz et al., 1995; Bringa et al., 1999a; Bringa et al., 1999).

5.2.1   Collisional Regime.   The transition from linear to non-linear regime occurs when the energy density in the cascade volume is of the order of the binding energy per atom, in which case sputtering yields are ~10-20. Enhancements, which are underestimated by LCC, are also produced by impact of ion clusters. Therefore, spike models have been applied to cluster ion sputtering of metals (Sigmund and Claussen, 1981). Heavy atomic projectiles (Dutkiewicz et al., 1995) and cluster ion impact (Andersen et al., 1998) can produce a hot region with a transient fluid at the center. Nuclear sputtering of condensed gas solids has been studied experimentally (David and Michl, 1989) and the transition from the LCC to the nonlinear regime has been described for one case (Pedrys, 1990).

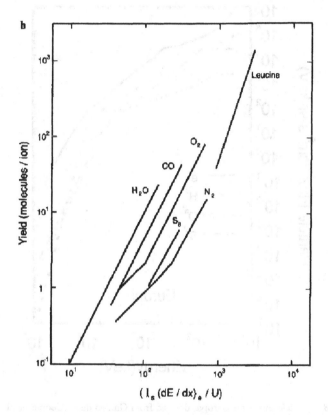

Figure 7 Experimental sputtering yields as a function of dE/dx, scaled with the binding energy U and the mean particle distance l, for several condensed gases (Johnson and Schou, 1993).

5.2.2 Electronic Excitations Regime.

In Fig. 7 it is seen that the sputtering for a number of low cohesive energy, condensed gas solids (O_2, N_2, CO and H_2O) is roughly quadratic in dE/dx at high excitation densities (high dE/dx). For the range of dE/dx shown, water ice does not exhibit the linear regime. This has only been observed at very low dE/dx using ~100 keV electrons (Heide, 1984). The yield for the amino acid leucine is also included in the figure. Over a narrow range of dE/dx it appears to be steeper, possibly cubic (Fenyö and Johnson, 1992).

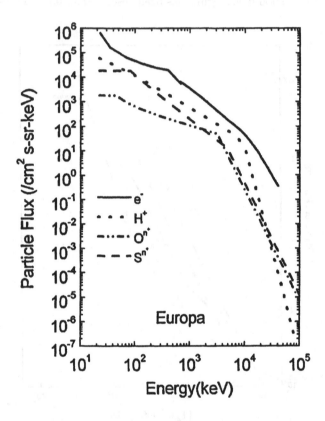

Figure 8 Particle fluxes on Europa, derived from Galileo data (Cooper et al., 2001).

A linear regime is observed in solid O_2. A single electronic recombination event decays as follows: $O_2^+ + e \rightarrow O+O+\Delta E$ (Johnson and Schou, 1993) with energy release, $\Delta E \approx 1 - 10$ eV shared between the atoms. Since $U \approx 0.08eV$, a single excitation can lead to sputtering. However, in water ice, $H_2O^+ + e \rightarrow H+OH+\Delta E$ or $H_2+O+\Delta E$ or $H_2O^++H_2O \rightarrow H_3O^++OH$, and $H_3O^+ + e \rightarrow H_2+OH+\Delta E$. Since $\Delta E \approx 1 - 5$ eV and the binding energy is much higher, $U \approx 0.5$ eV sputtering due to an individual excitation event is problematic. This is especially so since the lightest species carry off most of the energy resulting in hot H and H_2 that are seen in experiments (Sieger et al., 1997).

For weakly attached adsorbed water molecules other ejection processes are possible.

The energy spectra for heavy water (D_2O) ice from Brown et al. (1984) for 50 keV Ar^+ the Thompson spectrum agrees well with the experiment if the effective binding energy is approximately ten times smaller than the binding energy of water (~ 0.5 eV). When $(dE/dx)_{el}$ dominates[1], the low energy component grows, again suggestive of spike effects (Pedrys et al., 2000). As mentioned before, "Coulomb explosion" may contribute. At both low and high $(dE/dx)_e$ sputtering by this process varies as $(dJ/dx)^2$, where dJ/dx is the ionization per unit path length (Bringa and Johnson, 2002).

The yield is seen to increase with increasing angle to the normal, Θ. This occurs due to the increase of energy deposited near the surface up to a point ($\Theta \geq 60^{\circ}$) where projectile backscattering and surface roughness affect the yield. Angular data for sputtering of solid CO, O_2 and other condensed gas solids (Brown et al., 1984; Gibbs et al., 1988) can be well approximated as $Y(\Theta) = Y(0)(\cos\Theta)^{-1.6}$. MD simulations of ejection from a hot track also give the same dependence (Bringa and Johnson, 2001). Coincidentally, a thermal spike model for constant thermal diffusivity and narrow spike radius gives the same dependence at small angles (Johnson, 1989).

5.3. CHEMICAL EFFECTS

Incident particles and energetic photons not only cause sputtering but can also cause chemical alteration of the solid, a process called radiolysis or photolysis (Johnson and Quickenden, 1997). Therefore, even in a pure material a number of newly formed species can be ejected. The dominant ejecta are typically those with the lowest surface binding energy. For this reason ions, that have a strong attraction to the surface, are sputtered inefficiently even though they are studied most frequently since they are easy to detect. Radicals formed in repulsive decays can be ejected from the surface layer more easily than ions and can be an important component at low yields. Because their binding energies are still large, they have, typically, much lower yields than any close shell atoms or molecules that can be produced. Because the surface binding acts as a filter, the sputtering of ice by energetic ions is typically dominated by

---

[1]1.5 MeV He^+ in liquid water has $(dE/dx)_{el} = 18.2$ eV/A, $(dE/dx)_{nuc} = 0.02$ eV/A, and a penetration depth of 8.4 μm. 1.5 MeV Ar^+ in liquid water has $(dE/dx)_{el} = 83.5$ eV/A, $(dE/dx)_{nuc} = 7.4$ eV/A, and a penetration depth of 2.4 μm. 50 keV Ar^+ in liquid water has $(dE/dx)_{el} = 13.8$ eV/A, $(dE/dx)_{nuc} = 37.4$ eV/A, and a penetration depth of 10^{-1} μm. These values were calculated for liquid H_2O using SRIM 98 (Biersack and Eckstein, 1984).

H_2O, H_2 and O_2 with the latter becoming increasingly important with increasing temperature. Although the O_2 is not produced as efficiently as the radicals OH and H, it is very weakly bound to the solid and can diffuse along the ion track to the surface. Similarly, the sputtering of CO_2 and CO are both dominated by CO, O_2 and CO_2 with more refractory products such as carbon suboxides (C_xO_y) accumulating in the irradiated sample. In solid oxygen the dissociation of O_2 leads to the formation of ozone, which itself is readily dissociated releasing energy to the lattice (Fama et al., 2002).

Because the chemistry induced by energetic ions is qualitatively similar in most instances to that induced by photons and discussed extensively at this conference, it will be only touched on here. As in a space environment, irradiation in the presence of a vacuum interface will lead to the loss of the most volatile species (e.g., Na and K from most minerals, O_2 from a silicate, H_2 from water ice) and the most refractory species formed will accumulate (e.g., carbon chains in organic solids). In addition, segregation will occur under long term irradiation and defects will accumulate to form voids (interior surfaces). Radiation yields are often given as a G-value, the number of a particular radiation product per 100 eV deposited in the solid. Unfortunately, G-values vary slowly with ion energy and type and material temperature as they are affected by the local energy density.

6.   APPLICATIONS: PLANETARY

One of the most exciting areas of research in Planetary Science is the study of the sputtering and the chemistry induced in the surfaces of the icy moons of Jupiter and Saturn by the Jovian and Saturnian magnetospheric particle radiation. Observations of Io, Europa, Ganymede and Callisto by Galileo using the newly discovered telescope initiated enormous controversy and changed our image of the solar system over three centuries ago. Now observations of these moons using the instruments on the Galileo spacecraft and Hubble Space Telescope (HST) are again leading to radically new ideas on the origins of bodies in our solar system and, possibly, to new insights into the origins of biological activity. Objects in Saturn's magnetosphere are now being studied using HST and our understanding is also expected to grow dramatically when the CASSINI spacecraft arrives at Saturn in 2004.

6.1. SPUTTERING OF THE ICY GALILEAN SATELLITES

The new spacecraft and HST observations require an understanding of the changes induced in the surfaces of the Galilean satellites by the energetic ions and electrons trapped in the giant magnetosphere of Jupiter (Johnson et al., 2002). These energetic particles both alter the material and cause desorption of atoms and molecules, producing a tenuous atmosphere and ionosphere. Whereas the dominant surface material on the outer three Galilean moons is ice (Table 1), the inner moon, Io, is the most active volcanic object in the solar system. It has been totally dehydrated and has a surface coated in sulfur dioxide.

Also remarkable is the fact that one of the icy moons, Ganymede (Kivelson et al., 1996), has an intrinsic field and all of the icy moons have unusual conducting properties as indicated by the magnetometer measurements on the Galileo spacecraft. This has been attributed to the presence of underground, tidally heated oceans (Khurana et al., 1998). The magnitude of the conductivity deduced from the time variability of the local fields suggests the presence of underground 'salty' oceans. Supporting this hypothesis, the surface of Europa is chaotic with geologically young material on the surface. Recent infrared spectra are indicative of hydrated salts and organics (McCord et al., 1999), which would be consistent with material from an underground ocean. For this reason Europa is now considered to be an object on which biological materials might have evolved. Such an evolution, if it occurred, could be driven by the heat created by the tidal interaction with Jupiter or it could be driven by the energy of the Jovian magnetospheric particle radiation incident on to the surface (Cooper et al., 2001). A recent laboratory comparison of the shape of the water of hydration bands with the spectra obtained by the Galileo spacecraft suggests that the hydrated material might be frozen, hydrated H_2SO_4 (Carlson et al., 1999b). This material would be produced by the charged particle irradiation of sulfate salts, sulfur or SO_2 in an ice matrix (Johnson, 2001) so that a sulfur cycle is maintained by the incident radiation. Such activity produces oxidants such as SO_4^{-2}, H_2O_2 and O_2 which have potential importance for pre-biotic chemistry.

Using the data from the Energetic Particle Detector on the Galileo spacecraft, the ion and electron fluxes are given in Fig. 8. In addition, a lower energy component exists (Bagenal, 1994). Since the dominant volatile species on the Io is SO_2, with H_2O the dominant species on the other three moons, the composition of the plasma is predominantly H^+, O^{+z}, S^{+z}, with a smaller component of undissociated, molecular

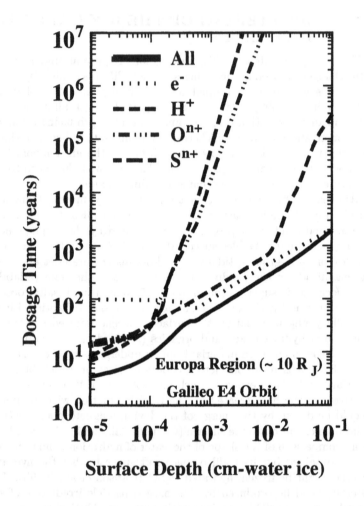

Figure 9 Time required to achieve a dose of 100 eV/molecule in water ice.

ions (SO_2^+, SO^+, NaX^+, etc) found, primarily, close to Io, the principal source of the plasma. Although the flux of ions and electrons is not large by laboratory standards, the plasma energy flux to the surfaces or atmospheres of Europa and Io is larger than the solar UV flux (Cooper et al., 2001).

An important issue for irradiation chemistry is the radiation dose vs. depth (Cooper et al., 2001). This is shown in Fig. 8 for Europa with the vertical axis the time required to achieve a dose of 100 eV per molecule in H_2O. This is a useful form since radiation chemists typical give yields as G-values. Geologist have suggested the youngest surface ages are $\lesssim 10^{6-7}$ years on Europa, during which time significant doses are acquired at all depths above ~ 10 g/cm^2, Fig 9. Since over most of the surface the solar photon penetration depths are much smaller than this, the optical layer (~ 1 μm) is primarily altered by the incident plasma.

Because any H_2 produced in irradiated ice diffuses efficiently to the surface and is lost to the vacuum, the irradiated ice is permanently altered. Therefore, there is a depletion of H and formation of trapped O and OH, so the penetrated layer is oxidizing, as well as having an enhanced D/H ratio. These radicals can react directly: OH+OH $\rightarrow$ H_2O_2 and OH+O$\rightarrow$ HO_2. Reimann et al. (1984) measured the fluence and temperature dependence for the production of O_2 from a thin ice sample in a vacuum exposed to energetic Ne$^+$. These ions were used to represent the energetic O$^+$ ions in the Jovian plasma. They found a correlation between the loss of H_2 and the production and loss of O_2. H_2 exhibited a prompt component followed by a gradually increasing component. The former comes from the surface and the latter depends on the film thickness for ions penetrating to the substrate. Reimann et al. noted an activation barrier 0.05-0.07eV and considered diffusion of the radicals produced. Sieger et al. (Sieger et al., 1997) found similar results for incident electrons. They concluded that a non-diffusing precursor is first formed (cross section $\sim 10^{-18} cm^2$) and a subsequent electron produces the observed O_2 (cross section $\sim 10^{-16} cm^2$) in a temperature independent process. Therefore, trapped O as H_2O - O , H_2O-O + e $\rightarrow$ H_2 + O_2 is a candidate and they suggest that the activation energy may be associated molecular orientation changes. Alternatively, a hot O atom reacting with a trapped OH + O(hot)$\rightarrow$H + O_2 may be a candidate.

Since energetic ion and electron irradiation produces H_2 and O_2, we predicted that the Europa has a tenuous O_2 atmosphere (Johnson, 1990). That is, the H_2 desorbed would readily escape from a moon's gravitational field but the heavier O_2 would not. In addition, the O_2, unlike H_2O, does not condense at the surface temperatures on these moons, $\gtrsim$80K. Remarkably an O_2 atmosphere was indirectly observed by HST both on Europa and Ganymede . In a parallel set of discoveries, a reflectance band in the visible, similar to that for solid O_2 (Calvin et al., 1996), was seen at low latitudes on Ganymede and, more recently, on Europa and Callisto. In addition, a likely radiation-induced UV feature associated with O_3 was seen on Ganymede (Noll et al., 1996) and subsequently on the icy satellites of Saturn.

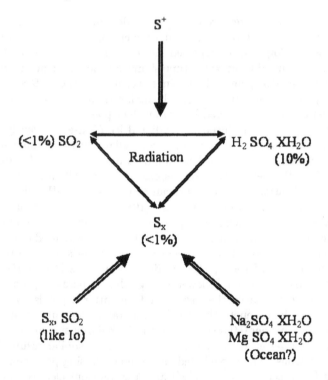

Figure 10   Sulfur cycle on Europa (Johnson, 2001).

The recent observation of peroxide on Europa gave a clearly identifiable radiolytic product (Carlson et al., 1999) This molecule, formed readily in the presence of a dilute sulfuric acid, may now be used as a kind of a dosimeter for surface radiation effects. Recently, Moore and Hudson(Moore and Hudson, 2000) showed that at the temperatures relevant to Europa an electron scavenger (like sulfuric acid) is needed to accumulate sufficient peroxide making the comparison with more distant icy objects more interesting and, possibly, allowing us to extract compositional information.

The exciting observation that the hydrated mineral bands seen by (McCord et al., 1999) are primarily radiolytically-formed, frozen, dilute sulfuric acid (Carlson et al., 1999b) was supported by the laboratory data on radiolysis of frozen sulfur/ water mixtures. Summarizing the available data, it was clear that the other radiolytic products, SO_2 and chain sulfur, are present. These are produced at the appropriate levels ($< 1\%$) to account for the visible reflectance in the chaos regions of

Europa and the depth of Europa's SO bands. Therefore, we proposed that, in the surface layer penetrated by the energetic ions and electrons, a radiolytic sulfur/ice cycle (Fig. 10) is maintained, with sulfur being cycled, rapidly on a geologic time scale, among three forms, independent of the original source of the surface sulfur. Therefore, a separate indicator is needed to determine the source of S at Europa. There is sufficient sulfur implanted to account for the observations(Johnson et al., 2002), however, in two recent papers (Johnson, 2000; Leblanc et al., 2002) it has been suggested that the radiolytic decomposition of $Na_2 SO_4 XH_2O \rightarrow Na + NaHSO_4 (X-1)H_2O \rightarrow 2 Na + H_2SO_4 (X-2)H_2O)$ could both initiate the sulfur cycle and produce the observed Na. Using this, the sodium observations at Europa were shown to be suggestive of an internal source of Na which apparently reaches the surface in the chaos areas, possibly salts or carbonates from Europa's underground ocean. CO_2 in ice is also seen, strengthening the picture that carbonates are being decomposed.

6.2.   SPUTTERING OF SATURN'S E-RING GRAINS

The role of radiation bombardment of icy surfaces is even more spectacular in Saturn's inner magnetosphere. There the small, bright icy moons of Saturn (Mimas, Enceladus, Tethys, Dione and Rhea) are also irradiated by the trapped plasma, as are the grains in the largest planetary ring in the solar system, the diffuse E-ring (Johnson et al., 1989). Whereas radiolytic surface chemistry is the dominant feature of the Jovian satellites, for the lower temperatures of the icy Saturnian satellites, sputtering of H_2O dominates, with smaller amounts of O_2 and H_2.

The presence of a giant, toroidal atmosphere of sputter-produced water-ice products at Saturn was expected. Remarkably, the OH component of that cloud was observed using Hubble Space Telescope. However, it was about thirty times more dense than predicted. In order to attempt to explain such densities we re-examined the sputtering of the icy surfaces, considering enhancements due to non-normal incidence, grain charging, and small grain size (discussed below), as well as a more careful extrapolation of the available data. We constructed the giant OH cloud by tracking the sputtered H_2O as it orbited Saturn, allowing for dissociation to form OH and then loss by ionization (Fig. 11). Using the best available data on the plasma and the most recent observations, we concluded that there must indeed be >10 times more surface available for sputtering than is seen optically, an exciting result (Jurac et al., 2001). This is most likely in the form of freshly produced small grains

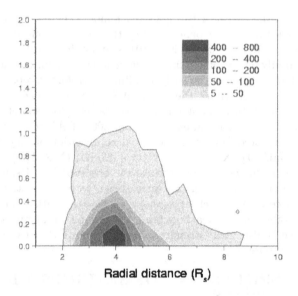

Figure 11   Neutral OH cloud for a large source at Enceladus (Jurac et al., 2001).

associated with Saturn's E-ring, either from activity on Enceladus or due to collisions of small, as yet unseen, objects. Cassini will have on it a good time of flight mass spectrometer capable of detecting sputtered molecular species in the plasma, so that for the first time in the outer solar system a secondary ion mass spectrometry (SIMS) experiment of the surfaces will be carried out, of the type performed routinely in the laboratory.

7.   APPLICATIONS: INTERSTELLAR MEDIUM

Dust in the interstellar medium (ISM) can exist in a variety of environments, from the hot intercloud medium to the cold neutral clouds. This dust contains most of the refractory material in the ISM and accounts for extinction, polarization, and scattering of light, as well as for certain absorption and emission features. The dust also plays an important role in determining the total energy balance, converting light to infrared emission and heating the clouds through photoelectron emission. It is estimated that $\sim 2/3$ of the grain mass in the ISM is tied up in grains with radius, a, less that $\sim 0.05\mu$m, accounting for the UV

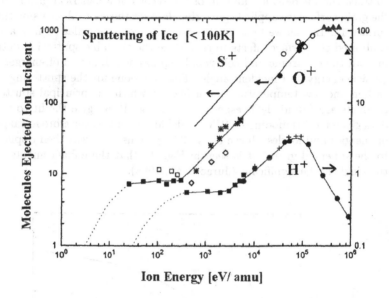

Figure 12 Sputtering yield of water ice from modified TRIM version (lines), calli-brated to experimental data shown as symbols

However, this is not a static environment since grains are being eroded (sputtered) by the hot gas atoms or the local plasma, by grain-grain collisions, and either directly or indirectly by the cosmic ray ions. This loss competes with grain growth. Here we review the data on the sputtering of small refractory or icy grains in a hot plasma formed in a shocked region as well as the sputtering of the icy mantles of larger grains by cosmic rays ions in molecular clouds. In Table II was shown typical energies of H^+ and He^{+z} in interstellar shocks. We first describe the enhanced sputtering of small grains by a plasma in the threshold region and we then consider spike models used for calculating the grain erosion.

7.1. ENHANCED SPUTTERING OF SMALL GRAINS

In the sputtering of small grains edge effects are dominant. It has been shown that weakly attached surface components are readily eroded and that the enhancement due to angular incidence is always important. In addition, energetic particles can penetrate the grain and sputter at the exit surface. Since the tracking of the energetic recoils in the grain is required, we created a MC BCA code by modifying the TRIM program to track the incident ion and all of the recoils in a spherical grain. At the grain surface we applied sputtering data for those recoils crossing the surface 12. In this we first test the modified code for a flat surface and calibrated the surface effects to reproduce the available sputtering data in both the electronic and collisional regimes as a function of energetic particle energy and incident angle. This was done for the sputtering of carbon and low temperature water ice, for which the principal sputter products are C and H_2O respectively. In Fig. 12 we give the fit to the data given by this modified TRIM model for water ice over a broad range of energies and angles (Jurac et al., 2001). This code was then applied to the spherical grain. It is seen in Fig. 13 that the enhancements in the yield can be significant (Jurac et al., 1998).

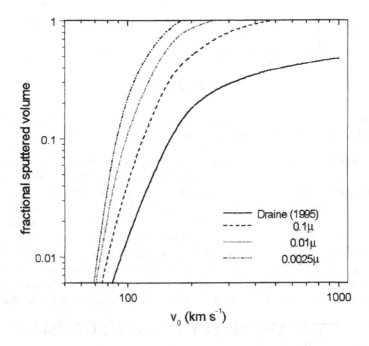

Figure 13 Fractional Sputtering yield from an interstellar shock with vellocity v_o for grains of radius $a = 0.1$, 0.01, and 0.0025 μm. The model from Draine (Draine, 1995) is also shown for comparisson.

7.2. "SPOT HEATING" AND "WHOLE GRAIN HEATING": TS MODELS

In the above, we neglected the total heat deposited, which is valid in many instances as both water and carbon are bound relatively tightly, but it clearly fails for very small grain size, large energy depositions and very volatile materials. Watson and Salpeter (WS) (Watson and Salpeter, 1972) proposed that small grains could be heated by incident cosmic ray (CR) bombardment. They evaluated the heating using a number of simplifying assumptions and calculated the evaporation rate of the grain material using a TS model. If heating and evaporation are localized in certain region of the grain they call it "spot heating", whereas if the energy deposited is spread nearly uniformly by the time sputtering occurs (as for small grains) they call it "whole grain heating". WS concluded that spot heating was negligible respect to whole grain heating because they consider only light CR ions. However, Léger, Jura and Omont (LJO) (Leger et al., 1985) showed that spot heating was dominant for volatile grains with radius $a > 0.25\mu m$ by CR Fe bombarding a CO ice mantle. Using the CR abundances and energy spectrum they estimated the evaporation rate for a CO mantle to be 70 molec/$(cm^2 s)$. Their analysis, however, relied on a number of weak hypotheses.

a) A hot region of 20-200 Å in radius is rapidly created in the solid. This may not be true for the 0.2 GeV/amu Fe they study because the size of the region where ionizations are created is of the order of few hundred Å's, but there is only ∼1 ionizations/Å along the path of the ion. Therefore the track is not continuous, the excitations are dilute and act independently, creating small regions, much hotter than expected if the energy is distributed in the whole track.

b) The temperature profile is assumed to have a fixed shape: first a Gaussian and then a rectangular profile. Because evaporation depends exponentially on temperature (see flux formula, Eq. 10) this affects the estimate.

c) In the case of a small region where the energy density is such that the mean energy per molecule is larger than the binding energy of those molecules, analytic spike models do not work, as described above.

d) Finite size effects are neglected. The collision cascade may not be fully contained within the grain as discussed above. The large pressure propagates radially so when it reaches the surface of a large grain it may be weak, but for a small grain it may produce spallation from the side.

Hasegawa and Herbst (HH) (Hasegawa and Herbst, 1993) pointed out that the grain goes through heating and cooling cycles. Assuming the mean temperature during heating is T_{eff}, the evaporation rate $\langle k_i \rangle$ for species i is:

$$\begin{aligned}
\langle k_i \rangle &\approx F\left(T_{eff}\right) k_i(T_{eff}, U_i) \approx 3.16 \ 10^{-19} \ k_i(70, U_i) \\
&= 3.16 \ 10^{-19} \ \nu_i \exp\left[-U_i / (k_B 70)\right]
\end{aligned} \quad (16)$$

They used $T_{eff} = 70$ K, and took the fraction of time the grain is hot as $F(T_{eff}) \approx t_{des}/t_{bombard}$, where t_{des} is the typical time for cooling due to desorbing species, $t_{des} \sim 10^{-5}$ s, and t_{des} is the typical time between CR bombardment by iron nuclei with sufficient dE/dx, $t_{des} \sim 3 \ 10^{13}$ s. $\nu(T_{eff})$ is an average desorption rate which depends on the binding energy of the species. This approach is convenient since it gives a general expression for the desorption rate which is only a function of the binding energy of the absorbed species, and it has been used in several papers. However, the approximation used for evaporation has the same problems mentioned above.

After a heavy CR passes through a large grain the following occurs (Bringa and Johnson, 2002b):

1) $10^{-17} - 10^{-13}$ s. Energy is deposited in a region with a radius of the order of 100 Å and is rapidly distributed by secondary electrons and hot recoils.

2) $10^{-13} - 10^{-11}$ s. Sputtering occurs due to kinetic energy transport following energetic energy release events giving Y_{col}. "Temperature" is not well defined during this non-equilibrium stage, but the kinetic energy of atoms can reach 10^4 K and hot ejecta can take away ~20% of the deposited energy.

3) $10^{-11} - 10^{-9}$ s. Early thermal sputtering, $Y_{th-early}$, occurs following decay of the energetic recoil cascades and due to lower energy, energy-release events. This comes mainly from an extended radial region around the ion track. The temperature profile in the grain is not uniform and temperature near the track can be few hundred K. Ejecta are colder and take away ~5% of the deposited energy.

4) $10^{-9} - 10^{-5}$ s. Temperature in the grain is roughly uniform and decreases gradually due to evaporative cooling giving $Y_{th-late}$.

For a small grain, the temperature may already be roughly uniform by 10^{-10} s, but will still be higher than ~100 K. Due to size effects the energy taken away in step 2 is larger than 20%. Radiative cooling may be neglected compared to evaporative cooling in all these stages for grains of radius $a \ll 0.1\mu$m. The total sputtering yield is:

$$Y = Y_{col} + Y_{th-early} + Y_{th-late} \quad (17)$$

LJO neglected Y_{col} and treat $Y_{th-early}$ roughly. HH neglect both Y_{col} and $Y_{th-early}$. The key point is that late thermal ejection can occur over relatively long times, giving a large contribution to the yield. However, the first two contributions may be as important as or even dominate the last. Taking this into account gives results in the high energy density regime within factors 2-10 of the LJO or HH estimates. In a recent paper the results from MD calculations and the extrapolation of the available laboratory data were used to estimate the fraction of CO and H_2O that will be seen in the gas phase in a low temperature molecular cloud due to desorption from grains by GeV CR-iron ions (Bringa and Johnson, 2002b). Fig. 14 shows snapshots from MD simulations of a grain hit by a cosmic ray. These simulations allow one to calculate $(Y_{col} + Y_{th-early})$, while $Y_{th-late}$ can be calculated following simple evaporation models. Although a number of the approximations in LJO and HH were incorrect, the ratios of gas-phase to condensed phase CO did not change considerably due to cancellation of errors.

8. SUMMARY

Energetic ions, electrons and photons in many regions of our solar system, in young stellar objects and in the interstellar medium impact and modify the imbedded surfaces. This leads to sputtering (desorption) as well as chemical alteration of the surfaces. This has been shown to be critical for understanding the reflectance spectra of the Galilean satellites of Jupiter (Johnson et al., 2002). X-ray-induced desorption may maintain gas phase volatiles in young stellar objects (Najita et al., 2001), heavy cosmic-ray ions (Bringa and Johnson, 2002b) and Lyman-α photons produced by cosmic rays protons may be responsible for maintaining the volatiles observed in the gas phase in cold molecular clouds, and energetic ions and electrons maintain a giant toroidal atmosphere in Saturn's inner magnetosphere 12.

In this chapter we briefly described the effect of radiation on materials in space. We primarily reviewed data for sputtering and, more importantly, the models needed to extrapolate that data or apply it to materials for which there is no data. It is important to remember that sputtering and the chemical changes induced in the material are intimately connected. First, reactions may be induced by implanted ions, as is the case for protons bombarding the lunar surface or sulfur ions bombarding Europa's surface. Second, the most volatile species in the material or the most volatile species formed by the radiation are preferentially ejected. This changes the composition of the material in the surface layer. This effect is striking demonstrated on Jupiter's moons.

Figure 14 Snapshots from MD simulations of CR-induced sputtering from a "CO-like" grain of radius $a = 0.013\mu$m. Upper pannel shows heating with a track of 50 Åradius, and lower pannel shows whole grain heating. Only two central slices are shown, few ps after the bombardment. Both cases assume a total energy deposition of 8 keV, consistent with a Fe CR of ≈ 2 GeV.

At Europa, oxygen, sodium and potassium atmospheres are maintained by decomposition of the surface (Johnson et al., 2002) and at Callisto a CO_2 atmosphere is likely to be due to decomposition of organics in the surface (Carlson, 1999c; Johnson, 2001). These surfaces have also been seen to contain trapped volatiles and refractory absorbing species produced by the radiation decomposition of the surface material. In the coming years, exciting spacecraft missions such as the Cassini mission and new space telescope data will considerably expand the observational data base. Understanding this data base will require a much better understanding of sputtering, radiolysis and photolysis for a much broader range of materials.

Acknowledgments

The work at Virginia was supported by the NSF Astronomy Divison and the Origins Program and NASA Headquarters. The work of E.M.B. was partially performed under the auspices of the U. S. Department of Energy by the University of California, Lawrence Livermore National Laboratory under Contract No. W-7405-Eng-48.

REFERENCES

Abraham, F.F., Broughton, J., Bernstein, N. and Kaxiras, E. (1999), *Phys. Rev. B*, **60**, 2391.
Allen, M.P. and Tildesley, D.J. (1987), *Computer Simulation of Liquids*, Clarendon Press, 1987.
Andersen, H.H. et al. (1998), *Phys. Rev. Lett.*, **80**, 5433.
Bagenal, F. (1994), *J. Geophys. Res.*, **99**, 11043.
Berger, M.J. and Paul, H. (1995), in ref. (IAEA-TECDOC799), p.415.
Bethe, H.A. (1930), *Ann. Phys.*, **5**, 325.
Betz, G. and Wien, K. (1994), *Int. J. Mass Spec. and Ion Proc.*, **140**, 1.
Biersack, J.P. and Eckstein, W.D. (1984), *Apply. Phys. A*, **34**, 73. (The latest version of "TRIM", SRIM 2000, can be downloaded for free in J.F. Ziegler's web site: http://www.srim.org/).
Biersack, J.P. and Haggmark, L.G. (1980), *Nucl. Instr. and Meth.*, **174**, 257.
Bloch, F. (1933), *Z. Phys.*, **81**, 363.
Bringa, E.M. and Johnson, R.E. (2000), *Surf. Sci.*, **451**, 108.
Bringa, E.M. and Johnson, R.E. (2001), *Nuc. Inst. Met. B*, textbf180, 99.
Bringa, E.M. and Johnson, R.E. (2002), *Phys. Rev. Lett.*, **88**, 165501.
Bringa, E.M. and Johnson, R.E. (2002b), *ApJ*, accepted. *Nucl. Instr. and Meth. in Phys. Res. B*, **193/194**, 370.
Bringa, E.M., Johnson, R.E. and Dutkiewicz, L. (1999a), *Nuc. Inst. Met. B*, **152**, 267.
Bringa, E.M., Johnson, R.E. and Jakas, M.M. (1999), *Phys. Rev. B*, **60**, 15107.
Bringa, E.M., Papaleo, R. and Johnson, R.E. (2002a), *Phys. Rev. B*, **65**, 094113.
Brown, W.L. et al. (1984), *Nucl. Instr. and Meth. B*, **1**, 307.
Calvin, W.M. (1999), *Planetary Report*, **XIX**, March/April, 8.
Calvin, W.M., Johnson, R.E. and Spencer, J.A. (1996), *Geophys. Res. Lett.*, **23**, 673.
Carlson, R.W., *Science*, **283**, 820.
Carlson, R.W. et al., *Science*, **283**, 2062.
Carlson, R.W., Johnson R.E. and Anderson, M.S. (1999), *Science*, **286**, 97.
CASP - Schiwietz, G. and Grande, P.L. (n.d.), CASP is available at the web site http://www.hmi.de/people/schiwietz/casp.html
Claussen, C. (1982), *Nucl. Instr. and Meth.*, **194**, 567.
Cobut V. et al. (1995), *Radiat. Phys. Chem.*, **51**, 229. Frongillo, Y. et al. (1998), *Radiat. Phys. Chem.*, **51**, 245.

Cooper, J.F., Johnson, R.E., Mauk, B.H., Garrett, H.B. and Gehrels, N. (2001), *Icarus*, **149**, 133.

David, D.E. and Michl, J. (1989), *Prog. Sol. Stat. Chem.*, **19**, 283.

Draine, B.T. (1995), *Astrophys. Space Sci.*, **233**, 111.

Dutkiewicz, Ł., Pedrys, R., Schou, J. and Kremer, K. (1994), *Europhysics Let.*, **27**, 323. Dutkiewicz, Ł., Pedrys R. and Schou, J. (1996), *Europhys. Lett.*, **36**, 301.

Dutkiewicz, Ł., Pedrys, R., Schou J. and Kremer, K. (1995), *Phys. Rev. Lett.*, **75**, 1407.

Eckstein, W.D. (1991), *Computer Simulations of Ion-Solid interaction*, Springer Verlag, Berlin.

Ercolesi, F. (n.d.), MD primer:
    http://www.sissa.it/furio/md/md/md.html. HTML or Postcript format.

Famá, M., Bahr, D.A., Teolis, B.D. and Baragiola, R.A. (2002), *Nucl. Instr. Methods. Phys. Res. B*, **193**, 775.

Fenyö, D. and Johnson, R.E. (1992), *Phys. Rev. B*, **46**, 5090.

Firsov, O.B. (1958), *Sov. Phys. JETP*, **6**, 534.

Fleischer, R.L., Price, P.B., Walker, R.M. and Hubbard, E.L. (1967), *Phys. Rev.*, **156**, 353.

Gibbs, K.M., Brown, W.L., Johnson, R.E. (1988), *Phys. Rev. B*, **38**, 11001.

Hasegawa, T. and Herbst, E. (1993), *MNRAS*, **261**, 83.

Heide, H.G. (1984), *Ultramicroscopy*, **14**, 271.

IAEA-TECDOC-799 (1995), *Atomic and Molecular data for radiotherapy and radiation research*, IAEA, Vienna.

Insepov, Z. and Yamada I. (1997), *Nucl. Instr. and Meth. B*, **121**, 44.

Jakas, M.M. (2000), *Rad. Effs. and Deffs. in Solids*, **152**, 157.

Jakas, M.M., Bringa, E.M. and Johnson, R.E. (2002), *Phys. Rev. B*, **65**, 165425.

Jakas, M.M. and Harrison, D. (1985), *Phys Rev. Lett.*, **55**, 1782.

Johnson, R.E. (1989), *J. de Physique C2*, **50**, 251.

Johnson, R.E. (1990), *Energetic Charged-Particle Interactions with Atmospheres and Surface*, Springer-Verlag, Berlin.

Johnson, R.E. (1993), in Baragiola, R.A. (ed.), *Ionization of Solids by heavy particles*, Plenum Press, p.419.

Johnson, R.E. (2000), *Icarus*, **143**, 429.

Johnson, R.E. (2001), in Dessler, R. (ed.), *Chemical Dynamics in Extreme Environments*, Adv. Ser. In Phys. Chem., World Scientific, Singapore, 11, Chap.8, p.390.

Johnson, R.E. et al. (1989), *Icarus*, **77**, 311.

Johnson, R.E. et al. (2002), in Bagenal, F. (ed.), *Jupiter: Planet, Satellites & Magnetosphere*, Univ. Arizona Press, Tucson.

Johnson, R.E. and Evatt, R. (1980), *Rad. Eff.*, **52**, 187.

Johnson R.E. and Liu, M. (1996), *J. Chem. Phys.*, **104**, 6041.

Johnson, R.E., Pirronello, V., Sundqvist, B.U.R., Donn, B. (1991), *ApJ*, **379**, L75.

Johnson, R.E. and Quickenden, T.I. (1997), *J. Geophys. Res.*, **102**, 10985.

Johnson, R.E. and Schou, J. (1993), in Sigmund, P. (ed.), *Fundamental Processes in The Sputtering of Atoms and Molecules*, Mat-fys. Medd. Dan. Vid. Selsk, **43**, Mem. Roy. Dan. Soc., Copenhagen, p.403.

Jurac, S., Johnson, R.E. and Donn, B. (1998), *ApJ*, **503**, 247.

Jurac, S., Johnson, R.E., Richardson, J.D. (2001), *Icarus*, **149**, 384.

Kelly, R. (1990), *Nucl. Instr. and Meth. B*, **46**, 441.

Khurana, K.K. et al. (1998), *Nature*, **395**, 777.

Kivelson, M.G. et al. (1996), *Nature*, **384**, 537.

Leblanc, F., Johnson, R.E. and Brown, W.L. (2002), *Icarus*, in press.

Leger, A., Jura, M. and Omont, A. (1985), *A&A*, **144**, 147.

Lesueur, D. and Dunlop, A. (1993), *Rad. Eff. and Def. in Solids*, **126**, 163.

Lindhard, J., Scharff, M. and Schiøtt, H.E. (1963), *K. Dan. Vidensk. Selsk. Mat. Fys. Medd.*, **33**, 14.

Madey, T.E., Johnson, R.E. and Orlando, T.M. (2002), *Surf. Sci.*, **500**, 838.

MARLOWE - M.T. Robinson's MARLOWE web page:
http://www.ssd.ornl.gov/Programs/MARLOWE/MARLOWE.html.

McCord, T.B. et al. (1999), *J. Geophys. Res.*, **104**, 11827.

Moore, M.H. and Hudson, R.L. (2000), *Icarus*, **145**, 282.

Movies: A number of results, including movies of MD simulations can be found at:
http://dirac.ms.virginia.edu/~emb3t/research/research.html.

Najita, J., Bergin, E.A., Ullom, J.N. (2001), *ApJ*, **561**, 880.

Noll, K.S. et al. (1996), *Science*, **273**, 341.

Paretzke, H.G., Goodhead, D.T., Kaplan, I.G. and Terrisol, M. (1995), in ref. (IAEA-TECDOC799), p.633.

Pedrys, R. (1990), *Nucl. Instr. and Meth. B*, **48**, 525.

Pedrys, R. et al. (2000), *Nucl. Instr. and Meth. B*, **164-165**, 861.

Rapaport, D.C. (1995), *The art of molecular dynamics simulation*, Cambridge Univ. Press.
http://www.cup.cam.ac.uk/
onlinepubs/ArtMolecular/ArtMoleculartop.html

Reimann, C.T. et al. (1984), *Surf. Sci.*, **147**, 227.

Ritchie, R.H. and Claussen, C. (1982), *Nucl. Instr. and Meth.*, **198**, 133.

Ryazanov, A.I., Volkov, A.E., Klaumünzer, S. (1995), *Phys. Rev. B*, **51**, 12107.

Sanche, L., Märk, T.D. and Hatano, Y. (1995),
in ref. (IAEA-TECDOC799), p.277. Herman, Z., Märk, T.D. and Sanche, L., same book, p.371.

Seiberling, L.E., Griffith, J.E. and Tombrello, T.A. (1980), *Rad. Eff.*, **52**, 201.

Shapiro, M.H. (1997), *Rad. Eff. and Defects in Solids*, **142**, 259.

Sieger, M.T., Simpson, W.C. and Orlando, T.M. (1997), *Phys Rev. B*, **56**, 4925.

Sigmund, P. (1969), *Phys. Rev.*, **184**, 383.

Sigmund, P. (1981), in Behrish, R. (ed.), *Sputtering by Particle Bombardment*, Top. Appl. Phys. **47**, Springer, Heidelberg, p.9.

Sigmund, P. and Claussen, C. (1981), *J. Appl. Phys.*, **52**, 990.

Sigmund, P. et al. (1989), *Nucl. Instr. and Meth. B*, **36**, 110.

Smith, R. ed. (1997), *Atomic & ion collision in solids and at surfaces*, Cambridge University Press.

Thompson, M.W. (1968), *Philos. Mag.*, **18**, 377.

Toulemonde, M., Dufour, Ch., Meftah, A. and Paumier, E., presented at the International Conference on Atomic Collisions in Solids, ICACS 18, Odense, Denmark.

Urbassek, H.M. (1987), *Nucl. Instr. and Meth. B*, **22**, 480.

Urbassek, H.M. (1997), *Rad. Effects and Defects*, **142**, 439.

Urbassek, H.M. and Michl, J. (1997), *Nucl. Instr. and Meth. B*, **122**, 427.

Watson, C. and Tombrello, T.A. (1985), *Rad. Eff.*, **89**, 263.

Watson, W.D. and Salpeter, E.E. (1972), *ApJ*, **334**, 771.

Zhigilei, L.V. and Garrison, B.J. (1999), in Diaz de la Rubia, T. et al. (eds.), *Multiscale Modelling of Materials*, Mat. Res. Soc. Symp. Proc., **538**, 491.

Ziegler, J.P., Biersack, J.P. and Littmark, U. (1985), *The stopping and range of ions in solids*, Pergamon Press, New York, U.S.A.

TRANS-NEPTUNIAN OBJECTS

Maria Antonietta Barucci and Jennifer Romon
DESPA, Observatoire de Paris, 5, place Jules Janssen 92195 Meudon Principal Cedex, France

Keywords: Solar System – minor bodies – TNOs – circumstellar disks

Abstract The presence, the physical and dynamical properties, the chemical composition and the possible origin of Trans Neptunians small bodies are reviewd. The enrichment of dust by collisions among TNOs in the Kuiper Belt is also discussed.

Introduction

The study of the structure and the history of our Solar System has been a central objective of astronomy over the last three centuries. Only until a few years ago, our Solar System was known to have nine planets on orbits ranging from 0.4 to 39 AU, numerous planetary satellites, the asteroid belt, located mainly between Mars and Jupiter, and the comets. The outer Solar System appeared to be empty, inhabited by the four giant planets, Pluto, and a small population of transient comets. After the discovery of Pluto in 1930, many astronomers tried to discover the planet X, even if they were thinking that it was too faint and too far to be discovered with the available telescopes. In the middle of this century, Edgeworth (1943, 1949) and Kuiper (1951) speculated that our Solar System could be surrounded by a disk of material left over from the formation of the planets. They suggested that the density of the Solar Nebula beyond Neptune was too small to have accreted a major planet, but the region could be populated by planetesimals. They speculated also that bodies could migrate inward from this region and they could be an alternative source of short-period comets (SPCs). Oort (1950) proved that long-period comets (LPCs) come from a spherical cloud (now known as the Oort Cloud) occupying the region from roughly 10,000 to 200,000 AU from the Sun. He assumed that SPCs were derived from LPCs,

V. Pirronello et al. (eds.), Solid State Astrochemistry, 395–416.

perhaps captured by giant planets. This assumption was cast in doubt by Whipple (1964) who also proposed a source of SPCs beyond Neptune. Simulations by Fernández (1980), Duncan et al. (1988), and Quinn et al. (1990), taking into account the low inclinations of Jupiter-family comets (JFCs), showed that their reservoir should be a belt or a disk just beyond Neptune.

The search for new objects in the outer solar system was impeded by their faintness. Kowal (1989), using the Palomar Schmidt, conducted photographic survey of about 6000 square degrees near the ecliptic, up to red magnitude 19.5, without finding any object. The high quantum efficiency of charge-coupled device (CCD) detectors was the turning point for this research, even if the first CCD surveys were negative (Luu & Jewitt 1988, Levison & Duncan 1990, Cochran et al. 1991, Tyson et al. 1992). On 1992 August 30, the first Trans-Neptunian object (named 1992 QB$_1$) was discovered by Jewitt & Luu (1993) using a 2048 × 2048 CCD on the University of Hawaii 2.2 m telescope at Mauna Kea Observatory. At this point, intensive surveys started and increased the number of discovered objects very rapidly. The discovery of these new objects and their study exploded in the last half decade. More than 300 bodies have been discovered so far, and more than 100,000 objects orbiting between 30 and 50 AU with diameter larger than 100 km have been estimated to exist on the base of sky survey statistics (Jewitt 1999).

We have now the observational evidence that our outer Solar System contains a vast population of icy bodies. These objects do not have a well defined name yet: some call them Trans-Neptunian objects (TNOs), some call them Kuiper Belt objects (KBOs), some others Edgeworth-Kuiper objects. These objects, so far from the Sun, contain the least processed material accessible to direct investigation and, as fossils of the protoplanetary disk, they can provide important information on the processes operating in the young Solar System. In this paper we give a review on what we know about the Kuiper Belt, its possible relationship with other small bodies in the Solar System, and the constraints it could give us about the other planetary systems.

1.   DETECTION TECHNIQUES OF TNOS

During the last year, more than 100 TNOs were discovered. Discovery rates since 1993 are shown on Figure 1. These numerous discoveries are the result of many specific observing programmes. The usual method to detect the TNOs is to acquire CCD images of a wide field over a time span of a few hours. Due to the Earth's revolution about the Sun, the TNOs are moving slowly (Fig. 2). Automatic methods are used

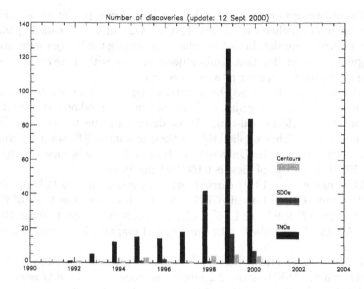

Figure 1 Discovery rates in the Trans-Neptunian population, Centaurs (see section 5), and scattered objects (SDOs, see section 3).

to detect these moving objects. Two points are critical regarding the efficiency of this method: the width of the field (a wider field allows to detect more objects) and the time span (a large time span, of a few days, allows to detect slower objects, i.e. farther ones).

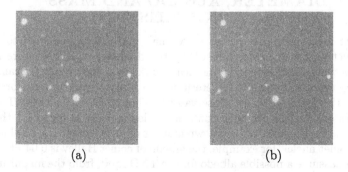

Figure 2 Images of TNO 1997 CU$_{29}$, taken at NTT (ESO, Chile). The time span between images (a) and (b) is about 1h10, i.e the object moved of 3.5″. The object is located at the center of both images.

Several groups are involved in Kuiper Belt survey programmes. There are different methods to detect TNOs. One of them, called 'pencil-beam' survey, consists in shifting and recombining the images using given angular rate and direction. Only objects moving with these angular rate and direction will appear as a point source.

Trujillo & Jewitt (1998) have developped a semi-automatic method to search for moving objects. Such automatic methods are required in order to perform real-time object detection, due to the high data collection rate. They applied this method to a large KBOs survey which was carried out at the University of Hawaii 2.2 m telescope (Jewitt et al. 1998). The field of view is 0.09 deg^2 per image.

Gladman et al. (1998) carried out observations at the Palomar 5 m and Canada-France-Hawaii (CFHT) 3.5 m telescopes. The fields of view used were 9.7' × 9.7' and 30' × 30', respectively. Objects from 10 to 100 AU can be searched. The limiting red magnitude reached is about 25 − 26.

Delsanti et al. (1999) began an observing programme with the Wide Field Imager (WFI) at the 2.2 meter telescope in La Silla Observatory (ESO, Chile). The objective of this survey is to be sensitive up to 300 AU, by observing the same field over a time span of three nights. Several adjacent fields of view of 34' × 33' allow to cover 1.25 deg^2 per night, and the upper limit magnitude is about 24. The objects are semi-automatically detected by their linear, slow motion. This survey has already allowed to discover about 30 objects.

To have a well determined orbit, many observations are at least necessary at three different oppositions.

2.   DIAMETER, ALBEDO AND MASS (POPULATION STATISTICS)

The brightness of a planet or a small body is proportional to the product of the geometric albedo by the diameter. As TNOs are not resolved and their albedo is practically unknown, their diameter cannot be determined. Attempts to determine the albedos have been performed using the Infrared Space Observatory (Thomas et al. 2000). These results are critical but they confirm the low albedo assumed for these objects. On the basis of the known albedos of few Centaurs (see Table 1) and comet nuclei (for example, the albedo of comet Halley is 0.04±0.01), we can assume a possible albedo for the TNOs and, from the magnitude, their diameters can be estimated using the equation:

$$p_R \cdot D^2 \cdot \Phi(\alpha) = 9 \cdot 10^{16} r^2 \cdot \Delta^2 \cdot 10^{0.4(m - m_R)}$$

where p_R is the geometric albedo, D the diameter (km), $\Phi(\alpha)$ the phase, m the apparent magnitude of the Sun, m_R the R magnitude of the object, r the heliocentric distance (in AU), and Δ the geocentric distance (in AU).

Assuming an albedo of 0.05, the diameter of these objects ranges between 740 km and 20 km for the fainter objects. The knowledge of the diameter distribution for the TNOs, as well as the total number of objects, is essential to estimate the mass of the population. On the basis of statistical surveys of the sky from ground based observations, it is possible to compute the Cumulative Luminosity Function (CLF) of the Trans-Neptunian population. The CLF is the number of TNOs per unit area of sky brighter than a given limit magnitude, measured as a function of the magnitude (Fig. 3). The CLF is a very important quantity because it reflects both the size distribution and the radial distance distribution in magnitude limit surveys. Fitting the surface density with the power law:

$$\log\left[S(m_R)\right] = \alpha(m_R - m_O)$$

where m_O is the red magnitude at which the surface density S reaches $1\,\mathrm{deg}^{-2}$. The CLF is well defined in the range of magnitude $20-26$, using only data obtained with CCD. Hubble Space Telescope measurements (Cochran et al. 1995) are available for fainter objects (28th magnitude), but these data are controversial. Least square fits to the CLF (Luu & Jewitt 1998) give $\alpha = 0.54 \pm 0.04$ and $m_O = 23.20 \pm 0.10$. Using a different method and different subsets of data, including HST measurements, Gladman et al. (1998) found $\alpha = 0.76 \pm 0.11$. Using the size distribution, Luu & Jewitt (1998) estimated the total mass at about $0.1 M_\oplus$ between 30 and 50 AU. This mass includes about 100,000 objects larger than 100 km. An extrapolation to small sizes indicates that there would be 10^{11} TNOs larger than 1 km diameter, whereas the extrapolation to large sizes seems to allow the existence of Pluto-size objects not yet discovered.

The surveys are generally focused to the ecliptical plane; the mean inclination observed is about 10° which has to be considered as a lower limit to the intrinsic inclination distribution due to the observational bias. In fact, objects of high orbital inclination spend a large fraction of each orbit far from the ecliptic and, therefore, are difficult to be detected. However most of the TNOs are so small that they will be impossible to be discovered by classical means. Only occultation detection can avoid this difficulty. In the next years, occultation programmes should provide a good prediction of the population (Roques & Moncuquet 2000). High speed photometry of a few well chosen stars would allow to better know

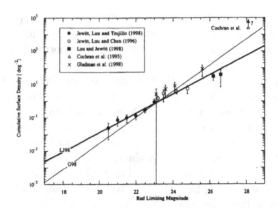

Figure 3 The plotted lines are the fitted cumulative luminosity functions (CLF), from Jewitt (1999).

the population of very small objects (e.g. objects of about 100 m radius at 40 AU). These observations will not allow to discover the occulting object because the orbit will remain unknown, but it will help to explore the population as a whole.

The Trans-Neptunian population which is known today contains too small mass. The surface mass density of the Solar System is known up to Neptune. If we extrapolate the surface mass density with a power law, the mass of the solar system should be $25M_\oplus$ between 30 and 50 AU (Fig. 4), which is much larger than the estimated one. Moreover, a minimum mass of $10M_\oplus$ is needed to explain Pluto-size objects growth. Probably more than 99% of the initial mass has been lost. Scattering by Neptune may have cleared part of the population, especially in the region close to the planet (Holman & Wisdom 1993). Collisions can have modelled the population especially inside 50 AU. Models of the collisional evolution of a massive Kuiper belt suggest that 90% of the mass inside 50 AU could have been lost due to collisions over the age of the Solar System (Stern & Colwell 1997, and Davis & Farinella 1998). The real loss mechanisms remain unidentified indeed.

3.   DYNAMICAL PROPERTIES

Three dynamical classes of Trans-Neptunian Objects can be defined:

The resonant objects: About 12% of the known TNOs are trapped in the 3:2 mean-motion resonance with Neptune (see Fig. 5). They have a semi-major axis of about 39.4 AU, large eccentricities and

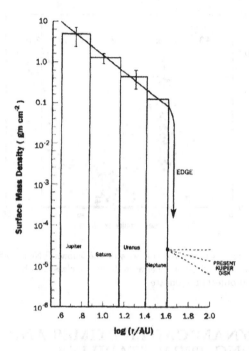

Figure 4 Non-volatile surface mass density of the Solar System vs. the heliocentric distance. The mass of each giant planet is spread out into a ring at its heliocentric distance (from Stern 1996a).

inclinations (up to 0.34 and 20° respectively). As these dynamical properties are the same as those of Pluto, the objects trapped in the 3:2 resonance are called Plutinos. A few objects are trapped in other resonances (2:1, 4:3, and 5:3).

The scattered disk: About 27 TNOs belong to the so-called scattered disk. These objects have very elongated orbits (0.34 < e < 0.94). To date, the discovered objects have a semi-major axis ranging from 51 to 675 AU.

The classical objects : The classical Kuiper Belt includes the objects between 41 and 47 AU. Most of them have nearly circular and low inclined orbits.

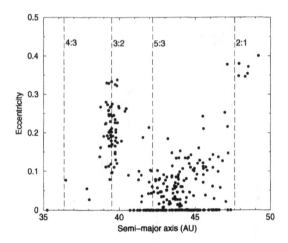

Figure 5 Eccentricity vs. semi-major axis of the known TNOs. The Plutinos are clearly shown near the 3:2 resonance, as well as the classical objects from 41 to 47 AU. The scattered objects are omitted.

3.1.   DYNAMICAL LIFETIMES AND LONG-TERM STABILITY

Duncan et al. (1995) proposed numerical simulations to study the long-term stability of TNOs. They calculated the dynamical lifetimes for 1300 low inclined orbits ($i = 1°$). Duncan et al. (1995) showed (Figure 1 of their paper) the stable and unstable zones they have found: for circular and low inclined initial orbits, there is a stable area between 36 and 40 AU, and complete stability beyond 42 AU; for more eccentric orbits and still low inclinations, the region before 42 AU is very unstable, except near resonances with Neptune.

The dynamical structure of the resonance 3:2 has been thoroughly studied. Morbidelli (1997), on the basis of numerical simulations of the diffusion speed, pointed out that instability timescales near the 3:2 resonance range from less than 1 Myr to more than 4 Gyr. The boundaries of the resonance are particularly unstable. Close to them, interactions with other resonances could generate dynamical chaos which would explain these instabilities.

Several studies about the orbital stability of bodies orbiting between the giant planets have also been carried out. For instance, Gladman & Duncan (1990) found that their dynamical lifetimes range from 10^5 to 10^7 years.

3.2. THE ORIGIN OF PLUTINOS: RESONANCE SWEEPING HYPOTHESIS

In order to explain the dynamical properties of the Plutinos, and particularly their great abundance, Malhotra (1995) suggested that an outward migration of Neptune's orbit could provide an efficient mechanism to sweep up many TNOs into Neptune's resonances. The following mechanism was proposed. During the last stages of planetary formation, a part of the residual icy planetesimals would have been ejected outward in order to form the Oort Cloud. These planetesimals would have been scattered by the giant planets. Angular momentum of the planets could have been changed by the back reaction of the planetesimal scattering. But the energy balance should differ from one planet to another. As the massive Jupiter provided all the energy to scatter the planetesimals, its orbital energy globally decreased, whereas Saturn, Uranus, and Neptune underwent an increase of their orbital energy. As a result their orbits expanded, while Jupiter's orbit decayed. Hahn & Malhotra (1999) performed numerical simulations of the migration of planets embedded in a massive debris disk, but they were unable to estimate accurately their magnitude and timescale. By analysing the orbital evolution of TNOs trapped in Neptune's resonances, Neptune orbital migration could be estimated. Indeed the eccentricity of a captured object increases, and the dispersion of the eccentricity depends on the magnitude of the orbital migration. From the observed eccentricities of the Plutinos, it has been infered that Neptune would have moved outward of about 9 AU. Hahn & Malhotra also found out that a substantial migration of Neptune over 10^7 years would require a debris disk of $30-100$ $M_\oplus$. However, many difficulties remain to fully explain the dynamical properties of the Kuiper Belt. For instance, the absence of observed TNOs between 35 and 39 AU - whereas this zone should be stable, according to Duncan et al. (1995) - could be explained by the outward migration of the 3:2 resonance that cleaned this region. But, in that case, the 2:1 resonance should have moved toward 47 AU and cleaned this region in the same way. So why do we observe objects there ?

4. COLLISIONAL PROCESSES AMONG THE TNOS

4.1. MODELS OF COLLISIONS

First studies on the collision rates among the TNOs pointed out that the collisions are an important mechanism to shape the belt (Stern 1995). Collisional evolution could even be dominant beyond 42 AU (at least until 60 AU) compared to dynamical evolution, as dynamical lifetimes in this region are longer than the age of the Solar System. Different authors (Stern 1995, Farinella & Davis 1997) found out that the large TNOs (diameter > 100km) are not much altered by collisional processes. The impacts are not energetic enough to disrupt them. However the energy could be sufficient to convert these large objects into rubble piles. The collisions undergone by the smaller objects (diameters < 50 − 100 km) would occur much more frequently. As a result, the size distribution of the small bodies would be close to equilibrium $(dN(r) \propto r^{-3.5})$ and would correspond to a collision cascade. Davis & Farinella (1997) estimated that about ten fragments (sizes ranging from 1 to 10 km) are produced every year near 40 AU. A few of them could be injected into strongly unstable regions and, escaping from the Trans-Neptunian region, could supply the short-period comets (Morbidelli 1997). Thus the SPCs could be collisional fragments, and they could have been altered in the interior of their parent bodies (compacting effect, heating). If this scenario is correct, the SPCs would not be primitive. Durda and Stern (2000) showed by numerical simulation that the TNOs surface is not primordial, but have been reworked by collisions. They showed that in the region 30-50 AU many craters can be produced on the surface over a period of 3.5 Gyr.

4.2.   DUST PRODUCTION

On the basis of his model of collisions in the Kuiper Belt, Stern (1996b) demonstrated that the collisions among the small TNOs are erosive, whereas the collisions with large objects can also result in accretion depending on their orbital parameters and their surface strength. He estimated the debris production rates as 3×10^{16} to 10^{19} g/yr. As in the case of the main asteroid belt, the mass distribution of these debris should follow a power law, and range from large fragments down to fine dust. Stern (1996b) also suggested that signatures of this dust production should be observable. He highlighted two types of detectable IR signatures:

- An IR background due to the thermal emission of fine dust;
- Local brightness variations of the background due to discrete dust clouds (resulting from recent collisions).

This population may also contribute to the Zodiacal Cloud[1], as proposed by Flynn (1996). Liou et al. (1996) studied the transport of the small Kuiper Belt dust particles (1 to 9 μm in diameter) to the inner Solar System. They considered the gravitational interaction due to the Sun and the planets, as well as solar radiation pressure, Poynting-Robertson and solar wind drag. They concluded that 20% of these dust grains could reach the Earth. But, collisions with interstellar grains could destroy a significant part of these dust particles, and thus reduce the contribution of the Kuiper Belt to the Zodiacal Cloud.

If the Kuiper Belt is a significant source to the Zodiacal Cloud, the KB grains might be collected in the Earth's stratosphere, at the same time as other IDPs (Interplanetary Dust Particles). Liou et al. (1996) pointed out that, when they reach the Earth, the KB grains have low velocities (due to low inclinations and eccentricities): this grain flux to the Earth is enhanced, and they have more chances to survive the atmospheric entry heating. It is much more difficult for cometary grains to reach the Earth without damage, due to their high relative velocity. Moreover, Flynn (1996) suggested that these grains could be distinguished from other IDPs as the duration of their space exposure is much longer than those of asteroidal or cometary IDPs.

5.   RELATIONSHIPS WITH OTHER SMALL BODY POPULATIONS

In 1950, Jan Oort, analysing the orbital parameters of some comets, suggested the existence of a reservoir of comets very far from the Sun. But this reservoir, the Oort Cloud, could not explain the dynamical characteristics of all the observed comets. In the 80's, numerical simulations confirmed that most short-period comets could not come from the Oort Cloud, but from a belt-like source (Duncan et al. 1988, Quinn et al. 1990). Levison & Duncan (1997) presented the numerical simulation of 2000 massless particles evolving from Neptune-encountering orbits for 1 Gyr. Their main result was that 30 % of these objects became visible comets (i.e. reached perihelion distance $q < 2.5$ AU) with orbital parameters typical of Jupiter-family comets (JFCs, periods < 20 years) and with a median transfer time of less than 100 Myr. The belt does not produce many Halley-type comets (periods between 20 and 200 years, high eccentricities, and high inclinations), which most likely come from the Oort cloud. They also found that about 5% of the population remain in the Kuiper Belt after the 1 Gyr-integration; these objects should have high eccentricities and could form a significant scattered population. This scenario would explain the origin of the short-period comets, as well as the existence of a scattered disk, as a result of planetary perturbations. The scattered objects we observe nowadays would be the signature of this scattering process which supplies the Jupiter-family comets.

Another scenario suggests that the scattered disk could be the remnant of a massive scattered disk of planetesimals resulting from Neptune's formation. Duncan & Levison (1997) estimated that, if it now contains 6×10^8 objects, the scattered disk could be the only source of Jupiter-family comets, without any contribution from the classical Kuiper Belt. They have also pointed out that 1% of the population remains in the scattered disk after a 4 Gyr-integration. As a result, the initial scattered disk should have contained about 6×10^{10} comets in order to supply the observed JFCs flux. Collisional processes inside the Kuiper Belt could also be suggested to explain the origin of the short-period comets (see section 4).

In addition to the short-period comets, two other populations could be related to the TNOs: Centaurs and irregular satellites.

The Centaurs. They are not a clearly defined population from a dynamical point of view, but they can be considered as a group of minor bodies orbiting between Jupiter (5.2 AU) and Neptune (30 AU) with short dynamical lifetimes (about 10^7 yr). They could have been ejected from the Kuiper Belt by planetary perturbations or mutual collisions. On the basis of numerical simulations, Levison & Duncan (1997) found that Centaur-like orbits could be a transition to the JFC-like orbits.

However the origin of the largest Centaurs remains unclear. They are too large to be collisional fragments. Purely dynamical processes could be involved but their dynamical lifetime is very short ($10^5 - 10^6$ yr, see Hahn & Bailey 1990), and a too large number of large TNOs would be required to supply the observed Centaurs. Non-disruptive collisions were also suggested as an alternative process.

The irregular satellites of the giant planets. The so-called regular satellites have nearly circular and low-inclined orbits, whereas the irregular satellites follow eccentric, highly inclined, and often retrograde orbits. Contrary to the regular satellites, supposed to be formed together with the central body, the irregular satellites are probably captured objects. Different scenarii were suggested to describe this process.

It is now well known that gravitational capture of small bodies by giant planets can occur. Dynamical studies of comets demonstrated that some members of the Jupiter family had been once temporarily captured as jovian satellites (Carusi et al. 1985). The most famous case is Shoemaker-Levy 9 which collided with Jupiter after being a jovian satellite during several revolutions. However, the probability of stable capture by purely gravitational processes is close to zero. A theory that would describe the permanent capture of a satellite should also consider an energy dissipation process, in addition to a possible temporary capture. Pollack et al. (1979) suggested that some satellites may have been captured by gas drag in protoplanetary nebula. They proposed that the prograde and retrograde satellites of Jupiter would be the result of the deceleration and fragmentation of two parent bodies which passed through the jovian nebula. Other satellites such as Phoebe (which follows a retrograde, outermost orbit about Saturn) could also be captured by gas drag. Such a process must have taken place shortly before the end of the planet formation.

The origin of the largest satellite of Neptune, Triton, has been particularly studied, and it could be a captured TNO. Goldreich et al. (1989) suggested that a collision between Triton and a satellite of Neptune would dissipate enough orbital energy to make possible the capture of Triton. The initially inclined and eccentric orbit of Triton would have been circularized by tidal effects due to Neptune. With this scenario, Goldreich et al. (1989) would explain the absence of a great number of satellites as about the other giant planets, as well as the highly eccentric orbit of Nereid that would have been ejected after many close approaches to Triton.

It is of great interest to study irregular satellites together with TNOs. As some irregular satellites could come from the Kuiper Belt, dynamics and physical/chemical properties of these both populations could help us to understand the origin of irregular satellites, as well as the history and evolution of the Trans-Neptunian Objects.

6.   COMPOSITION

Spectroscopy is the mainly used technique to investigate the chemical properties of the bodies in our Solar System, from main planets to comets and asteroids. It is the only way to study the chemical composition of their atmosphere or their surface. Unfortunately, spectroscopy cannot be applied to most of the TNOs. Even with the largest telescopes available today (8-10 meter class), spectra can only be obtained for the objects brighter than 20th magnitude. In addition to the problems encountered by all astronomers studying faint objects, we have to take into account several difficulties. As they belong to the Solar System, the TNOs are moving objects relatively to the stars (about $3.5''$ per hour, see Fig. 2). For this reason, we need to take into account the spatial variation of the sky background, in addition to its temporal variation. The TNOs rotate around their spin axis (measurements of rotational periods are only available for a few objects, see Table 1). Depending on the time of the observations, we will observe one side of the object or the other, and we will probably obtain quite different spectra as both sides of the object could not be the same. But the main problem is that we usually need long exposure time to get high S/N spectra, which means to get a mean spectrum over a rotating surface. Thus, spectral features will be smoothed; for example, a small area of pure water ice could not be detected. For this reason, it is not suitable to co-add spectra obtained at different observing times, and we cannot get very high S/N spectra.

Table 1 Physical properties of Centaurs (upper rows) and TNOs (lower rows): in case several estimations were published, we took the mean of these different values. References: 1. Bus et al. 1989; 2. Luu & Jewitt 1990; 3. Dahlgren et al. 1991; 4. Marcialis & Buratti 1993; 5. Buie & Bus 1992; 6. Hoffmann et al. 1993; 7. Brown & Luu 1997; 8. Davies et al. 1998; 9. Romanishin & Tegler 1999; 10. Hainaut et al. 2000.

Object	Period (hrs)	Albedo	Diameter (km)	References
2060 Chiron	5.918 ± 0.002	0.15 ± 0.05	180 ± 10	1, 2, 3, 4
5145 Pholus	9.980 ± 0.004	0.04 ± 0.03	190 ± 22	5, 6
8405 Asbolus	8.90 ± 0.04	–	–	7, 8
10199 Chariklo	–	0.045 ± 0.010	302 ± 30	
1994 TB	~ 6	–	–	9
1994 VK_8	~ 10	–	–	9
1995 QY_9	~ 8	–	–	9
1996 TO_{66}	6.25 ± 0.03	–	–	10

Today, less than ten objects, including Centaurs, could be studied using spectroscopic techniques. The spectra that have already been published will be described below.

We need accurate data on a large sample of objects in order to investigate the whole population. To obtain data on the largest number of TNOs, photometric techniques are used. To date, visible colours (B, V, R, and I magnitudes) have been measured for about 30 objects, and near-IR colours (J, H, and K) for only few of them. Even if broad-band colours can be obtained for much more objects than spectra, observations and data reduction are still very difficult. As mentionned above, the TNOs are moving objects, and thus, the exposure time cannot be longer than $30 - 45$ minutes to avoid trailing effect. With a 4-meter class telescope, we can observe objects up to $23 - 24^{\text{th}}$ magnitude, whereas most of the objects are fainter than mag. 23. Careful observations and reductions are necessary to minimize potential systematic errors due to faint background-source contamination. Detailed study of a data reduction method can be found in Romon (1999).

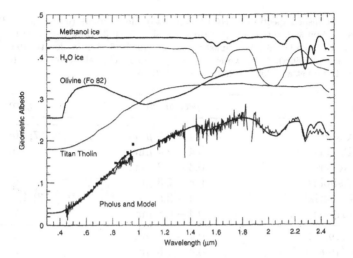

Figure 6 Spectrum of Centaur Pholus: observations and model (from Cruikshank et al. 1998).

The physical and chemical properties of the TNOs are still poorly known due the difficulties mentionned above. Spectral observations are the best way to study their composition, especially in the near-infrared filters ($1 - 2.4$ μm) where the ices signatures are most diagnostic. Up to now, near-infrared spectroscopy has been attempted using Keck (Hawaii) and VLT (ESO, Chile). Infrared spectra of four TNOs and four Centaurs have been published (see Table 2). Some other spectra have been obtained in the visible range. This spectral range is not expected to show strong signatures, but the slope of the spectra is very useful to constrain the composition of the surface.

Chiron and Pholus are the brightest objects and their spectra are of good quality. Pholus which is a very red object, seems to be composed of organic compounds such as methanol ice, water ice and silicates (Cruikshank et al. 1998). Its spectrum (Fig. 6) shows absorption features of water ice at 1.5 and 2 μm, an absorption at 2.3 μm maybe due to the presence of light organic compounds in solid state (methanol ice), while a very broad feature between 0.6 and 1.6 μm seems to be due to silicates. The presence of complex organic compounds (tholin-type material) on the surface has been also suggested.

Chiron, which is the most famous Centaur, is the only one now recognized to be a comet as cometary activity has been detected. It shows

Table 2 Published spectra of TNOs (lower rows) and Centaurs (upper rows). References: 1. Luu & Jewitt (1996); 2. Brown et al. (1997); 3. Brown et al. (1998); 4. Brown & Koresko (1998); 5. Cruikshank et al. (1998); 6. Luu & Jewitt (1998); 7. Barucci et al. (1999a); 8. Brown et al. (1999); 9. Barucci et al. (2000a); 10. Brown (2000); 11. Luu et al. (2000); 12. Brown et al. (2000)

Object	Spectral range	References
2060 Chiron	$0.4 - 2.5$ μm	7; 10; 11
5145 Pholus	$0.4 - 2.4$ μm	5
7066 Nessus	$0.4 - 0.8$ μm	
8405 Asbolus	$0.4 - 2.3$ μm	9
10199 Chariklo	$0.4 - 2.5$ μm	3; 4; 7; 10
10370 Hylonome	$0.4 - 0.8$ μm	7
15789 (1993 SC)	$0.4 - 2.4$ μm	2
15874 (1996 TL_{66})	$0.4 - 2.5$ μm	6
1996 TO_{66}	$1.4 - 2.5$ μm	8
2000 EB_{173}	$1.4 - 2.5$ μm	12

a featureless spectral behaviour or spectra with water ice depending on the heliocentric distance. Asbolus were observed by Barucci et al. (2000a) and by Brown (2000), and the near-infrared spectra seemed to be nearly featureless. Observations made by the Hubble Space Telescope showed a very inhomogeneous surface (Kern et al. 2000). Its spectrum varies from a nearly featureless continuum to complicated absorption due probably to a mixture of ices. Kern et al. suggested that it has a very bright crater on one side of its surface covered some peculiar icy material. Chariklo shows spectra with absorption due to water ice which may be in amorphous state.

1996 TL_{66} shows flat featureless spectra similar to that of dirty water ice. TNO 1993 SC shows a spectrum (Fig. 7), although heavily smoothed, with absorption features very similar to those found on Triton or Pluto, particularly at 2.2 μm that suggested the presence of hydrocarbon ices such as CH_4, C_2H_2, C_2H_4 or C_2H_6. The red continuous reflectance suggests also the presence of more complex hydrocarbons.

Even if the available spectra are of poor quality, they show a wide diversity of surfaces among the objects. This diversity is confirmed by photometric observations of a larger sample of TNOs. Optical (BVRI) and near-infrared (JHK) colours have been obtained since 1996 (see eg. Jewitt & Luu 1998; Barucci et al. 2000b; Davies et al. 2000). These

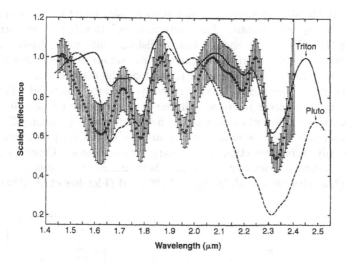

Figure 7 Spectrum of TNO 1993 SC: observations and comparison with spectra of Pluto and Triton (from Brown et al. 1997).

colours exhibit a wide diversity ranging from neutral to very red, implying different surface composition from one object to another. Centaurs and TNOs appear to have very similar spectral characteristics; this supports the hypothesis that Centaurs are ejected from the Kuiper Belt by planetary scattering, and represent transitional objects between TNOs and short-period comets. On the basis of B, V, and R colours, Tegler & Romanishin (1998) suggested that TNOs and Centaurs could be split into two groups: the first one containing neutral objects and the other one containing very red objects. Analysing a larger data sample, Barucci et al. (1999b, 2000b) found a very complex and non homogeneous population.

Red colour can be due to surfaces containing carbon-rich compounds (e.g. organics and tholins), and dark albedo can be due to ice surfaces exposed to cosmic radiation. Laboratory measurements show that a simple ice mixture bombarded by energetic particles can cause surface darkening and modify the chemical structure. The differences found among TNO surfaces may originate from compositional variation in the TNO population or from time-dependent irradiation processes. As the TNOs are supposed to have formed in the same region of the solar system, at low temperature (~ 50 K), it is difficult to explain a wide diversity among their initial composition. On the basis of a simple model, Luu

& Jewitt (1996) first suggested that the diversity could be due to collisions. Taking into account that collisional evolution is very important
among the TNOs, they showed that collisions could reveal fresh material from the interior and thus explaining the colour diversity depending
on the quantity of fresh material present on the surface. As a result,
a TNO with large areas of fresh, recently excavated material will appear neutral, whereas red objects have not undergone a recent collision.
Cometary activity could also explain the colour diversity among TNOs.
It has already been detected on Centaur Chiron at its perihelion. It
is the only direct detection of cometary activity among Centaurs and
TNOs. Cometary activity could have been indirectly detected around
1996 TO$_{66}$ (Hainaut et al. 2000) and 1994 TB (Fletcher et al. 2000).

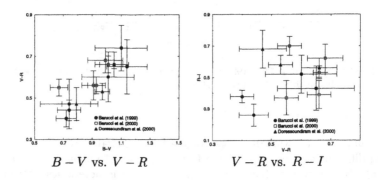

$$B - V \text{ vs. } V - R \qquad V - R \text{ vs. } R - I$$

Figure 8   Colour distribution of TNOs

7.   CIRCUMSTELLAR DISKS

Circumstellar disks which occur around nearby main sequence stars
are called "second generation" disks because the dust they contain is
produced by collisions among larger bodies such as comets and asteroids.
The prototypes of such disks are those observed around β Pictoris and
Vega. Recent high-resolution images showed the existence of structures
in these disks such as rings and central gaps. For example, images of the
dust near ϵ Eridani, obtained around 850 μm by Greaves et al. (1998),
show a dust ring which peaks at 60 AU and could extend from 35 to
75 AU. These structures would indicate the presence of larger bodies
such as planetesimals or forming planets. Imaging of β Pic, which is
probably the best-studied of the circumstellar dust disks, also showed
that the disk presents a finite thickness (Kalas & Jewitt 1995). This
would imply the presence of 1000 km bodies in the disk.

We also got other evidences of the presence of planetary bodies in these disks. Silicates have been detected at 10 μm in the spectrum of β Pic; this emission is very close to the one found in cometary spectra (Knacke et al. 1993). Redshifted temporal variations of the β Pic disk spectra could be attributed to bodies falling down to the star and evaporating (Beust et al. 1998).

Circumstellar disks have been compared to the Kuiper Belt and the solar system zodiacal dust. The small-grain component of the Kuiper Belt have not yet been observed, but it might exist as shown by collisional models. In case that the initial Kuiper Belt had been more massive than the actual one, it would have looked like the observed second generation disks. Artymowicz (1997) compared β Pic and the solar system zodiacal dust. He noticed similarities as well as differences. The differences could be due to an early evolutionary stage of β Pic disk.

The study of the circumstellar disks could help us to understand the history of the Kuiper Belt, whereas the Kuiper Belt could help to investigate the circumstellar disks. For example, Liou & Zook (1999) suggested that the presence of giant planets in circumstellar disks could affect the disk morphology in the same way as our jovian planets affect the Kuiper Belt interplanetary particles. Thus, the study of the Kuiper Belt is closely linked to the circumstellar disks research field.

8.   FORMATION AND EVOLUTION

Planetesimals formed in the outer part of the solar nebula followed various paths toward the present distribution of matter in the solar system. Cruikshank et al. (1997) suggested several possible pathways starting from the interstellar medium (Fig. 9). The planetesimals remaining in the trans-Neptunian region, after the completion of planet formation, constitute the Kuiper Belt. The evolution of objects from the Kuiper belt to the Centaur population is also well detailed in Fig. 9. During the planet formation, when the outer Solar System reached a configuration close to the present state (the four giant planets which accreted most of their mass, and the nebular gas already dispersed), the remaining population of icy planetesimals were scattered to wider orbits. Due to gravitational effects, planetesimals were scattered inward and outward. Planetesimals condensing in the Solar Nebula beyond about 20 AU should retain a large amount of volatile and organic material; only a small part of this material could have been lost by evaporation during collisions. After the formation of Neptune, TNOs changed from intensive accretion to intensive collisions, resulting in erosion and mass depletion of the population. They have been chemically processed through interactions with cosmic rays and solar ultraviolet radiation (space weathering mechanisms). Collisions could refresh intensively the surface material especially on the inner part of the Kuiper Belt.

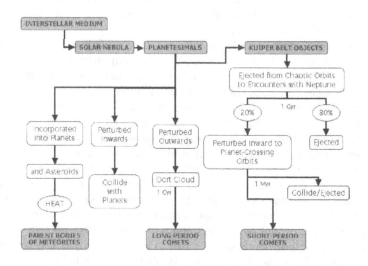

Figure 9   Evolution scenario proposed by Cruikshank et al. (1997).

Our knowledge of the Trans-Neptunian population is at present still confused. The study of this population which represents the remnant planetesimals, composed of the oldest and least processed planetary materials, will allow to have more precise information on the frontiers of our Solar System.

Notes

1. The Zodiacal Cloud is made by interplanetary dust particles that are concentrated in the ecliptical plane.

REFERENCES

Artymowicz, P. (1997), *An. Rev. of Earth and Pl. Sci.*, **25**, 175.

Barucci, M.A., de Bergh, C., Cuby, J.-G., le Bras, A., Schmitt, B. and Romon, J. (2000a), *A&A*, **357**, L53.

Barucci, M.A., Doressoundiram, A., Tholen, D., Fulchignoni, M., Lazzarin, M. (1999b), *Icarus*, **142**, 476.

Barucci, M.A., Lazzarin, M. and Tozzi, G.P. (1999a), *AJ*, **117**, 1929.

Barucci, M.A., Romon, J., Doressoundiram, A., Tholen, D.J. (2000b), *AJ*, **120**, 496.

Beust, H., Lagrange, A.-M., Crawford, I.A., Goudard, C., Spyromilio, J., Vidal-Madjar, A. (1998), *A&A*, **338**, 1015.

Brown, M.E. (2000), *AJ*, **119**, 977.

Brown, M.E., Blake, G.A. and Kessler, J.E. (2000), *ApJ*, **543**, L163.

Brown, M.E. and Koresko, C.D. (1998), *ApJ*, **505**, L65.

Brown, W.R., Cruikshank D., Pendleton, Y. (1999), *ApJ*, **519**, L101.

Brown, W.R., Cruikshank, D., Pendleton, Y., Veeder G. (1997), *Science*, **276**, 937.

Brown, W.R., Cruikshank, D., Pendleton, Y., Veeder, G. (1998), *Science*, **280**, 1430.

Brown, W.R. and Luu, J.X. (1997), *Icarus*, **126**, 218.

Buie, M.W. and Bus, S.J. (1992), *Icarus*, **100**, 288.

Bus, S.J., Bowell, E., Harris, A.W. and Hewitt, A.V. (1989), *Icarus*, **77**, 223.

Carusi, A., Kresak, L., Perozzi, E. and Valsecchi, G.B. (1985), *Long-term evolution of short-period comets*, Adam Hilger Ltd. Bristol and Boston.

Cochran, A.L., Cochran, W.D. and Torbett, M.V. (1991), *BAAS*, **23**, 131.

Cochran, A.L., Levison, H.F., Stern, S.A. and Duncan, M.J. (1995), *ApJ*, **503**, L89.

Cruikshank, D.P. (1997), in Pendleton, Y.J. and Tielens, A.G.G.M. (eds.), *From Stardust to Planetesimals*, ASP conf. Series, vol.122, p.315.

Cruikshank, D.P., Roush, T.L., Barthdomew, M.J., Geballe, T.R., Pendleton, Y.J., White, S.M., Bell, J.F. III, Davies J.K., Owen, T.C., de Berh, C., Tholen, D.J., Bernstein, M.P., Brown, R.H., Tryka, K.A. and Dalle Ore, C.M. (1998), *Icarus*, **135**, 389.

Dahlgren, M., Fitzsimmons, A., Lagerkvist, C.-I. and Williams, I.P. (1991), *MNRAS*, **250**, 115.

Davies, J.K., Green, S., McBride, N., Muzzerall, E., Tholen, D., Whiteley, R.J., Foster, M.J. and Hillier, J.K. (2000), *Icarus*, **146**, 253.

Davies, J.K., McBride, N., Ellison, S.L., Green, S.F. and Ballantyne, D.R. (1998), *Icarus*, **134**, 213.

Davis, D.R. and Farinella, P. (1997), *Icarus*, **125**, 50.

Delsanti, A.C., Hainaut, O.R., Boehnhardt, H., Dalahodde, C.E., Sekiguchi, T. and West, R.M. (1999), *BAAS*, **31**, 1115.

Doressoundiram, A., Barucci, M.A. and Romon, J. (2000), *BAAS*, **32**, 2110.

Duncan, M.J. and Levison, H. (1997), *Science*, **276**, 1670.

Duncan, M.J., Levison, H.F. and Budd, S. M. (1995), *AJ*, **110**, 3073.

Duncan, M.J., Quinn, T. and Tremaine, S. (1988), *ApJ*, **328**, L69.

Durda, D.D. and Stern, S.A. (2000), *Icarus*, **145**, 220.

Edgeworth, K.E. (1943), *J. of British Astron. Soc.*, **53**, 181.

Edgeworth, K.E. (1949), *MNRAS*, **109**, 600.

Farinella, P. and Davis, D.R. (1996), *Science*, **273**, 938.

Fernández, J. A. (1980), *MNRAS*, **192**, 481.

Fletcher, E., Fitzsimmons, A., Williams, I.P., Thomas, N. and Ip, W.-H. (2000), in Fitzsimmons, A. (ed.), *Minor Bodies in the outer Solar System*, Springer, in press.

Flynn, G.J. (1996), in Gustafson, B.A.S. and Hanner, M.S. (eds.), *Physics, Chemistry, and Dynamics of Interplanetary Dust*, ASP Conf. Series, vol.104, p.171.

Gladman, B. and Duncan, M. (1990), *AJ*, **100**, 1680.

Gladman, B., Kavelaars, J.J., Nicholson, P.D., Loredo, T.J. and Burns, J.A. (1998), *AJ*, **116**, 2042.

Goldreich, P., Murray, N. Longaretti, P.Y. and Banfield, D. (1989), *Science*, **245**, 500.

Greaves, J.S., Holland, W.S., Moriarty-Schieven, G., Jenness, T., Dent, W.R F., Zuckerman, B., McCarthy, C., Webb, R.A., Butner, H.M., Gear, W.K. and Walker, H.J. (1998), *ApJ*, **506**, L133.

Hahn, G. and Bayley M.E. (1990), *Nature*, **348**, 132.

Hahn, J.M and Malhotra, R. (1999), *AJ*, **117**, 3041.

Hainaut, O.R., Dalahodde, C.E., Boehnhrdt, H., Dotto, E., Barucci, M.A., Meech, K.J., Bauer, J.M., West, R.M. and Doressoundiram, A. (2000), *A&A*, **356**, 1076.

Hoffmann, M., Fink, U., Grundy, W.M. and Hicks, M. (1993), *JGR*, **98 E4**, 7403.

Holman, M. and Wisdom, J. (1993), *AJ*, **105**, 1987.

Jewitt, D. (1999), *Annu. Rev. Earth Planet. Sci.*, **27**, 287.

Jewitt, D. and Luu, J. (1993), *Nature*, **362**, 730.

Jewitt, D. and Luu, J. (1998), *AJ*, **115**, 1667.

Jewitt, D., Luu, J. and Trujillo, C. (1998), *AJ*, **115**, 2125.

Kalas, P. and Jewitt, D. (1995), *AJ*, **110**, 794.

Kern, S.D., McCarthy, D.W., Buie M.W., Brown R.H., Campins, H. and Rieke, M. (2000), *ApJ*, **542**, L155.

Knacke, R.F., Fajardo-Acosta, S.B., Telesco, C.M., Hackwell, J.A., Lynch, D.K. and Russell, R.W. (1993), *ApJ*, **418**, 440.

Kowal, C. (1990), *Icarus*, **77**, 118.

Kuiper, G.P. (1950), in Hynek, J.A. (ed.), *Astrophysics*, New York, McGraw-Hill, p.375.

Levison, H. and Duncan, M. (1990), *AJ*, **100**, 315.

Levison, H. and Duncan, M. (1997), *Icarus*, **127**, 13.

Liou, J.-C. and Zook, H.A. (1999), *AJ*, **118**, 580.

Liou, J.-C., Zook, H.A., and Dermott, S.F. (1996), *Icarus*, **124**, 429.

Luu, J.X. and Jewitt, D. (1988), *AJ*, **95**, 1256.

Luu, J.X. and Jewitt, D. (1990), *AJ*, **100**, 913.

Luu, J.X. and Jewitt, D. (1996), *AJ*, **112**, 2310.

Luu, J.X. and Jewitt, D. (1998), *ApJ*, **502**, L91.

Luu, J.X., Jewitt, D.C. and Trujillo, C. (2000), *ApJ*, **531**, L151.

Malhotra, R. (1995), *AJ*, **110**, 420.

Marcialis, R.L. and Buratti, B.J. (1993), *Icarus*, **104**, 234.

Morbidelli, A. (1997), *Icarus*, **127**, 1.

Oort, J.H. (1950), *Bull. Astr. Inst. Netherlands*, **11**, 91.

Pollack, J.B., Burns, J.A. and Tauber, M.E. (1979), *Icarus*, **37**, 587.

Quinn, T., Tremaine, S. and Duncan, M.J. (1990), *ApJ*, **355**, 667.

Romanishin, W., Tegler, S.C., Levine, J. and Butler, N. (1997), *AJ*, **113**, 1893.

Romon, J. (1999), *Objets de Kuiper: Analyse des sources d'erreurs sur la magnitude*, Rapport de stage de DEA, Université Paris XI.

Roques, F. and Moncuquet, M. (2000), *Icarus*, **147**, 530.

Stern, S.A. (1995), *AJ*, **110**, 856.

Stern, S.A. (1996a), in Retting, T.W. and Hahn, J.M. (eds.), *Completing the inventory of the Solar System*, ASP Conf. Series, vol.107, p.209.

Stern, S.A. (1996b), *A&A*, **310**, 999.

Stern, S.A. and Colwell, J.E. (1997), *ApJ*, **490**, 879.

Tegler, S.C. and Romanishin, W. (1998), *Nature*, **392**, 49.

Thomas, N., Eggers, S., Ip, W.-H., Lichtenberg, G., Fitzsimmons, A., Jorda, L., Keller, H.U., Williams, I.P., Hahn, G. and Rauer, H. (2000), *ApJ*, **534**, 446.

Trujillo, C. and Jewitt, D. (1998), *AJ*, **115**, 1680.

Tyson, J.A., Guhathakurta, P., Bernstein, G. and Hut, P. (1992), *BAAS*, **24**, 1127.

Whipple, F.L. (1964), *Proc. Nat. Acad. Sci.*, **51**, 711.

CHEMICAL MODELS: WHERE TO START FROM?

Osama M. Shalabiea

Astronomy Program, SEES, Seoul National University, Seoul 151-742, Korea;
and Department of Astronomy, Cairo University, Cairo, Egypt

Keywords: Interstellar medium - clouds - abundances - molecular processes - dust

Abstract   Chemical models are sensitive to the adopted set of initial input parameters. One of these crucial parameters is the set of the adopted gas-phase
elemental abundances with which the reactions networks start to operate. Based on the most recent observations and theoretical data, we
investigate the effect of initial abundance variation on both gas-phase
and gas-grain chemical models. At early-time stages less than 1 Myr,
there is little difference between results with different initial [C]/[O]
ratios. This holds for gas-phase and gas-grain models. At later evolutionary time or in the steady state, the result of the gas-grain model
shows little or no difference due to variations of the initial [C]/[O] ratios.
On the contrary, at the late or at steady state times, the abundances
of chemical species using gas-phase models are very sensitive to any
variation of the initial [C]/[O] ratios.

Introduction

During the last decade, chemical models have progressed noticeably.
Many of the key rates of reactions have been improved in the gas-phase
Le Teuff et al. (2000). Also gas-grain interactions have gotten more
attention, showing their crucial role in chemical models (Hasegawa et al.
1992, Willacy & Williams 1993, Shalabiea & Greenberg 1994, Shalabiea
et al. 1998, Willacy & Millar 1998, Ruffle & Herbst 2000,2001).

After years of extensive, theoretical, experimental, and observational
studies, interstellar chemical modeling still has many puzzling problems.
One of these problems is the set of the adopted gas-phase elemental
abundances. The direction in which each reaction network proceeds
strongly dictated by variation of initial abundances. There are three

V. Pirronello et al. (eds.), Solid State Astrochemistry, 417–429.
© 2003 *Kluwer Academic Publishers. Printed in the Netherlands.*

parts to this problem of particular interest. First is the carbon-to-oxygen ratio [C]/[O], second is the sulfur depletion and the third is the depletion of heavy elements.

Several authors have studied its effects on gas-phase models (e.g. Watt 1985; Terzieva & Herbst 1998). However, the effects of varying the elemental abundances on gas-grain models have not yet been investigated.

The main objectives of the present work are (1) to investigate how sensitive the gas-grain models are to initial elemental abundance variation compared with the gas-phase models, and (2) to study under which initial conditions and sets of elemental abundance, the results can agree with observations.

In Section 1 constraints on the initial elemental abundances are briefly given. The chemical models and initial conditions we have used are described in Section 2. The results and discussion are presented in Section 3 and the conclusions are given in Section 4.

1.   CONSTRAINTS BASED ON OBSERVATIONS AND GRAIN MODELS

The observed abundances of different atomic elements in the interstellar medium (ISM) tend to be lower than those adopted as the cosmic (solar) reference standard.

The C, O, N, S, and metals such as (Fe, Si, Mg) are among the most abundant elements, which commonly condense to form the solids of interstellar grains. The level of their depletion is dependent on the physical condition in their environment. Shalabiea & Greenberg (1994) discussed in some detailed the justifications for the assumptions used in interstellar cloud chemical evolution.

The *real* depletion of the initial elements still has some uncertainties. However, based on recent observations from the Goddard High Resolution Spectrograph (GHRS) (Cardelli et al. 1996; Savage & Sembach, 1996; Sofia et al. 1997,1998; Genaciński, 2000) and theoretical grain models Li & Greenberg (1997), they are well constrained within the following ranges: carbon $(3.55\text{-}1.4 \times 10^{-4})$, oxygen $(7.41\text{-}2.94 \times 10^{-4})$ with a ratio of [C]/[O] from 0.21 to 0.54, while for sulfur it can vary from its solar value (1.62×10^{-5}) to the depleted value 3.23×10^{-6}. For the heavy elements, the use of high-metal abundances seem to meet both the observational and theoretical constrains. These constraints upon the initial elemental abundance are discussed in more details (Shalabiea, 2001). Using the above mentioned constraints, their effect and crucial role on chemical models are investigated in this work.

2. CHEMICAL MODELS AND INITIAL CONDITIONS

Two chemical models have been used: one is purely the gas-phase while the second is characterized by combining grain interactions with the gas-phase chemistry.

The chemical evolution of interstellar clouds using these two models is investigated under fixed physical parameters. For dense interstellar clouds the following initial conditions have been used: a cloud total number density n_0 $(n(H) + 2n(H_2))=2\times10^4$ cm^{-3}, $T_{gas}=T_{grain}=10K$, and a visual extinction Av=10 mag. The standard cosmic ray ionization rate $\zeta=1.3\times10^{-17}s^{-1}$ has been used. The enhanced gas-grain network from the so-called unmodified model (Shalabiea et al. 1998) has been used. It contains 656 species and 4742 reactions.

Table 1a Adopted elemental abundances relative to total hydrogen for the standard model

Element	[X]/[H]
He	1.40E-1
N	7.94E-5
O	3.02E-4
C$^+$	1.38E-4
S$^+$	3.23E-6
Si$^+$	1.74E-6
Fe$^+$	1.74E-7
Na$^+$	2.00E-7
Mg$^+$	1.07E-6

Table 1b [C]/[O] ratio and their initial abundances

[C]/[O]	[C]/[H]	[O]/[H]
0.2	6.01E-5	3.02E-4
0.46	1.38E-4	3.02E-4
0.8	2.45E-4	3.02E-5

We have listed the standard set of the adopted elemental abundances in Table 1a. We assumed that C, S, other heavy elements are initially ionized and most of hydrogen is in the molecular form. For metals we have used the so-called "high-metal" elemental abundances. Under this set of initial condition, there is no bistable solution (see Shalabiea & Greenebrg 1995 and Lee et al. 1998).

3.   RESULTS AND DISCUSSION

To examine the sensitivity of varying the [C]/[O] ratio, the results are given at three different ratios which are 0.2, 0.46 (standard) and 0.8. This ratio is varied between the lowest and highest values we may expect. The relative abundances of C, O and [C]/[O] ratios are listed in Table 1b.

Since ages of the interstellar dense molecular clouds are not certain (van Dishoeck et al. 1993), we have presented the results at four times of evolution: 0.1, 0.5, 1 and 10 Myr. The results for some key species are shown in figures 1-5. Gas-phase models (a) are presented at the top of the figures while gas-grain models (b) are presented at the bottom.

Results for two simple diatomic molecules C_2 and SO are presented in Figures 1 and 2, respectively. In Fig. 1a there is a clear difference between the low (0.2) and high (0.8) [C]/[O] ratios. As expected the higher carbon the higher C_2. At early time that difference is only about one order of magnitude (1.15×10^{-9} to 3.0×10^{-8}). With time evolution the difference increases to reach about four orders of magnitudes (8.89×10^{-13} - 3.7×10^{-9}) at late time, which is indeed significant. On the contrary the results for gas-grain models, presented at Fig. 1b, show smaller differences (within two orders of magnitude) for C_2 in comparison with the gas-phase results in Fig. 1a. Its evolution is steadier with time than that of the gas-phase model.

It is clear from Figures 1 and 2 that even a small variation in the [C]/[O] ratio can severely alter the abundances of C_2 and SO. At early time, most of the parent (destroyer) molecules have similar or slightly different abundances in both gas-phase and gas grain models. For example at 0.1 Myr, O abundances are 2.6×10^{-4} and 1.6×10^{-4} for gas-phase and gas-grain models respectively, while its abundance at 10 Myr are 1.8×10^{-4} and 2.0×10^{-6} for models (a) and (b), respectively. The same is true for other neutral species such as nitrogen whose abundances at early time (0.1 Myr) are 7.9×10^{-5} and 4.8×10^{-5} for models (a) and (b) respectively, while its abundance at late time (10 Myr) is 3.4×10^{-5} and 7.2×10^{-7} for models (a) and (b). This is about 2 orders higher in model (a) than model (b). Therefore the C_2 abundances in the models (a) and (b) are close at early time. On the contrary, at late times, model (b) has a higher C_2 abundance than model (a) due to the lower abundances of its main destructive species O and N. In comparison with observations for TMC-1, where the C_2 abundance is 5×10^{-8}, the best fit of these

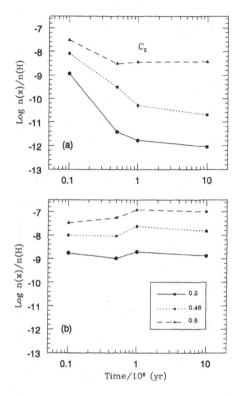

Figure 1 Time evolution of the relative abundances of C_2. (a) for gas-phase and (b)for gas-grain models for three values of [C]/[O] ratio (0.2 (solid line), 0.46 (long-dashed line) and 0.8 (dotted line)

results is for a [C]/[O] ratio 0.46 at a time of 1 Myr for the gas-grain model. For gas-phase model, at the same time and for the same [C]/[O] ratio, the C_2 is much lower than observation. Its abundance can only fit with observation for high [C]/[O] ratio of 0.8.

The results for SO are shown in Figure 2 for models (a) and (b) respectively. As for C_2, the differences of the SO abundances in model (a) are large between the [C]/[O] ratio 0.2 and 0.8. For models (b) this difference between the low and high [C]/[O] ratio of SO abundance is much smaller than the difference in model (a). It is within an order of magnitude. At early time, the SO abundance in model (b) is only two times higher than its abundance in model (a). This is mainly due to the high abundance of its precursor species O_2 and the low abundance of its destructive species C^+ in model (b). On the contrary, at late time

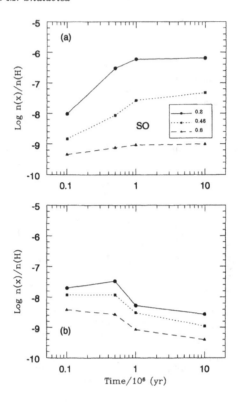

Figure 2   Same as Fig. 1 but for SO.

its abundance in model (a) is higher by about two orders of magnitude than in model (b). This is because its main production via the reactions of $HSO^+ + e$ is diminished to be only 2% instead of 42% at early time in model (b). The reaction $(S+O_2)$ contributes 44% in its formation in model (a) at late time. This is due to the high gas-phase abundance of elemental O in model (a). In model (b) most of the oxygen is depleted on the grain surfaces in the form of CO, CO_2, and H_2O solid species. Therefore, SO becomes low with time evolution for model (b).

The observed abundances for SO in TMC-1 and L134N are 5×10^{-9} and 2×10^{-8} respectively. If we try to fit such observational values with the models, for model (b) in Fig. 2a, it is only at 0.5 Myr for 0.8 [C]/[O] ratio for TMC-1, while for L134N agreement can be achieved at 1 Myr for a [C]/[O] ratio of 0.46. The results shown for models (b) are in agreement with SO observational abundance in both TMC-1 and L134N for a [C]/[O] ratio of 0.46 at 1 and 0.5 Myr respectively.

For these simple diatomic molecules, we can conclude that gas-grain models are much less sensitive to the [C]/[O] variations than are gas-phase models.

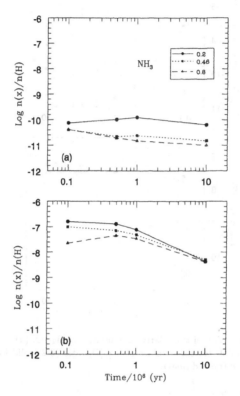

Figure 3 Time evolution of the relative abundances of NH₃. At the top gas-phase models (a) are presented while the gas-grain models (b) at the bottom. The results are given at three values of [C]/[O] ratio, 0.2 (solid line), 0.46(long-dashed line) and 0.8 (dotted line)

To examine the role of [C]/[O] variations on a bit more complex species which does not include C and O atoms, we present results for NH₃ in Figure 3. It shows that NH₃ follows the same trend where the effect due to the [C]/[O] variation is very small for both of the models (a) and (b). However the results for models (b) is higher than those of models (a) by orders of magnitudes. The main formation route for NH₃ is via $NH_4^+ + e \longrightarrow NH_3 + H$. Its destruction routes are via its reaction with the Si^+, S^+, H_3^+, C^+, H^+ and He^+ ions, and neutral-neutral reactions with CN.

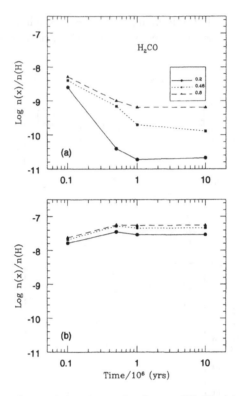

Figure 4 Time evolution of the relative abundances of H₂CO. (a) for gas-phase and (b)for gas-grain models for three values of [C]/[O] ratio (0.2 (solid line), 0.46 (long-dashed line) and 0.8 (dotted line))

In model (b), ammonia is formed mainly on grain surfaces then desorbed to the gas phase, while its destruction routes are via its accretion onto grain surfaces, and reactions with ion species (H_3^+, He^+, Si^+, H^+) and neutral- neutral reaction with CN.

From these routes of NH_3 formation and destruction, it is clear that NH_3 has very little change due to the variation of carbon-to-oxygen ratio. This is mainly because neither carbon, nor oxygen, nor their bearing molecular species are involved into its formation or destruction.

In comparison with observations, the results of models (b) are in agreement with the TMC-1 abundance (10^{-8}) at about 5 Myr. Mean while in comparison with L134N, where the fractional abundance of ammonia is 2×10^{-7}, the best agreement requires an early time of 0.5 Myr and

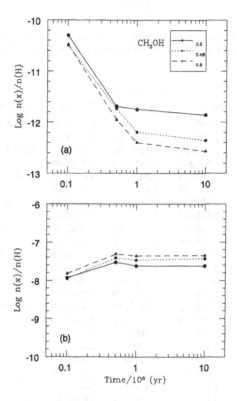

Figure 5 Same as Fig. 4 but for CH₃OH.

[C]/[O] 0.46. If this is the case then NH$_3$ could be a good indicator for the age of these molecular clouds.

The results for more complex species such as H$_2$CO and CH$_3$OH are shown in Figures 4 and 5 respectively. For model (a), the difference between the 0.2 and 0.8 results is two orders of magnitude at late time, while at early time its abundance is almost the same. Since at early time most of the H$_2$CO parent species are of similar abundances, there is a little difference between the different ratios of [C]/[O]. When [C]/[O] becomes high, a natural consequence is that the H$_2$CO abundance increases. Thus the difference between the low and the high [C]/[O] ratio at late time becomes around two orders of magnitude, as shown in Fig. 4a. Unlike the (a) model, the gas-grain model (b) shows a little sensitivity to the variations of the [C]/[O] ratio. The difference in that later case is almost negligible.

Table 2 Fractional abundances with respect to total H for two different ratios of [C]/[O] (0.2, 0.8) at three different evolutionary times (0.5, 1, 10 Myr) for gas-phase chemical model

		Gas-phase model			Observations[a]	
	[C]/[O]	0.5	1	10 (Myr)	TMC-1	L134N
CO	0.2	6.08E-05	6.090E-05	6.090E-05	1.6E-04	1.8E-04
	0.8	2.23E-04	2.324E-04	2.337E-04		
OH	0.2	1.90E-08	2.100E-08	2.237E-08	6.0E-7	1.6E-7
	0.8	5.896E-09	5.970E-09	6.878E-09		
CN	0.2	2.79E-10	1.543E-10	8.622E-11	6.0E-08	1.0E-09[c]
	0.8	1.45E-07	1.010E-07	7.481E-08		
NO	0.2	2.62E-08	2.905E-08	2.943E-08	< 6.0E-08	1.2E-07
	0.8	5.82E-09	7.513E-09	7.838E-09		
C_2H	0.2	3.32E-12	4.519E-13	2.999E-13	1.0E-07	1.0E-08[c]
	0.8	2.14E-08	8.543E-09	7.472E-09		
C_2O	0.2	3.12E-15	4.231E-16	2.671E-16	1.2E-10[b]	..
	0.8	4.93E-13	2.485E-13	2.290E-13		
HCN	0.2	4.65E-10	3.107E-10	1.465E-10	4.0E-08	1.4E-08[c]
	0.8	1.60E-08	7.684E-09	4.956E-09		
C_2S	0.2	2.14E-11	1.687E-12	8.803E-13	1.6E-08	1.2E-09
	0.8	4.69E-10	3.807E-10	9.456E-09		
H_2S	0.2	1.30E-11	1.102E-11	3.460E-12	<1.0E-09	1.6E-09
	0.8	5.77E-12	5.309E-12	2.567E-12		
OCS	0.2	4.94E-09	5.121E-09	4.943E-09	4.0E-09	4.0E-09
	0.8	5.76E-09	4.168E-09	3.929E-09		
SO_2	0.2	1.18E-07	2.732E-07	3.150E-07	<2.0E-09	6.0E-09[c]
	0.8	3.18E-11	3.548E-11	3.906E-11		
HC_3N	0.2	1.08E-12	7.027E-14	1.993E-14	1.2E-07[b]	8.0E-10[c]
	0.8	2.00E-09	6.531E-10	4.274E-10		
C_4H	0.2	1.66E-13	1.191E-14	6.846E-15	8.0E-07[c]	2.0E-09
	0.8	4.76E-10	1.057E-10	7.486E-11		
C_6H	0.2	1.71E-17	8.319E-19	1.030E-22	4.0E-10[b]	..
	0.8	1.30E-13	8.096E-15	8.096E-15		
C_3H_2	0.2	2.29E-13	1.602E-14	9.797E-15	8.0E-09[b]	4.0E-09
	0.8	7.69E-10	2.343E-10	1.855E-10		
CH_2CO	0.2	1.58E-12	9.633E-13	7.784E-13	1.2E-09[b]	<1.4E-09
	0.8	5.85E-11	3.315E-11	2.642E-11		
HCO^+	0.2	1.64E-10	1.825E-10	1.769E-10	1.6E-08	1.6E-08
	0.8	2.09E-10	2.007E-10	1.906E-10		
$N2H^+$	0.2	2.51E-12	6.409E-12	2.268E-11	1.0E-09	1.2E-09[c]
	0.8	2.03E-13	5.037E-13	3.442E-12		
HCS^+	0.2	7.96E-12	1.224E-11	1.258E-11	6.0E-10[b]	1.2E-10
	0.8	9.11E-11	6.545E-11	7.468E-11		

[a] Ohishi et al. (1992) unless otherwise noted.
[b] Ohishi & Kaifu (1998).
[c] Dickens et al. (2000).

Table 3 Fractional abundances with respect to total H for two different ratios of [C]/[O] (0.2 and 0.8) at three (0.5, 1.0 and 10 Myr) different evolutionary times for gas-grain model

	[C]/[O]	0.5	Gas-grain 1	10 (Myr)	Observations TMC-1	L134N
CO	0.2	9.746E-06	2.788E-06	1.505E-06	1.6E-04	1.6E-04
	0.8	3.313E-05	8.332E-06	4.166E-06		
OH	0.2	1.599E-07	4.072E-07	6.120E-07	6.0E-07	1.6E-07
	0.8	6.142E-08	1.456E-07	1.967E-07		
CN	0.2	1.786E-08	1.296E-08	8.034E-09	6.0E-08	1.0E-09[c]
	0.8	1.829E-07	9.677E-08	5.694E-08		
NO	0.2	2.083E-07	1.729E-07	1.816E-07	< 6.0E-08	1.2E-07
	0.8	5.117E-08	4.762E-08	4.714E-08		
C_2H	0.2	3.052E-10	4.151E-10	2.700E-10	1.0E-07	1.0E-08[c]
	0.8	2.618E-08	2.541E-08	1.724E-08		
C_2O	0.2	6.056E-14	3.799E-14	1.889E-14	1.2E-10[b]	..
	0.8	2.202E-12	1.155E-12	7.097E-13		
HCN	0.2	9.945E-09	6.800E-09	3.926E-09	4.0E-08	1.4E-08[c]
	0.8	3.938E-08	3.169E-08	2.117E-08		
C_2S	0.2	4.829E-12	6.693E-13	2.109E-13	1.6E-08	1.2E-09
	0.8	4.805E-10	6.026E-11	1.715E-11		
H_2S	0.2	7.663E-09	7.269E-09	7.337E-09	<1.0E-09	1.6E-09
	0.8	4.805E-10	6.026E-11	1.715E-11		
OCS	0.2	3.134E-10	3.058E-11	6.460E-12	4.0E-09	4.0E-9
	0.8	1.279E-09	8.995E-11	1.358E-11		
SO_2	0.2	2.405E-09	4.811E-10	3.221E-10	<2.0E-09	6.0E-09[c]
	0.8	5.639E-11	2.111E-11	1.073E-11		
HC_3N	0.2	1.419E-10	7.382E-11	4.260E-11	1.2E-08[b]	8.0E-10[c]
	0.8	8.986E-09	6.948E-09	5.212E-09		
C_4H	0.2	2.540E-12	1.114E-12	4.038E-13	8.0E-07[c]	2.0E-09
	0.8	2.059E-09	1.151E-09	6.140E-10		
C_6H	0.2	4.735E-15	3.209E-15	1.557E-15	4.0E-10[b]	..
	0.8	1.247E-12	7.147E-13	4.209E-13		
C_3H_2	0.2	1.568E-11	9.964E-12	5.136E-12	8.0E-09[b]	4.0E-09
	0.8	1.992E-09	1.262E-09	1.196E-08		
CH_2CO	0.2	1.373E-12	6.368E-13	5.586E-13	1.2E-09[b]	<1.4E-09
	0.8	2.929E-11	1.448E-11	1.036E-11		
HCO^+	0.2	1.223E-10	7.555E-11	6.644E-11	1.6E-08	1.6E-08
	0.8	3.071E-10	1.973E-10	1.929E-10		
$N2H^+$	0.2	1.768E-11	3.354E-11	4.839E-11	1.0E-09	1.2E-09[c]
	0.8	8.644E-12	2.425E-11	4.371E-11		
HCS^+	0.2	4.321E-12	2.074E-12	1.470E-12	3.0E-10[b]	1.2E-10
	0.8	3.783E-11	1.753E-11	1.061E-11		

[a] Ohishi et al. (1992) unless otherwise noted.
[b] Ohishi & Kaifu (1998).
[c] Dickens et al. (2000).

In comparison with observation, the results for models (b) seem a bit higher than that of TMC-1 and L134N (2×10^{-8}). On the other hand, the results of H_2CO in models (a) are low by three orders of magnitude for [C]/[O] ratio 0.2. Even for the highest [C]/[O] ratio (0.8) they are still ten times lower than the observed value.

The observed abundances of CH_3OH for TMC-1 (3×10^{-9}) and L134N (5×10^{-9}) are higher than the values of model (a) in Fig. 5a, by more than 4 orders of magnitude. For model (a), the difference of the CH_3OH results between the high [C]/[O] (0.8) and the low one (0.2) reaches an order of magnitude at late time. This difference is almost negligible at early time ($\leq 10^5$ Myr). Its main formation route in models (a) is via, $CH_5O^+ + e \longrightarrow CH_3OH + H$, while its destruction comes via reactions with H_3^+, H^+, He^+, and photon induced by cosmic rays. In the gas-grain model (b) the abundances are higher than the observed values by one order of magnitude. This could be due to either its high formation rate on the grain surfaces and/or its efficiency of desorption. From these two examples of complex molecules, we still find a crucial role for the variation of the [C]/[O] elemental abundance in model (a). Model (b), on the other hand, is relatively insensitive to the variation of the [C]/[O] ratio. These variations are almost negligible for gas-grain models if the [C]/[O] ratio varies only from 0.4 to 0.5 as suggested by Pratap et al. (1997) or from 0.4 to 0.8 by Bergin et al. (1997) in studies with gas-phase models. The results for more gas-phase species are listed in Tables 2 and 3 for gas-phase and gas-grain models respectively. For comparison the observed abundances in TMC-1 and L134N are presented. It is useful in understanding at which ratio and evolution time, results are in agreement with observation for gas-phase and gas-grain model. We listed the results at the two extreme values of [C]/[O] (0.2 and 0.8) at three evolutionary times. The results in Table 2 show the high sensitivity of molecular abundances to the variation of [C]/[O] ratio, in particular for carbon-bearing molecules. In Table 3 the effect is minimized by mutual gas-grain interaction. Although there are still some differences in the gas-grain, they are much lower than those in the gas-phase model.

4.   CONCLUSIONS

This work *confirms* the vital role of the adopted set of the gas-phase elemental abundance for chemical models, especially for gas-phase models. On the contrary, variations of the carbon-to-oxygen elemental abundances in gas-grain chemical models show little or no effect. For both models, early-time (0.1 Myr) results show less sensitivity to changes in the [C]/[O] ratio than those at mid (1.0 Myr) or late time (10 Myr) of evolution. For further studies of the importance of adopted set of initial elemental abundance on chemical models, we conclude that gas-grain model, which includes both gas and solid-phase chemistry, provides a more complete basis for comparison with observations.

Acknowledgments

I acknowledge the NATO support of my visit to Erice, Italy. I am very grateful to Prof. Eric Herbst for providing me the chemical network of his group at the Ohio State University and for many helpful suggestions. I further thank Prof. S.S. Hong and for stimulating discussions.

REFERENCES

Bergin, E.A., Goldsmith, P.F., Snell, R.L., Langer, W.D. (1997), *ApJ*, **482**, 285.
Cardelli, J.A., Meyer, D. M., Jura, M. and Savage, B.D. (1996), *ApJ*, **467**, 334.
Dickens, J. E., Irvine, W. M., Snell, R. L. et al. (2000), *ApJ*, **542**, 870.
Genaciński, P. (2000), *Acta Astron.*, **50**, 133.
Hasegawa, T. I., Herbst, E. and Leung, C. M. (1992), *ApJSS*, **82**, 167.
Le Teuff, Y. H., Millar, T. J., Markwick, A. J. (2000), *AAS*, **146**, 157.
Lee, H.-H, Roueff, G., Pineau Des Forêts, G. et al. (1998), *A&A*, **334**, 1047.
Li, A. and Greenberg, J.M. (1997), *A&A*, **323**, 566.
Ohishi, M., Irvine, W.M. and Kaifu, N. (1992), in Singh, P.D. (ed.), *Astrochemistry of Cosmic Phenomena*, Kluwer, Dordrecht, p. 171.
Ohishi, M. and Kaifu, N. (1998), in *Chemistry and Physics of Molecules and Grains in Space*, Faraday Discussions No. 109, The Faraday Division of the Royal Society of Chemistry, London, p.205.
Pratap, P., Dickens, J. E., Snell, R. L. et al. (1997), *ApJ*, **486**, 862.
Ruffle, D. and Herbst, E. (2000), *MNRAS*, **319**, 837.
Ruffle, D. and Herbst, E. (2001), *MNRAS*, **322**, 770.
Savage, B.D. and Sembach, K.R. (1996), *Ann. Rev. A&A*, **34**, 279.
Shalabiea, O.M. (2001), *A&A*, **370**, 1044.
Shalabiea, O.M., Caselli, P., Herbst, E. (1998), *ApJ*, **502**, 652.
Shalabiea, O.M. and Greenberg, J.M. (1994), *A&A*, **290**, 266.
Shalabiea, O.M. and Greenberg, J.M. (1995), *A&A*, **296**, 779.
Sofia, U. J., Cardelli, J. A., Guerin, K. P., Meyer, D. M. (1997), *ApJ*, **482**, L105.
Sofia, U. J., Fitzpatrick, E., Meyer, D. M. (1998), *ApJ*, **504**, L47.
Terzieva, R. and Herbst, E. (1998), *ApJ*, **501**, 207.
van Dishoeck, E. F., Black, G. A., Draine, B. T., Lunine, J. I. (1993), in Levy, E.H. and Lunine, J.I. (eds.), *Protostars and Planets III*, Univ. of Arizona press, p.163.
Watt, G. D. (1985), *MNRAS*, **212**, 93.
Wilacy, K. and Millar, T. (1998), *MNRAS*, **298**, 562.

THE CHEMICAL EVOLUTION OF TMC-1
Possible Role of Gas-Grain Chemistry

M. S. El-Nawawy
Department of Astronomy, Cairo University, Egypt.

Keywords: contraction - TMC-1: chemistry - grain: surface chemistry

Abstract: The chemical evolution in the core of contracting I.S. cloud is studied to understand the observed structure in TMC-1. The results fit mostly with the observations. The possible role of gas-grain chemistry is touched.

INTRODUCTION

The interstellar cloud chemistry has been treated recently via: (1)- pseudo time-dependent models, in which the variations of abundances as a function of time are followed through solving a coupled system of stiff, non-linear, first-order, ordinary differential equations. In these models molecular abundances evolve under fixed physical conditions (e.g. Langer et al. 1984; Herbst & Leung 1986; Millar 1990; Nejad et al. 1990; Williams & Hartquist 1991); and (2)- evolutionary models, in which all the different physical parameters will not be longer constant. Pseudo time-dependent models can fit the observed high abundances of the most molecular species in TMC-1 only at early times, $t \sim 10^5$ yr (van Dishoeck 1995 and Millar et al. 1989). In this sense the evolutionary model should be near to the real situation in the star forming regions. In the second class one has to study the chemistry during contraction (El-Nawawy et al. 1997, hereafter EHM). In addition, study of chemistry during contraction would help us in simulating the mutual effects between chemistry and dynamics. Howe et al. (1996) studied the observed cores in TMC-1, concentrating on sulphur species and using collapse models.

V. Pirronello et al. (eds.), Solid State Astrochemistry, 431–439.
© 2003 *Kluwer Academic Publishers. Printed in the Netherlands.*

Amin and El-Nawawy (1997) integrated the chemical rate equations and the MHD equations simultaneously and EHM studied the chemical evolution in the central core of contracting interstellar clouds. The chemical evolution is studied for various physical and chemical conditions, including the effects of varying the cosmic ray ionization rate, in order to understand the observed structures in TMC-1 and the extended ridge cloud in Orion. Their results give good agreement with observations, much better than the pseudo time-dependent models. So it appears clear that the process of star formation does have significant and potentially observable effects on the chemistry of the gas. This point will be highlighted in a forthcoming paper.

El-Nawawy (2000) extended the study of the chemistry of TMC-1 including larger network that contains reactions of Deuterium and PAH, and using the UMIST rate file RATE95 (Millar et al. 1997). His results in model 3 fit well with the observations. Some species have enhanced values more than the observations, such as CS, C_2S, C_3S and C_2H. Such enhancement possibly returns to the assumed initial elemental abundances, which is an important factor in chemical modelling of the interstellar clouds. Some other species show minor discrepancies such as OH, OCS, H_2S and C_3H_4. The problem of these molecules is expected to be solved in terms of the gas-grain interactions (e.g. Nejad et al. 1994). It is aimed in the present work to extend the previous study of the chemical evolution to interpret the more detailed observed chemical structures in TMC-1 and to through some light on the possible role of grain surface chemistry in it.

1.   THE MAGNETO-HYDRO-DYNAMIC EQUATIONS (MHD)

The contraction of an interstellar cloud in star forming regions is studied by solving the equations of MHD. Assuming that the cloud is of density ρ, velocity v, gas pressure p, gravitational potential Ψ, temperature T, magnetic field strength B and current density J, then the desired MHD equations, in their general form, are:

The continuity equation,

$$\frac{\partial \rho}{\partial t} + \nabla \cdot (\rho v) = 0 \qquad (1)$$

The momentum equation,

$$\frac{\partial (\rho v)}{\partial t} + \nabla \cdot \rho vv = -\nabla p - \rho \nabla \Psi + (J/c) \times B \qquad (2)$$

The gravitational potential Ψ is given by solving Poisson's equation,

$$\nabla^2\Psi = 4\pi\rho G \qquad (3)$$

where G is the gravitational constant. Equations (1 - 3) were solved using the finite difference method given by El-Nawawy et al. (1992). The pressure at the cloud boundary was considered uniform and constant. The magnetic term in Equation (2) is neglected in the present work. The original grid was divided into shells with equal spacing. The coordinates were constructed to move approximately with the matter. Therefore the central core is contracting in size and increasing in mass. We have assumed that the cloud is in spherical symmetric, then one can follow the collapse only in one dimension. We have divided the cloud model into 100 shells. The spacing between shells is an important factor has to be chosen carefully to be sure that the gravitational potential is calculated with sufficient accuracy (El-Nawawy and Aiad 1988). The division of the cloud into 100 shells fulfills this condition.

The chemical kinetic equations were integrated as a function of time. The abundance of the different species at time step $t = t_0 + \Delta t$ is provided from that at the previous time step. The corresponding density at the same time is taken from the contraction equations. Our procedure does simultaneous integration of the chemical rate equations and the hydrodynamic equations. The schematic diagram describing the procedure used to follow the chemistry and hydrodynamics can be found in Amin and El-Nawawy (1997).

2.   THE CHEMICAL SCHEME USED TO STUDY TMC-1

We have considered the network used by EHM and developed by El-Nawawy (2000) by the addition of detailed chemistries of PAH and Deuterium. The UMIST rate file RATE95 (Millar et al. 1997) has been used to build the desired network. The reactions of PAH are taken from Flower & Pineau des Forets (1995), while those of Deuterium are taken from Caselli et. al. (1998). Finally the reaction network contains 1333 reactions involving 293 species. For ion-dipolar molecule collisions, long-range attractive forces lead to rate coefficients which posses an inverse temperature dependence; consequently, these reaction rates increase highly at low temperature. The influence of these enhanced ion-dipolar rate coefficients has been studied by Herbst & Leung (1986) and Millar & Herbst (1990). The enhanced rates used and calculated by EHM have been used in the present calculation in what we call the fast reaction ratefile and give it the name ratefile2, while ratefile1 is given to that one without the above mentioned enhancements.

We have studied the contraction from low density, n = 10 cm^{-3}, and up to n = 10^5 cm^{-3}. The temperature changes during contraction, in terms of n and Av, according to equation (6) in EHM, where the visual extinction Av is calculated by integrating the column densities over the cloud shells. The initial temperature is assumed to be 100 K. A restriction is placed on the temperature not to fall below 10 K. The cloud mass is taken to be 5×10^4 solar mass. This mass is representative of the parent molecular cloud out of the dense formed cores.

3.   THE CHEMICAL EVOLUTION OF TMC-1

We have studied the contraction of five models. In the third model, ratefile2 has been used, while in the rest of the models, we have used ratefile1. The initial elemental abundances are those of model 4 in EHM with decreasing the fractional abundance of oxygen to 1.8×10^{-4} in models 1 and 2 and to 1.1×10^{-4} in the models 3-5. The initial abundance of S = 4×10^{-7} in model 1 and 2×10^{-7} in the rest of models. The observations of TMC-1 are taken from Hiraahara et al. (1992), Ohishi et al. (1992) and Ohishi and Kaifu (1998). In our calculations, the rate of ionization by cosmic ray is taken with the value 3×10^{-17} s^{-1}. The results of the models are given in Table 1.

The results of model 1 predict good agreement with observations. There are discrepancies in the fractional abundances of OH, H$_2$S, OCS, C$_3$H$_4$ and HC$_9$N by about one order of magnitude. In addition the resulting values of CS, C$_2$S and C$_3$S in model 1 are enhanced more than the observations by more than one order of magnitude. This enhancement resulting from using high initial value of sulphur as mentioned above. The calculations given by model 2 carry more discrepancies than given by model 1. The calculated fractional abundances of the molecular species OH, H$_2$S, C$_5$H, CH$_3$C$_3$N, C$_2$H$_3$CN, HC$_7$N, HC$_9$N, CH$_3$CHO, C$_3$H$_4$ and HC$_5$N differ from their observations by values ranging between one-half and two orders of magnitude. In model 3, the rate of depletion of Oxygen is enhanced and surprisingly the results fit with the observations of TMC-1 much better than that given by models 1 and 2. Only the calculated fractional abundances of H$_2$S, OCS and OH differ, in model 3, from their observed values by about one order of magnitude. Such final results would evaluate the importance of studying the chemical evolution during contraction where the chemistry is evolving with the physical parameters during the collapse process.

The ion density comes out from the three models with an average value of 3.5×10^{-8}. It is expected that the problem of OH and H$_2$S hence OCS could be treated in terms of gas-grain chemistry. From the other side, we still need to look for modifications in the present rate file, particularly there are many reactions need to be tested at low temperature (10 K). We still have uncertainty in many branching ratios and the extensions of neutral-neutral reactions are waiting new developed experimental works.

4.   POSSIBLE ROLE OF DUST IN STAR FORMING REGIONS

Although gas phase models of the chemistry of interstellar clouds have enjoyed a fair degree of success in explaining observed abundances of gas phase molecules (Herbst and Leung 1989, 1990, EHM), it has long been known that they could not explain the large abundance of molecular hydrogen. Early work by Salpeter and collaborators (e.g. Hollenbach and Salpeter 1971) showed that recombination of two hydrogen atoms on the surfaces of interstellar grains could efficiently form H_2. Formation of heavier molecular species on grain surfaces also undoubtedly occurs, but unlike H_2 these molecules cannot evaporate efficiently at low prevailing temperatures.

Significant surface abundances of species such as H_2O, CO, CH_3OH and CH_4 have been detected in a variety of cool regions in front of infrared continuum sources (Lacy et al. 1991, Whittet et al. 1991, Knacke and Larson 1991, Tielens et al. 1991, Grim et al. 1991). Even a rather quiescent region such as Taurus has enough background infrared sources to permit detection of surface CO and H_2O. Although it can be argued that the observed surface molecules are produced at least in part via gas-phase deposition, the large abundances of molecules such as water and methanol would appear to exceed the most optimistic gas phase calculated results. Recently, it is confirmed that in star-forming regions and hot cores, gas-grain interactions actively modify the gas phase chemistry by evaporation and photochemistry (Millar et al. 1991).

Depending on the results shown in models 1-3, we have assumed an ejection of some OH and H_2S from grain surface into the gas phase in model 4. The rate of OH evaporation from grain surface into the gas is assumed to be $2.5 \times 10^{-28} n^2 \sqrt{T} \sim 10^{-10}$ times the rate of H_2 formation and that of H_2S is $1.0 \times 10^{-31} n^2 \sqrt{T} \sim 10^{-12}$ times the rate of H_2 formation. The calculated fractional abundance of H_2S resulting from model 4 fit well with the observations, confirming the importance of gas-grain chemistry. Comparing models 3 and 4 (without and with inclusion of gas-grain chemistry, respectively), one can find that the injection of H_2S will enhances the abundance of CS by at least one order (two orders from the observed value).

The calculated fractional abundances of C_2S, C_3S, H_2CS and HCS+ are approximately similar in the two models. The abundance of OCS given by model 4 is about half its observed value. The fractional abundance of OH fit with the observations only at a later time of contraction ($t = 8.76$ Myr), where the density has increased to about 9.0×10^4 cm^{-3}.

In model 5, we have followed another chemical scheme to study gas-grain interactions. We have included freezing of neutrals on grain surface and desorption by H_2 formation, Cosmic ray heating and explosive desorption (Shalabiea & Greenberg 1994). The network contains 245 gas species and 102 grain species. The grain rate file is taken from Hasegawa et al. (1992) and Hasegawa & Herbst (1993). The network contains 347 species in 3012 reactions (611 reactions for grain surface chemistry). This model is so complex that the computations require so huge time. To reduce the time of computation in this case, I have studied the chemistry in pseudo time-dependent model. The results of model 5 show that the calculated fractional abundances of some species are very low in comparison with observations, particularly CH_2CN and HCOOH. The calculated fractional abundances given by model 5 for C_2H, C_5H, H_3C_3N, HCS^+, H_2CN^+ are less than the observations by about one order of magnitude, and that of CH_3CHO, C_3H_4, HC_5N and H_3C_4N are less than their observed values by two orders. Many other species comes with relative values (compared to observations) lower by about 0.5 to 0.25. Although, the fractional abundances of OH and most of the sulphur species CS, C_2S, C_3S, H_2S and OCS given by model 5 fit approximately to their observed values, confirming the role of grain surface chemistry in the evolution of these species. Another point has to be noticed, that the electron density given by model 5 is smaller than the values given by the rest of the models. This result is consistent with the conclusion of Bergin & Langer (1997) that the electron density will decrease as we include the grain surface chemistry in our scheme of chemistry.

Table 1. The fractional abundance as given by models 1-5 in comparison with observations of TMC-1.

Species	Observations	Model 1	Model 2	Model 3	Model 4	Model 5
C_2	2.5(-8)	1.3(-8)	1.2(-8)	1.8(-8)	2.1(-8)	4.8(-9)
CH	1.0(-8)	1.0(-8)	1.0(-8)	1.2(-8)	1.2(-8)	3.0(-9)
CN	1.5(-8)	2.1(-8)	1.2(-8)	4.3(-8)	3.4(-8)	9.2(-9)
CO	4.0(-5)	5(-5)	4.5(-5)	5.8(-5)	4.3(-5)	4.0(-5)
NO	1.5(-8)	1.2(-8)	1.4(-8)	8.4(-9)	8.0(-9)	1.3(-8)
OH	1.5(-7)	3.0(-8)	5.0(-8)	3.0(-8)	3.0(-8)	1.2(-7)
SO	2.5(-9)	5.0(-9)	5.0(-9)	3.1(-9)	2.8(-9)	4.0(-9)
C_2H	2.5(-8)	1.2(-7)	6.0(-8)	1.4(-7)	1.5(-7)	6.0(-9)
CS	5.0(-9)	1.5(-7)	6.0(-8)	1.6(-8)	3.1(-7)	5.0(-9)
C_2S	4.0(-9)	1.2(-8)	3.4(-9)	1.3(-8)	2.3(-8)	2.0(-9)
C_3S	5.0(-10)	5.0(-9)	1.2(-9)	6.1(-9)	6.1(-9)	3.0(-10)
OCS	1.0(-9)	1.6(-10)	9.6(-10)	2.0(-10)	5.0(-10)	1.1(-9)
H_2S	2.5(-10)	1.3(-11)	3.0(-11)	2.3(-11)	3.3(-10)	8.0(-11)
C_2O	3.0(-10)	2.5(-10)	2.8(-10)	3.6(-10)	3.6(-10)	2.0(-11)
HCN	1.0(-8)	1.1(-8)	9.0(-9)	6.2(-8)	1.1(-7)	8.0(-9)
HNC	1.0(-8)	1.2(-8)	8.0(-9)	3.6(-8)	6.2(-8)	5.0(-9)
SO_2	< 5.0(-10)	2.0(-10)	1.0(-10)	1.3(-11)	1.3(-11)	2.8(-9)
C_3H	2.5(-10)	1.4(-8)	3.0(-9)	1.9(-8)	2.0(-8)	2.5(-10)
C_3N	5.0(-10)	6.0(-9)	1.0(-9)	1.5(-8)	1.4(-8)	1.2(-10)
C_3O	5.0(-11)	3.4(-11)	3.0(-11)	7.4(-11)	6.5(-11)	1.0(-12)
H_2CO	1.0(-8)	4.0(-8)	1.4(-8)	1.1(-7)	1.5(-7)	1.0(-8)
H_2CS	1.5(-9)	1.7(-9)	1.4(-9)	2.7(-9)	6.7(-9)	3.6(-10)
NH_3	1.0(-8)	2.0(-8)	1.1(-8)	3.1(-8)	3.0(-8)	9.0(-8)
CH_2CN	2.5(-9)	4.0(-9)	1.0(-9)	4.9(-8)	7.9(-8)	-------
CH_2CO	5.0(-10)	3.0(-9)	4.9(-9)	4.9(-9)	6.9(-9)	1.0(-20)
C_3H_2	5.0(-9)	5.0(-9)	5.0(-9)	8.2(-9)	4.8(-9)	1.0(-9)
C_4H	1.0(-8)	2.0(-7)	2.0(-7)	6.2(-7)	5.0(-7)	4.0(-9)
HCOOH	1.0(-10)	3.0(-9)	1.3(-9)	2.6(-9)	2.6(-9)	1.0(-25)
HC_3N	3.0(-9)	3.0(-8)	3.0(-9)	8.5(-8)	5.0(-8)	1.4(-9)
CH_3CN	5.0(-10)	5.0(-10)	2.0(-10)	1.5(-8)	1.5(-8)	1.4(-10)
C_4H_2	4.0(-10)	2.0(-8)	1.2(-8)	6.8(-8)	2.2(-8)	4.2(-10)
C_5H	1.5(-10)	1.6(-10)	1.6(-10)	1.4(-9)	1.4(-9)	1.5(-11)
CH_3OH	1.0(-9)	2.0(-9)	5.0(-10)	4.5(-9)	2.4(-9)	1.0(-9)
CH_3CHO	3.0(-10)	3.2(-10)	3.0(-10)	5.0(-10)	6.5(-10)	1.0(-12)
H_3C_3N	1.0(-10)	9.0(-11)	1.0(-11)	7.2(-10)	1.4(-9)	1.0(-11)
C_3H_4	3.0(-9)	9.0(-11)	3.0(-11)	3.9(-10)	5.5(-10)	7.0(-11)
C_6H	5.0(-11)	1.0(-8)	5.0(-11)	3.2(-8)	3.1(-8)	5.0(-11)
HC_5N	1.5(-9)	1.5(-9)	2.5(-10)	1.4(-8)	2.4(-8)	3.0(-11)
H_3C_4N	2.5(-10)	1.0(-10)	1.5(-12)	2.6(-10)	2.7(-10)	8.6(-12)
C_5H_4	1.0(-10)	1.9(-10)	1.1(-10)	1.6(-9)	2.5(-9)	5.0(-11)

a(-b) means a x 10^{-b}.

Table 1 (continued). The fractional abundance as given by models 1-5 in comparison with observations of TMC-1.

Species	Observations	Model 1	Model 2	Model 3	Model 4	Model 5
HC$_7$N	5.0(-10)	1.1(-10)	2.0(-11)	9.9(-10)	1.4(-9)	--------
HC$_9$N	1.5(-10)	1.6(-11)	2.5(-12)	1.9(-10)	3.0(-10)	--------
HCO$^+$	4.0(-9)	7.0(-9)	7.0(-9)	5.9(-9)	5.8(-9)	3.7(-9)
HCS$^+$	3.0(-10)	6.0(-10)	1.6(-10)	6.0(-10)	6.0(-10)	3.0(-11)
H$_2$CN$^+$	1.0(-9)	2.0(-10)	2.5(-10)	1.4(-10)	1.5(-10)	1.5(-10)
N$_2$H$^+$	2.5(-10)	1.8(-10)	2.1(-10)	1.6(-10)	1.6(-10)	1.2(-10)
DCO$^+$	4.0(-10)	4.1(-10)	4.2(-10)	4.0(-10)	4.2(-10)	4.1(-10)
PAH$^-$	----------	4.2(-10)	4.3(-10)	4.1(-10)	4.1(-10)	----------
E	----------	3.9(-8)	4.0(-8)	3.0(-8)	3.0(-8)	1.4(-8)

a(-b) means a×10^{-b}.

5.   CONCLUSIONS

Study of TMC-1 using only gas phase chemistry and following the chemical evolution during real contraction of the interstellar cloud model, initial C/O ~ 0.66 and rate of ionization by cosmic ray equals 3×10^{-17} s^{-1}, provided us with the most suitable theoretical model which can fit mostly well with the observations of TMC-1. H$_2$S, OH and C$_3$H$_4$ show about one order of magnitude of discrepancy. Decreasing C/O more than the value mentioned above in the initial model will not help us in predicting the observations, particularly the large molecules.

We have developed two chemical models including the grain surface chemistry. It appears that the problem of H$_2$S and OH can be solved by gas-grain models. But again there are some other disagreements, indicating that we still need much efforts to reduce the uncertainties in the data files for the grain surface reactions. The electron density decreases in the gas-grain models more than given by those of pure gas phase.

REFERENCES

Amin, M.Y. and El-Nawawy, M.S. (1997), Earth Moon Planets, 75, 25.
Bergin, E.A. and Langer, W.D. (1997), ApJ, 486, 316.
Caselli, P., Walmsley, C.M., Terzieva, R. and Herbst, E. (1998), ApJ, 499, 234.
El-Nawawy, M.S. (2000), in Chem1, Cairo university, Astrochemistry, 1.
El-Nawawy, M.S. and Aiad, A. (1988), Proc. Math. Phys. Soc. Egypt., 65, 85.
El-Nawawy, M.S., Aiad, A. and El-Shalaby, M.A. (1988), MNRAS, 232, 908.
El-Nawawy, M.S., Ateya, B.G., Aiad, A. (1992), Ap&SS, 190, 257.
El-Nawawy, M.S., How, D. A. and Millar, T.J. (1997), MNRAS, 292, 481.
Flower, D.R. and Pineau des Forets, G. (1995), MNRAS, 275, 1049.

Grim, R.J.A. et al. (1991), *A&A*, **243**, 473.

Herbst, E. and Leung, C.M. (1986), *MNRAS*, **222**, 689.

Herbst, E. and Leung, C.M. (1989), *ApJSS*, **69**, 271.

Herbst, E. and Leung, C.M. (1990), *A&A*, **233**, 177.

Hasegawa, T.I. and Herbst, E. (1993), *MNRAS*, **261**, 83.

Hasegawa, T.I., Herbst, E. and Leung, C.M. (1992), *ApJSS*, **82**, 167.

Hirahara, Y. et al. (1992), *ApJ*, **394**, 539.

Hollenbach, D.J. and Salpeter, E.E. (1971), *ApJ*, **163**, 155.

Howe, D.A., Taylor, S.D. and Williams, D.A. (1996), *MNRAS*, **279**, 143.

Knacke, R.F. and Larson, H.P. (1991), *ApJ*, **367**, 162.

Lacy, J.H. et al. (1991), *ApJ*, **376**, 556.

Langer, W.D. et al. (1984), *ApJ*, **277**, 581.

Millar, T.J., Bennett, A. and Herbst, E. (1989), *ApJ*, **340**, 906.

Millar, T.J., Farquhar, P.R.A., Willacy, K. (1997), *A&AS*, **121**, 139.

Millar, T.J. and Herbst, E. (1990), *A&A*, **231**, 466.

Millar, T.J., Herbst, E., Charnley, S.B. (1991), *ApJ*, **369**, 147.

Nejad, L.A.M., Hartquist, T.W. and Williams, D.A. (1994), *Ap&SS*, **220**, 261.

Ohishi, M., Irvine, W.M. and Kaifu, N. (1992), in Singh, P.D. (ed.), *Astrochemistry of Cosmic Phenomena*, Kluwer, Dordrecht, p.171.

Ohishi, M. and Kaifu, N. (1998), *Faraday Discussions*, **109**, 205.

Shalabiea, O. M. and Greenberg, J. M. (1994), *A&A*, **290**, 266.

Tielens, A.G.G.M., Tokunaga, A.T., Geballe, T.R. and Baas, F. (1991), *ApJ*, **381**, 181.

van Dishoeck, E.F. (1995), in Winnewisser, G. and Pelz, G.H. (eds.), *Physics and Chemistry of Interstellar Molecular Clouds*, Springer, Berlin, p.225.

Whittet, D.C.B. et al. (1991), *A&A*, **251**, 524.

Williams, D.A. and Hartquist, T.W. (1991), *MNRAS*, **251**, 351.

THE MODEL OF CHEMICAL COMPOSITION OF HALLEY DUST PARTICLES

B.M. Andreichikov, G.G. Dolnikov
Space Research Institute Russian Academy of Sciences, Moscow, Russia

YU.P. Dikov
Institute of Ore Deposits, Petrography, Mineralogy and Geochemistry Russian Academy of Sciences, Moscow, Russia

Keywords: Halley – dust particles

Abstract: In this short paper we present a quick view of the composition of the nucleus of comet Halley deduced by the analysis of the dust particles detected by VEGA and GIOTTO spacecraft.

The dust particles in the coma of comet Halley were studied in situ by various instruments onboard the VEGA-1,2 and GIOTTO spacecrafts in 1986 (see Table 1 for details). The time-of-flight mass spectrometer (PUMA-1,2 and PIA) measured the composition of Halley dust particles in the coma of the comet at a distance from the nucleus ranging from 600 to ~180,000 km.

Table 1. Chronology and some details of the three close encounters with comet Halley.

Spacecraft	Distance to nucleus (km) min	Relative velocity (km/c)	Distance to Sun (a.u.)	Date of maximum rapprochement
VEGA-1	8889	79.2	0.79	6 March 1986
VEGA-2	8030	76.8	0.83	9 March 1986
GIOTTO	596	68.4	0.90	14 March 1986

V. Pirronello et al. (eds.), Solid State Astrochemistry, 441–446.

Estimates of the dust-to-gas ratio in the coma of comets range from 0.5 to 2; current best estimates for Halley tend toward the latter value (McDonnell et al., 1991). In this short report we present the inferred composition of the nucleus of Halley as it can be deduced only from solid particles in the assumption that the elemental composition of dust is representatative of the nuclear one. To obtain reliable statistical result, we use only data of Puma-1 (long/wide mode) from Mukhin et al. (1991). In this case we have 1236 good quality spectra of icy and rocky particles with different mass (255 particles have mass $>5 \cdot 10^{-12} \div \sim 5 \cdot 10^{-13}$ g, 365 particles - $\sim 10^{-13} \div \sim 10^{-14}$ g and 616 particles - $\sim 3 \cdot 10^{-15} \div \sim 8 \cdot 10^{-17}$ g). The average atomic abundances of the elements in Halley's dust are shown in Table 2.

Table 2. Average atomic abundances of the elements in Halley's dust (Jessberger et al., 1988; Mukhin et al., 1991), Sun and carbonaceous (CI) chondrites (Anders and Grevesse, 1989).

Element	H	C	N	O	Mg	Si	S	Ca	Fe
Dust[a]	2025	814	42	890	100	185	72	6.3	52
Dust[b]	2937	548	74	491	100	94	65	14	88
Sun	3290000	1178	367	2795	100	109	58	7	97
CI	492	71	6	712	100	93	48	6	84

[a]Jessberger et al. (1988); [b]Mukhin et al. (1991)

Data from Jessberger et al. (1988) given in this table are based on a total of 79 high quality mass-spectra from PUMA-1 without calculating any weight factor, although it is obvious the small particles have less CHON elements.

Spectra are divided on 11 weight groups, as in Mukhin et al. (1991), and we calculated average abundances in every group. The sum of amplitudes (A) of the main 16 elements in a spectrum is normalized to 100% (i.e. $\Sigma A_i = 100\%$, where i – a element in a spectrum). For each element we have

$$A_i\% = A_i / \sum A_i \tag{1}$$

And the average abundance in every weight group (j) is

$$A_{ij}\% = \frac{1}{n} \sum_{j=1}^{n} A_i\%_j \tag{2}$$

The resulting average abundance for each element versus the mass of dust particles (from $8 \cdot 10^{-17}$ to $5 \cdot 10^{-12}$ g) is shown in Figure 1.

The main comet-forming elements are H, C, O, N, Mg, Si, S and Fe (it was already noted by us in Andreichikov et al., 1997). From Fig. 1 it is clear that the content of the hydrogen decreases and carbon increases as a function of the weight of particles. The ratio H/C for heavy grains (~5) falls to ~0.5

for light grains. The correlation between these elements allows to expect that the main organic components of heavy particles are hydrides (mainly methane and silane), while the mineral part is mainly represented by hydrates of the other elements. In light grains a dominant element is carbon, and the mineral component increases because of the growing content of Mg, Si and refractory elements. Variations of the relative abundances of elements are not monotonic (particularly for H/C ratio). This reflects a transition from aliphatic to aromatic hydrocarbons (Andreichikov et al., 1998a). According to Andreichikov et al. (1998b) the 11 weight groups of dust particles came mainly from 3 parts of cometary body: core, mantle and crust, in a sort of layered structure,

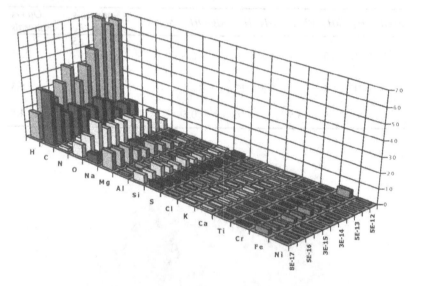

Figure 1. The dependence of the elemental composition of Halley dust upon the weight of particles.

and each shells consists of two main components - hydrides (basically hydrocarbons) and minerals.

The main component of the primary material of the core consists of hydrides - CH_4 (59,5 vol.%), SiH_4 (10%). The mineral part consists of hydroxides (mainly - $Mg(OH)_2$ -6,5%), sulphides (mainly - FeS), oxides. Comet material evolves with decomposition of $Mg(OH)_2$, oxidation of SiH_4 (3, 4) and decomposition of methane.

$$2Mg(OH)_2 + SiH_4 = MgSiO_3 + MgO + 4H_2 \uparrow \quad (3)$$
$$2Mg(OH)_2 + SiH_4 = Mg_2SiO_4 + 4H_2 \uparrow \quad (4)$$

$$2Mg\,(OH)_2 + SiH_4 = Mg_2Si + 4H_2O \uparrow \qquad (5)$$

Thus the comet is a differentiated body. The processes of the silane oxidation can be of the source of the energy for differentiation. The inferred composition of the core is given in Table 3 and in Figure 2, that of the mantle in Table 4 and in Figure 3, while the chemical composition of the crust is given in Table 5 and in Figure 4.

Table 3. The chemical composition of the core of the nucleus of Halley comet (vol.%).

Shell	CH₄	SiH₄	RNH₂	NH₃	H₂0	Mg(OH)₂	MgSiO₃	Σ minerals of Fe	Others minerals
0	59.5	10	-	6.5	0	6.5	-	9.65 (pyrhotine - 5.0)	7.85
1	51	8	-	9.5	13	11.5	-	1.7 (troilite)	5.3
2	63	0	3.5	-	16	0.5	11	1.2 (troilite)	4.8

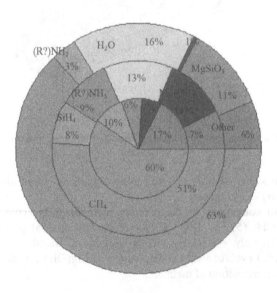

Figure 2. The core of the nucleus of Halley comet.

Table 4. The chemical composition of the comet mantle (vol.%).

Shell	C_nH_m	(R?)NH$_2$	MgSiO$_3$	Σ minerals Fe	Other minerals
3	62 C$_4$H$_{10}$ butane	5	17	3.5 (troilite)	12.5
4	52 C$_4$H$_8$ butilen	8	17	4.5 (pyrrotine)	18.5
5	54 C$_4$H$_6$ butadiene	8	18	3.6 (pyrrotine)	16.4

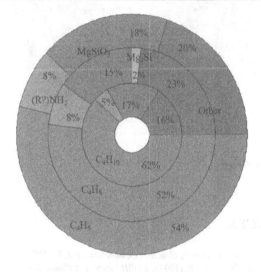

Figure 3. The mantle of the nucleus of Halley comet.

Table 5. The chemical composition of the comet crust (vol.%).

Shell	CnHm	C	(R?)NH$_2$	MgSiO$_3$	Mg$_2$Si	Σ minerals Fe	Others minerals
	38 C$_5$H$_6$ (cyclopentadien)	-	12	14	7	6.8 (troilite)	12.5
	47.5 C$_5$H$_6$ (cyclopentadien)	-	8.5	11	7	5.8 (troilite)	20.2
	35 (aromatic hydrocarbons)	-	12	21	9	3.5 (troilite)	19.5
	51 (aromatic hydrocarbons)	-	6	6	12	5.4 (Fe,Ni)nSm	19.6
	22 (aromatic hydrocarbons)	25	6	-	16.5	8 (Fe,Ni)nSm	22.5

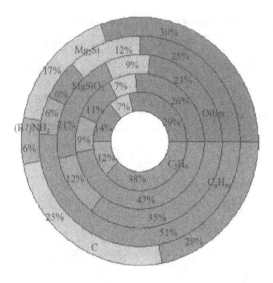

Figure 4. The comet crust.

REFERENCES

Anders, E. and Grevesse, E. (1989), *Geochim.Cosmochim.Acta,* **53**, 197.

Andreichikov, B.M., Dikov, Yu.P. (1997), *XXVIII Lunar and Planet Sci.Conf.,* Houston: LPI, NASA, p.43.

Andreichikov, B.M., Dolnikov, G.G., Dikov, Ju.P (1998a), *XXIXI Lunar and Planet Sci.Conf.,* Houston: LPI. NASA, abstr.1051.

Andreichikov, B.M., Dolnikov, G.G., Dikov, Ju.P (1998b), *XXIXI Lunar and Planet Sci.Conf.,* Houston: LPI, NASA, abstr.1052.

Jessberger, E.K., Christoforidis, A. and Kissel, J. (1988), *Nature,* **332**, 691.

McDonnell, J.A. M., Lamy, P. L. and Pankiewicz, G.S. (1991), in Newburn, R. L. Jr., Neugebauer, M. and Rahe, J. (eds.), *Comets in the Post-Halley Era*, Kluwer Academic Publishers, Dordrecht, p.1043.

Mukhin, L., Dolnikov, G., Evlanov, E., Fomenkova, M., Prilutsky, O. and Sagdeev, R. (1991), *Nature,* **350**, 480.

List of participants

Santi AIELLO
Dipartimento di Astronomia e
Fisica dello Spazio,
Largo Fermi 5,
50125 Firenze, ITALY

Louis J. ALLAMANDOLA
NASA Ames Research Center,
MS 245-6, Moffet Field,
CA 94034, UNITED STATES

Giuseppe BARATTA
Osservatorio Astrofisico,
Viale A. Doria 6,
95125 Catania, ITALY

Antonella BARUCCI
DESPA,
Observatoire de Paris – Meudon,
5 Place Jules Janssen,
92195 Meudon Cedex, FRANCE

Gozen BEREKET
Chemistry Department,
Osmangazi University,
26040 Eskisehir, TURKEY

Alexei BEREZHNOI
Lunar and Planetary Department,
Sternberg State Astronomical
Inst.,
Moscow University,
Universitetskij prosp. 13,
Moscow, 119899, RUSSIA

Ofer BIHAM
Racah Institute of Physics,
Hebrew University,
Jerusalem 91904, ISRAEL

Ferdinando BORGHESE
Dipartimento di Fisica della
Materia e Tecnologie Avanzate,
Università di Messina,
Salita Sperone 31,
98166 Messina, ITALY

Sergio BRANCIAMORE
Dipartimento di Chimica,
Università di Firenze,
Via G. Capponi 9,
50100 Firenze, ITALY

Eduardo M. BRINGA
Engineering Physics and
Astronomy Department,
University of Virginia,
Charlottesville,
VA 22901, UNITED STATES

John BRUCATO
Osservatorio Astronomico di
Capodimonte,
via Moiariello 16,
80131 Napoli, ITALY

Pietro CALANDRA
Dipartimento di Chimica-Fisica,
Università di Palermo,
Viale delle Scienze, Parco
d'Orleans II,
90128 Palermo, ITALY

Stephanie CAZAUX
Kapteyn Astronomical Institute,
PO Box 800,
9700 AV Groningen,
NETHERLANDS

Gennady DOLNIKOV
IKI Profsojuznaja 84/32,
Moscow, 117810, RUSSIA

Francois DULIEU
Université de Cergy-Pontoise,
umr 8588 du CNRS,
95806 Cergy Cedex, FRANCE

Irina EGOROVA
Dept of Astronomy,
Odessa State University,
Shevchenko Park,
Odessa 270014, UKRAINE

Pascale EHRENFREUND
Leiden Observatory,
Postbus 9513, Niels Bohrweg 2,
2300 RA Leiden,
NETHERLANDS

Mohamed S. EL-NAWAWY
Astronomy Dept.,
Faculty of Science,
Cairo University,
Giza, EGYPT

Dirk FABIAN
Astrophysical Institute,
University Jena,
Schillergässchen 3,
D – 07743 Jena, GERMANY

Jean Hughes FILLION
Université de Cergy-Pontoise,
umr 8588 du CNRS,
95806 Cergy Cedex, FRANCE

Helen FRASER
Leiden Observatory,
P.O. Box 9513,
NL-2300 RA Leiden,
NETHERLANDS

Jan FULARA
Institute of Physics,
Polish Academy of Sciences,
Al.. Lotnikow 32-46,
PL-02-668 Warszawa, POLAND

Gazinur GALAZUTDINOV
Special Astrophysical
Observatory,
Nizhnij Arkhyz, Zelenchukskaya,
Karachaevo-Cherkesia 369167,
RUSSIA

Mikhail GERASIMOV
IKI, Profsojuznaja 84/32,
Moscow, 117810
RUSSIA

Piotr GNACINSKI
Institute of theoretical Physics and
Astrophysics,
University of Gdansk,
ul. Wita Stwosza 57,
80-952 Gdansk, POLAND

J. Mayo GREENBERG
Leiden Observatory,
Postbus 9513, Niels Bohrweg 2,
2300 RA Leiden,
NETHERLANDS

Eugenia HALMAGEAN
National R&D Institute of
Microtechnologies,
Str. Erau Ianco Nicolae nr.32 B,
72996 Bucharest, ROMANIA

Thomas HENNING
Astrophysical Institute,
University Jena,
Schillergässchen 3,
D - 07743 Jena, GERMANY

Eric HERBST
Department of Physics,
The Ohio State University,
174 W. 18th Ave,
Columbus, OH 43210-1106,
UNITED STATES

Sasha HONY
Astronomical Institute,
University of Amsterdam,
Kruislaan 403,
1098 SJ Amsterdam,
NETHERLANDS

Antonella IATÌ
Dipartimento di Fisica della
Materia e Tecnologie Avanzate,
Università di Messina,
Salita Sperone 31,
98166 Messina, ITALY

Grzegorz KACZMARCZYK
Torun Center for Astronomy,
Nicholas Copernicus University,
Gagarina 11,
Pl – 87 – 100 Torun, POLAND

Yeghis KEHEYAN
Istituto di Chimica Nucleare,
CNR, Area della Ricerca di
Roma,
00016 Monterotondo Stazione,
ITALY

Ciska KEMPER
Astronomical Institute,
University of Amsterdam,
Kruislaan 403,
1098 SJ Amsterdam,
NETHERLANDS

Jacek KRELOWSKI
Torun Center for Astronomy,
Nicholas Copernicus University,
Gagarina 11,
Pl – 87 – 100 Torun, POLAND

Aurelie LE BRAS
DESPA,
Observatoire de Paris – Meudon,
5 Place Jules Janssen,
92195 Meudon Cedex, FRANCE

Jean Louis LEMAIRE
Université de Cergy-Pontoise,
umr 8588 du CNRS,
95806 Cergy Cedex, FRANCE

Aigen LI
Department of Astrophysical
Sciences,
Princeton University,
Peyton Hall, Princeton,
NJ 08544-1001,
UNITED STATES

Giulio MANICÒ
Dipartimento di Metodologie
Fisiche e Chimiche,
Università di Catania,
Viale A. Doria 6,
95125 Catania, ITALY

Andrzej MEGIER
Torun Center for Astronomy,
Nicholas Copernicus University,
Gagarina 11,
Pl – 87 – 100 Torun, POLAND

Vito MENNELLA
Osservatorio Astronomico di
Capodimonte,
via Moiariello 16,
80131 Napoli, ITALY

Guillermo MUNOZ CARO
Leiden Observatory,
P.O. Box 9513,
NL-2300 RA Leiden,
NETHERLANDS

Jaroslaw NIRSKI
Torun Center for Astronomy,
Nicholas Copernicus University,
Gagarina 11,

Pl – 87 – 100 Torun, POLAND

Fabio PICHIERRI
Dipartimento di Fisica,
Università di Trieste,
Via Valerio 2,
34127 Trieste, ITALY

Valerio PIRRONELLO
Dipartimento di Metodologie
Fisiche e Chimiche per
l'Ingegneria,
Università di Catania,
Viale A. Doria 6,
95125 Catania, ITALY

James PIZAGNO
Astronomy Department,
Ohio State University,
174 W. 18th Ave, Columbus,
OH 43210, UNITED STATES

Marina PROKOPJEVA
Astronomical Institute,
Saint Petersburg University,
Bibliotechbaya pl.2,
Petrodvoretz Saint Petersburg,
198904, RUSSIA

Giuseppe RAGUNÌ
Dipartimento di Metodologie
Fisiche e Chimiche,
Università di Catania,
Viale A. Doria 6,
95125 Catania, ITALY

Mark G. RAWLINGS
Observatory, P.O. Box 14,
Taehtitorninmaki,
University of Helsinki,
FIN-00014 Helsinki,
FINLAND

Steven RODGERS
Space Science Division,
M/S 245-3,
NASA Ames Research Center,
Moffett Field, CA 94035,
UNITED STATES

Kurt ROESSLER
Institut fur Nuklearchemie,
Forshungszentrum Juelich,
D – 52425 Juelich, GERMANY

S. ROMANKOV
Institute of Physics and
Technology,
480082 Almaty 82,
KAZAKHSTAN

Jennifer ROMON
DESPA,
Observatoire de Paris – Meudon,
5 Place Jules Janssen,
92195 Meudon Cedex,
FRANCE

Joe E. ROSER
Physics Department,
Syracuse University,
201 Physics Building,
Syracuse, NY 13244,
UNITED STATES

Miroslaw SCHMIDT
N. Copernicus Astronomical
Center,
Rabianska 8,
Pl - 87-100 Torun, POLAND

Osama M. SHALABIEA
Astronomy Dept.,
Faculty of Science,
Cairo University,

Giza, EGYPT

Valery I. SHEMATOVICH
Institute of Astronomy,
Russian Academy of Sciences,
48 Pyatnitskaya St.,
Moscow, 109017, RUSSIA

Chuanjian SHEN
Leiden Observatory,
P.O. Box 9513,
NL-2300 RA Leiden,
NETHERLANDS

Muriel SIZUN
Laboratoire des Collisions
Atomiques et Moleculaires,
Bat.351, Universite Paris-Sud,
F-91405 Orsay Cedex,
FRANCE

Bertrand STEPNIK
Institut d'Astrophysique Spatiale,
Bat. 121, Universite Paris-Sud,
91405 Orsay Cedex,
FRANCE

Achim TAPPE
Onsala Space Observatory,
S-43992 Onsala,
SWEDEN

Gianfranco VIDALI
Physics Department,
Syracuse University,
201 Physics Building,
Syracuse, NY 13244,
UNITED STATES

Tomasz WESELAK
Torun Center for Astronomy,
Nicholas Copernicus University,

Gagarina 11,
Pl – 87 – 100 Torun, POLAND

David A. WILLIAMS
Department of Physics and
Astronomy,
University College London,
Gower Street,
London WC1E6BT,
UNITED KINGDOM

Bogdan WSZOLEK
Torun Center for Astronomy,
Nicholas Copernicus University,
Gagarina 11,
Pl – 87 – 100 Torun, POLAND

Laimos ZACS
VIRAC Akademijas Laukums 1,
LV-1050 Riga, LATVIA

Index

454

Species Index